This is Volume 129 in
MATHEMATICS IN SCIENCE AND ENGINEERING
A Series of Monographs and Textbooks
Edited by RICHARD BELLMAN, *University of Southern California*

The complete listing of books in this series is available from the Publisher upon request.

Probability Methods
for Approximations in Stochastic Control
and for Elliptic Equations

Probability Methods for Approximations in Stochastic Control and for Elliptic Equations

Harold J. Kushner
DEPARTMENTS OF APPLIED MATHEMATICS
AND ENGINEERING
BROWN UNIVERSITY
PROVIDENCE, RHODE ISLAND

ACADEMIC PRESS New York San Francisco London 1977
A subsidiary of Harcourt Brace Jovanovich, Publishers

ACADEMIC PRESS, INC.
111 Fifth Avenue, New York, New York 10003

United Kingdom Edition published by
ACADEMIC PRESS, INC. (LONDON) LTD.
24/28 Oval Road, London NW1

Library of Congress Cataloging in Publication Data

Kushner, Harold Joseph, Date
Probability methods for approximations in stochastic control and for elliptic equations.

(Mathematics in science and engineering ; v. 129)
1. Control theory. 2. Differential equations, Elliptic. 3. Approximation theory. 4. Probabilities. I. Title. II. Series.
QA402.3.K79 519.2 76-13942
ISBN 0–12–430140–1
AMS (MOS) 1970 Subject Classifications: 93E20, 93E25, 65M10, 60B10, and 90C40

PRINTED IN THE UNITED STATES OF AMERICA

To
Linda, Diana, and Nina

Contents

Chapter 3. Markov Chains and Control Problems with Markov Chain Models

Chapter 4. Elliptic and Parabolic Equations and Functionals of Diffusions

Chapter 5. A Simple Application of the Invariance Theorems

Chapter 6. Elliptic Equations and Uncontrolled Diffusions

Chapter 7. Approximations for Parabolic Equations and Nonlinear Filtering Problems

Chapter 8. Optimal Stopping and Impulsive Control Problems

Chapter 9. Approximations to Optimal Controls and Nonlinear Partial Differential Equations

Chapter 10. Approximations to Stochastic Delay Equations and to Diffusions and Partial Differential Equations with Reflecting Boundaries

Preface

This book deals with a number of problems concerning approximations, convergence, and numerical methods for stochastic control problems, and also for degenerate elliptic and parabolic equations. The techniques that are developed seem to have a broader applicability in stochastic control theory. In order to illustrate this, in Chapter 11 we give a rather natural approach to the formulation and proof of the separation theorem of stochastic control theory, which is more general than the current approaches in several respects.

The ideas of the book concern a number of interesting techniques for approximating (cost or performance) functionals of diffusions and optimally controlled diffusions and for approximating the actual diffusion process, defined by stochastic differential equations of the Itô type, both controlled and uncontrolled. Since many of the functionals that we seek to compute or approximate are actually weak solutions of the partial differential equations (i.e., the weak solution can be represented as a functional of an associated diffusion), the techniques for approximating the weak solutions are closely related to the techniques for approximating the diffusions and their functionals. Also, the form of the partial differential equation that is (at least formally) satisfied by a functional of interest actually suggests numerical methods for the probabilistic or control problem.

We develop numerical methods for optimal stochastic control theory and prove the required convergence theorems. Neither for this nor for any of the other problems do we require that the cost or optimal cost functions be smooth or satisfy any particular partial differential equation in any particular sense. Nor do we require, a priori, that the optimal control exist. Existence is a by-product of our method. The numerical techniques are intuitively reasonable, admit of many variations and extensions, and seem to yield good numerical results.

The main mathematical techniques are those related to the use of results in the theory of weak convergence of a sequence of probability measures. The technique seems to provide a point of view that not only suggests numerical methods but also unites diverse problems in approximation theory and in stochastic control theory. The ideas of weak convergence theory are being used more and more frequently in various areas of applications. But this book and previous papers by the author and some of his students seem to be the only currently available works dealing with applications to stochastic control theory or to numerical analysis. The proofs are purely probabilistic. Even when dealing with numerical methods for partial differential equations, we make no explicit smoothness assumptions and use only probabilistic methods and assumptions.

Chapter 1 discusses some of the necessary probabilistic background, including such topics as the Wiener process, Markov processes, martingales, stochastic integrals, Itô's lemma, and stochastic differential equations. It is assumed, however, that the reader has some familiarity with the measure theoretic foundations of probability. In Chapter 2, we describe the basic ideas and results in weak convergence theory, at least insofar as they are needed in the rest of the book.

The computational methods of the book are all equivalent to methods for computing functionals of finite Markov chains or for computing optimal control policies for control problems with Markov chain models. Many efficient computational techniques are available for these problems. In particular, the functionals for the uncontrolled Markov chains are all solutions to finite linear algebraic equations. The Markov chain can arise roughly as follows. We start with the partial differential equation that, at least formally, is satisfied by a functional of the diffusion, and we apply a particular finite difference approximation to it. If the approximation is chosen carefully (but in a rather natural way), then the finite difference equation is actually the equation that is satisfied by a functional of a particular Markov chain, and we can immediately get the transition probabilities for the chain from the coefficients in the finite difference equation. The local properties of this chain are very close to the local properties of the diffusion, in the sense that there is a natural time scaling with which we interpolate the chain into a continuous parameter process, and the local properties of the interpolation and diffusion are close in certain important respects. Also, the functional of the Markov chain, which is the solution to the approximating equation, is similar in form to a "Riemann sum" approximation to the original functional of the diffusion.

At this point, the theory of weak convergence comes in, and we show that the functional of the chain does indeed converge to the desired functional of the diffusion as the difference intervals go to zero. Similarly, the approximation to the weak sense solution to the partial differential equation converges to the weak sense solution. The interpolation of the chain also converges (in a suitable sense) to a solution to the stochastic differential equation. Of course, the finite difference algorithm is classical. But neither the convergence proofs nor the conditions for convergence are classical. Also, the method can handle a much broader class of functionals than those that may possibly solve some partial differential equation.

It is not necessary that we use finite difference methods; their use does, however, yield an automatic way of generating a family of approximating chains, whether or not the functional is smooth. However, many types of approximations are usable, provided only that they yield the correct limiting properties. Indeed, this versatility is one of the strong points of the approach.

Approximating with Markov chains (whether or not we use classical finite difference techniques) allows us to use our physical intuition—to guide us in the choice of a chain, or in the selection of a computational procedure for solving the equation for the functional of the chain. Our sense of the "dynamics" of the process plays a useful role and can assist us in the selection of procedures that converge faster.

In the case of the optimal control problem, we start by approximating the nonlinear (Bellman) partial differential equation, which is formally satisfied by the minimal cost function. With a suitable choice of the approximation, the discrete equations are just the dynamic programming equations for the minimal cost function for the optimal control of a certain Markov chain. Again, there are many types of useful approximating chains. This nonlinear partial differential equation, or optimal control, case is much more difficult than the uncontrolled or linear partial differential equation case. However, the ideas of weak convergence theory again play a very useful role. Under broad conditions, we can show that the sequence of optimal costs for the controlled chain converge to the optimal cost for the controlled diffusion. Indeed, it can even be shown that the (suitably interpolated) chains converge, in a particular sense, to an optimally controlled diffusion.

In Chapter 3, we give the required background concerning the equations satisfied by various functionals of Markov chains, both controlled and uncontrolled. Our method is able to treat optimal control problems with various types of state space constraints. However, this often requires a linear programming (rather than a dynamic programming) formulation, and this is also discussed in Chapter 3.

Chapter 4 discusses the relations between diffusion processes and elliptic and parabolic partial differential equations, both nondegenerate and degenerate and linear and nonlinear. Proofs are not given. The representation of the solutions of the linear equations in terms of path functionals of the diffusion is discussed, as well as the relation between certain nonlinear equations and optimal stochastic control problems. Chapter 5 is an introduction to the techniques and results of the sequel. In order to illustrate some of the simpler ideas, the techniques of weak convergence theory are applied to a simple two-point boundary value problem for a second order differential equation.

In Chapter 6, we begin the systematic exploitation and development of the ideas. The motivation for the types of approximations is given, and the approximation of a variety of functionals of uncontrolled diffusion and linear elliptic equations is treated. We also show how to approximate an invariant measure of the diffusion by an invariant measure of an approximating chain, discuss the use of the approximations for Monte Carlo, and give some numerical data. The approximations that are explicitly discussed are derived by starting with finite difference techniques; all of them yield Markov chain approximations to the diffusion. However, it should be clear from the development that many other methods of approximation can be handled by the same basic techniques. The general approach taken here should motivate and suggest other methods with perhaps preferable properties for specific problems.

Chapter 7 deals with the parabolic equation and with the probabilistic approach to approximation and convergence for explicit and implicit (and combined) methods. Furthermore, approximations to a (currently much studied) class of nonlinear filtering problems are discussed. Some numerical data concerning approximations to an invariant measure are given.

In Chapter 8, we begin the study of nonlinear partial differential equations and approximations to optimal control problems, in particular to the optimal stopping and impulsive control problems. The discretizations of the optimization problems for the diffusion yield similar optimization problems on the approximating Markov chains. We are able to prove that the approximations to the optimal processes and cost functions actually converge to the optimal processes and cost functions, respectively. The study of nonlinear partial differential equations and optimal control problems continues in

Chapter 9, where a variety of approximations and control problems are discussed. In order to show that the limiting cost functionals are truly minimal (over some specified class of control policies) and that the limiting processes have the probabilistic properties of the optimally controlled diffusion, a number of techniques are developed for approximating arbitrary controls and for proving admissibility or existence. It is expected that many aspects of the general approach will be quite useful in other areas of stochastic control theory. Additional numerical data appear in Chapters 8 and 9. Again, it must be emphasized that much more work needs to be done (investigating various types of approximations) in order to understand fully which types of approximations are preferable and why.

In Chapter 10, we treat two types of extensions of the ideas in Chapters 6 and 7. First, approximations to stochastic differential difference equations and to path functionals of such processes are developed. Then, we discuss the problem of diffusions that are reflected from a boundary and the corresponding partial differential equations with mixed Neumann and Dirichlet boundary conditions.

It is hoped that the book will help open wider the door to an interesting direction of research in stochastic control theory. Similar techniques can be applied to the problem where the stochastic differential equation has a "jump term" and the partial differential equations are replaced by partial differential integral equations. See, e.g., [K12] or the forthcoming work of Kushner and DiMasi [K16].

Numbering. Example. Theorem 3 of Section 2 of Chapter 5 will always be referred to as Theorem 5.2.3. Equation (3) of Section 2 of Chapter 5 will be referred to as Eq. (2.3) in Chapter 5 and as Eq. (5.2.3) in other chapters.

Acknowledgments

A number of people have helped the author with various parts of the manuscript. He would especially like to thank Giovanni DiMasi for his reading of some of the more critical parts, and Sandra Spinacci and Katrina Avery for their excellent typing of the several drafts. It is also a pleasure to acknowledge the support of many research projects, over several years, by the Office of Naval Research, the Air Force Office of Scientific Research, and the National Science Foundation, which led directly to the ideas published in this book.

CHAPTER 1

Probability Background

This chapter contains a brief survey of several of the more important concepts and results in probability theory that will be used in the rest of the book. Most of the concepts are by now fairly standard in work in stochastic control theory and in many other areas of applications, for they are basic in the construction and study of continuous parameter models. But it is worthwhile to include them here in the interest of self-containment and in order to introduce them in the form in which they will be used.

The chapter is not intended to be an introduction to probability or stochastic process theory. It is generally assumed in this book that the reader has some familiarity with probability theory, especially with the concepts of Markov chains, stochastic convergence, conditional expectation, and the separability and measurability of processes. The requisite background is contained in many excellent books, e.g., Breiman [B4], Gikhman and Skorokhod [G2], Neveu [N1], or Wong [W2] (the first sixty or so pages in the latter reference).

Section 1.1 deals with the Wiener process, which plays a fundamental role in the study of diffusion Markov processes and continuous martingales.

Section 1.2 gives a few results on martingales. Section 1.3 deals with Markov processes and Section 1.4 with the basic definitions and properties of stochastic integrals (integrals with respect to a martingale process). Finally, in Section 1.5, the topic of stochastic differential equations is developed.

1.1 The Wiener Processes

1.1.1 Definitions

Let $(\Omega, \mathscr{B}, P)$ denote a probability space; i.e., the set Ω is the space of "elementary" outcomes, $\mathscr{B}$ is a σ-algebra of subsets of Ω, and P is a probability measure on $\mathscr{B}$. Let $\{w(t), \infty > t \geq 0\}$ denote a family of real-valued random variables defined on the space. Generally, the generic variable ω of the set Ω will not be explicitly written. However, when there is a greater possibility of ambiguity, the ω will be included. Also, $w(\omega, t)$ will denote the value of the random variable $w(t)$ at ω, and $w(\omega, \cdot)$ will denote the path corresponding to ω. The entire process $\{w(t), \infty > t \geq 0\}$ will be denoted by $w(\cdot)$. *The process* $w(\cdot)$ *is said to be a standard Wiener process if it satisfies* (a) *and* (b):

(a) it is a separable process†;
(b) it has independent, stationary, and normally distributed increments with zero mean value and $w(0) = 0$.

By (b) we mean: for each integer n and set of real numbers $0 \leq t_1 \leq t_2 \leq \cdots \leq t_n$, the random variables

$$\{w(t_{i+1}) - w(t_i), 1 \leq i < n\}$$

are mutually independent with zero mean and $w(t_{i+1}) - w(t_i)$ has a normal distribution with variance $t_{i+1} - t_i$ and $w(0) = 0$. The term "standard" will not be used. A "nonstandard" Wiener process satisfies (a) and (b), except that the variance can depend on time.

Any process satisfying properties (a) and (b) has continuous paths w.p. 1. Let $\mathscr{B}(R)$ and $\mathscr{B}(R^+)$ denote the Borel fields over the real line R and over the positive part of the real line $R^+ = [0, \infty)$, resp. By virtue of the continuity w.p. 1, there is a *measurable version* of $w(\cdot)$ in the sense that there is a version for which, for each set $A \in \mathscr{B}(R)$, we have

$$\{\omega, t : w(\omega, t) \in A\} \in \mathscr{B} \times \mathscr{B}(R^+).$$

If we are interested only in a Wiener process or in functionals of a Wiener process, then the sample space Ω can be chosen to be $C[0, \infty)$, the space of

† All processes used in the book will be assumed to be separable.

continuous functions on $[0, \infty)$. Then each ω corresponds to a continuous function on $[0, \infty)$ and conversely, and $\mathscr{B}$ is the smallest σ-algebra on Ω which contains all the open sets in the topology of uniform convergence on finite intervals.

A vector-valued Wiener process is defined to be a vector of mutually independent real-valued Wiener processes.

There are several equivalent definitions of the Wiener process. One (in terms of martingale concepts) that is particularly convenient in applications will be given in Section 4.3. The normality of the distribution of the increments is implied by the independence and stationarity of the increments if we add the condition: for each $t \geq 0$ and $\varepsilon > 0$,

$$\lim_{s \downarrow 0} \frac{P\{|w(t+s) - w(t)| \geq \varepsilon\}}{s} = 0$$

(see Breiman [B4, Chapter 12]).

If $Y(t)$, $t \geq 0$ is a sequence of random variables on $(\Omega, \mathscr{B}, P)$ and for each $t \geq 0$, $\{Y(s), s \leq t\}$ is independent of $w(u) - w(t)$, $u \geq t$, we say that $Y(\cdot)$ is *nonanticipative with respect to* $w(\cdot)$, or simply *nonanticipative* if the Wiener process is obvious.

Suppose that $\mathscr{B}_t$ is the minimal σ-algebra over which $w(s)$, $s \leq t$, is measurable, and $\mathscr{B} = \bigcup_{t \geq 0} \mathscr{B}_t$, and that P is the measure induced on $\mathscr{B}$ by $w(\cdot)$. Then $(\Omega, \mathscr{B}, P)$ may be called a *Wiener process space*.

1.1.2 Stopping Times

Let $(\Omega, \mathscr{B}, P)$ denote a probability space, and let $\mathscr{B}_t$ and $Y(t)$, $\infty > t \geq 0$, denote a nondecreasing sequence of sub-σ-algebras of $\mathscr{B}$ $(\mathscr{B}_{t+s} \supset \mathscr{B}_t, s \geq 0)$ and a sequence of random variables defined on the probability space. We say that the sequence $\{\mathscr{B}_t\}$ and process $Y(\cdot)$ are *adapted* [*or that* $Y(\cdot)$ is adapted to $\{\mathscr{B}_t\}$] if $Y(t)$ is $\mathscr{B}_t$ measurable for each $t \geq 0$. $\mathscr{B}_t$ may be larger than the minimal σ-algebra over which $\{Y(s), s \leq t\}$ is measurable.

Let $\{\mathscr{B}_t\}$ denote a nondecreasing sequence of sub-σ-algebras of $\mathscr{B}$. The nonnegative real-valued function τ is called a *stopping time* with respect to $\{\mathscr{B}_t\}$ if it is defined on a set of positive probability and $\{\tau \leq t\} \in \mathscr{B}_t$ for each t. (As a convention, if such a random variable τ is not defined at some ω, we set it equal to ∞ there. Later, we will deal with stopping times which may actually take infinite values.)

There is a natural set of events $\mathscr{B}_\tau \subset \mathscr{B}$ which is associated with a stopping time τ. By definition, a set A is in $\mathscr{B}_\tau$ if and only if

$$A \cap \{\tau \leq t\} \in \mathscr{B}_t$$

for each $t \geq 0$. The events in $\mathscr{B}_\tau$ "occur" prior to (or at) time τ. Indeed, $\mathscr{B}_\tau$ is a σ-algebra. If $\mathscr{B}_t$ is the minimal σ-algebra over which a process $Y(s)$, $s \leq t$, is measurable, then (loosely speaking) whether or not any particular subset of Ω is in $\mathscr{B}_\tau$ can be determined by simply observing $\{Y(s), s \leq \tau\}$.

Stopping times arise frequently in stochastic process theory. They are often the moments at which specific events occur. Often we are concerned with studying the quantities $Y(\tau)$, or the process (or some functional of it) $Y(\cdot)$ only up until some event has occurred. In addition, the stopping times are frequently introduced as purely (and useful) technical devices to be used in the course of a proof.

Suppose that the process $Y(\cdot)$ has right continuous paths w.p. 1 and is adapted to $\{\mathscr{B}_t\}$, and τ is a stopping time with respect to $\{\mathscr{B}_t\}$. Then $Y(\tau)$ is a random variable and is $\mathscr{B}_\tau$ measurable (Neveu [N1, Chapter III.6]).

Examples of Stopping Times

Let $Y(\cdot)$ take values in a Euclidean space M and have right continuous paths w.p. 1, and let $\mathscr{B}_t$ denote the smallest σ-algebra over which $Y(s)$, $s \leq t$, is measurable.† Let A denote a Borel set in M, and $Y[s, t]$ the range of $Y(u)$, $s \leq u \leq t$, and define the function of ω

$$\tau = \inf\{t : Y(t) \in A\} = \sup\{t : Y[0, t] \in M - A\}$$

(the first entrance time into A; if $Y(t) \notin A$, all $t < \infty$, define the $\inf\{\ \}$ to be ∞). Then (Dynkin [D4, Chapter IV.1]) τ is a stopping time if A is closed and either A or $M - A$ is compact, or if A is open and $M - A$ is compact.

1.1.3 Properties of the Wiener Process

Owing to its fundamental role in the construction and study of Markov processes with continuous paths, the Wiener process and several of its properties will be used frequently in the sequel.

(a) (Breiman [B4, Chapter 12]) Almost every path is nowhere differentiable.

(b) For each $T > 0$, let T_n denote an arbitrary partition of the interval $[0, T]$ of the form

$$T_n = \{t_i^n, i \leq n : 0 = t_1^n \leq \cdots \leq t_n^n = T\}.$$

Define $|T_n| = \sup_i |t_{i+1}^n - t_i^n|$. Then we have the "quadratic" variation formula (Doob [D2, Chapter 8])

$$\lim_{|T_n| \downarrow 0} \sum_i |w(t_{i+1}^n) - w(t_j^n)|^2 = T \qquad \text{w.p. } 1, \tag{1.1}$$

† We will often use the notation $\mathscr{B}(Y(s), s \leq t)$ for such a σ-algebra.

provided that T_{n+1} includes the points $\{t_i^n\}$ of T_n. In any case (1.1) holds in mean square.

Both (a) and (b) suggest that the path is rather wild locally. This is corroborated by the law of the iterated logarithm (c) (Breiman [B4, Chapter 12]):

$$\text{(c)}\quad \overline{\lim_{t\downarrow 0}} \frac{w(t)}{(2t\log\log(1/t))^{1/2}} = 1 \qquad \text{w.p. } 1.$$

Property (c) implies that [recall that $w(0) = 0$] in each interval $[0, s]$, $s > 0$, $w(\cdot)$ takes positive and negative values infinitely often. It is also true that the location of a zero of $w(\cdot)$ is the limit from the right of an infinite sequence of zeroes and, generally, if τ is such a zero, then the shifted function $w_\tau(\cdot)$, defined by $w_\tau(t) = w(t + \tau) - w(\tau)$, takes positive and negative values each infinitely often in any interval $[0, s]$, $s > 0$. To see this, let t_0 denote any positive number and define

$$\tau = \inf\{t \geq t_0 : w(t) = 0\}, \qquad \mathscr{B}_t = \mathscr{B}(w(s),\ s \leq t).$$

Then τ is a stopping time relative to $\{\mathscr{B}_t\}$, and it can be shown (Breiman [B4, Chapter 12]) that $w_\tau(\cdot)$ is a standard Wiener process. The assertion follows, since (c) holds for $w_\tau(\cdot)$.

1.2 Martingales

1.2.1 Definition and Martingale Inequalities

Let $(\Omega, \mathscr{B}, P)$ be a probability space and $\{\mathscr{B}_t\}$ a nondecreasing family of sub-σ-algebras of $\mathscr{B}$. For any σ-algebra $\mathscr{A} \subset \mathscr{B}$ and vector-valued random variable Y, we use either notation $E_{\mathscr{A}} Y$ or $E[Y \mid \mathscr{A}]$ for the conditional expectation of Y given $\mathscr{A}$ (assuming that it exists). Suppose that the real-valued process $Y(\cdot)$ is adapted to $\{\mathscr{B}_t\}$. Then $\{Y(t), \mathscr{B}_t, \infty > t \geq 0\}$ is a *martingale*, if for each real $t \geq 0$, $s \geq 0$,

$$E|Y(t)| < \infty, \qquad E[Y(t+s) \mid \mathscr{B}_t] = Y(t) \quad \text{w.p. } 1. \tag{2.1}$$

We may simply say that $Y(\cdot)$ is a martingale if the $\{\mathscr{B}_t\}$ sequence is obvious. Martingale processes play a very basic role in the subjects dealt with in this book, but will not be used in any particularly deep way. They will be used mainly to obtain upper bounds for certain probabilities, and to show that some of our derived processes are related in a specific way to a Wiener process and, ultimately, to a stochastic differential equation. We will merely state some of the properties which will be used.

For any $T \geq 0$, $\varepsilon \geq 0$ (Doob [D2, Chapter 7])

$$P\left\{\sup_{T \geq t \geq 0} |Y(t)| \geq \varepsilon\right\} \leq E|Y(T)|^2/\varepsilon^2, \tag{2.2}$$

$$E \sup_{T \geq t \geq 0} |Y(t)|^2 \leq 4E|Y(T)|^2. \tag{2.3}$$

Both (2.2) and (2.3) hold on $[0, \infty)$ if $E|Y(T)|^2$ is replaced by $\lim_{T \to \infty} E|Y(T)|^2$.

A martingale $\{Y(t), \mathscr{B}_t, \infty > t \geq 0\}$ is said to be *square integrable* if $E|Y(t)|^2 < \infty$ for all $t \geq 0$. If the martingale is square integrable and continuous, then there is a *unique continuous nondecreasing process* $A(\cdot)$ with $A(0) = 0$ so that (Wong [W2, p. 166]; Kunita and Watanabe [K14]; Meyer [M3, Chapter 7]) if we define $M(t) = Y^2(t) - A(t)$ then

$$\{M(t), \mathscr{B}_t, \infty > t \geq 0\}$$

is a martingale.

The function $A(\cdot)$ may be written as $\langle Y \rangle$, with values $\langle Y \rangle_t$. For an arbitrary $T > 0$, let T_n denote a partition of the interval $[0, T]$, where

$$T_n = \{t_i^n : 0 = t_1^n \leq t_2^n \leq \cdots \leq t_n^n = T\}.$$

Then

$$\lim_{|T_n| \to 0} \sum_{i=1}^{n-1} |Y(t_{i+1}^n) - Y(t_i^n)|^2 = A(T), \tag{2.4}$$

where the limit is in probability. The function $A(\cdot)$ is constant over an interval only if the martingale is constant over that interval.

If $\{Y(t), \mathscr{B}_t, \infty > t \geq 0\}$ is a vector-valued martingale (i.e., each component of $Y(\cdot)$ is a martingale with respect to $\{\mathscr{B}_t\}$), which is continuous and square integrable, then there is a unique continuous nonnegative definite matrix-valued process $A(\cdot)$, so that $A(0) = 0$, $A(\cdot)$ is nondecreasing (in the sense of nonnegative definite matrices) and $\{M(t), \mathscr{B}_t, \infty > t \geq 0\}$ is a matrix-valued martingale, where

$$M(t) = Y(t)Y'(t) - A(t).$$

We may use the symbols $\langle Y \rangle$ and $\langle Y \rangle_t$ for $A(\cdot)$ and $A(t)$, resp. Let $\langle Y_i, Y_j \rangle$ denote the i, jth component of the matrix $\langle Y \rangle$. Then $Y_i(t)Y_j(t) - \langle Y_i, Y_j \rangle_t$ is a martingale with respect to $\{\mathscr{B}_t\}$. Also, (2.4) holds if the square $|Y(t_{i+1}^n) - Y(t_i^n)|^2$ is replaced by the outer product

$$(Y(t_{i+1}^n) - Y(t_i^n))(Y(t_{i+1}^n) - Y(t_i^n))'$$

(Kunita and Watanabe [K14, Section 1]). $A(\cdot)$ is known as the *quadratic variation* of $Y(\cdot)$, owing to the property (2.4). If $E\{[Y_i(t + s) - Y_i(t)] \times [Y_j(t + s) - Y_j(t)] \mid \mathscr{B}_t\} = 0$ w.p. 1, for all $s \geq 0$, $t \geq 0$, then $\langle Y_i, Y_j \rangle = 0$.

If, for each $t \geq 0$, $\mathscr{B}_t \supset \mathscr{B}(w(s), s \leq t)$ and $\{w(u) - w(t), u \geq t\}$ is independent of $\mathscr{B}_t$, then $\{w(t), \mathscr{B}_t, \infty > t > 0\}$ is a martingale, and (2.4) is a generalization of (1.1). Indeed, if $\{Y(t), \mathscr{B}_t, \infty > t \geq 0\}$ is a continuous vector-valued martingale with *quadratic variation* $A(t) = tI$, where I is the identity matrix, then $Y(\cdot)$ is a Wiener process. This important property will be proved in Section 1.4.3, which also contains additional material on martingales.

1.3 Markov Processes

1.3.1 Definition. Homogeneous Markov Process

Suppose that $\{X(t), \infty > t \geq 0\}$ is a sequence of vector-valued random variables which is defined on some probability space. Roughly speaking, the process is a Markov process if, knowing the present state, then knowledge of how the path evolved up to its present value yields no additional information concerning the future evolution of the path. Intuitively, for any set Γ in the state space of the process we would have

$$P\{X(t+s) \in \Gamma \mid X(s), s \leq t\} = P\{X(t+s) \in \Gamma \mid X(t)\} \qquad \text{w.p. 1.}$$

The following more exact definition is specialized to our needs. A less restrictive definition and a rather thorough study of Markov processes appears in Dynkin [D4].

Let $(\Omega, \mathscr{B})$ denote a measure space and S a Borel set in some Euclidean space. The set S will be the *range* (or *state space*) of the process, and $\mathscr{B}(S)$ will denote the Borel field over S. Let $P_x\{\cdot\}$, $x \in S$, denote a family of probability measures on $(\Omega, \mathscr{B})$, let $\{\mathscr{B}_t, \infty > t \geq 0\}$ denote a family of nondecreasing sub-σ-algebras of $\mathscr{B}$, and let $X(\cdot)$ be an S-valued random process on $(\Omega, \mathscr{B}, P_x)$, for each $x \in S$. Let $P(\cdot, \cdot, \cdot)$ be a function from $S \times [0, \infty) \times \mathscr{B}(S)$ to $[0, 1]$, such that $P(\cdot, t, \Gamma)$ is $\mathscr{B}(S)$ measurable for each $t \geq 0$ and $\Gamma \in \mathscr{B}(S)$, $P(x, t, \cdot)$ is a probability measure on $\mathscr{B}(S)$ for each $t \geq 0$ and $x \in S$, and $P(x, 0, \Gamma) = I_\Gamma(x)$, where $I_\Gamma(\cdot)$ is the indicator function of the set Γ [$I_\Gamma(x) = 1$ if $x \in \Gamma$ and $= 0$ otherwise].

The collection $(\Omega, \mathscr{B}, \mathscr{B}_t, P_x, X(\cdot), x \in S)$ [which we often refer to merely as $X(\cdot)$] is a *homogeneous Markov process*, with *transition function* $P(\cdot, \cdot, \cdot)$, if $X(t)$ is $\mathscr{B}_t$ measurable for each $t \geq 0$ and (3.1) holds for each $x \in S$, $t \geq 0$ and $\Gamma \in \mathscr{B}(S)$.

$$P_x\{X(t+s) \in \Gamma \mid \mathscr{B}_t\} = P(X(t), s, \Gamma) \qquad \text{w.p. 1.} \tag{3.1}$$

Equation (3.1) is called the *Markov property*. It is a precise statement of the intuitive notion of the first paragraph, which said roughly that it is of no value to know the values of $X(u)$, $u < t$, to predict the value of $X(t+s)$, if the value of $X(t)$ is known. According to (3.1), the left-hand side of (3.1) equals (w.p. 1) $P_{X(t)}\{X(s) \in \Gamma\}$. If (3.1) holds, then $P(\cdot, \cdot, \cdot)$ must satisfy the *Chapman–Kolmogorov equation* (3.2)

$$P(x, t+s, \Gamma) = \int_S P(x, t, dy)P(y, s, \Gamma) \tag{3.2}$$

for all $x \in S$, $s, t \geq 0$, and $\Gamma \in \mathscr{B}(S)$. The quantity $P(x, t, \Gamma)$ is the probability that if $X(0) = x$, then $X(t) \in \Gamma$.

Let $\mathscr{B}_t(x) \equiv \mathscr{B}(X(s), s \leq t)$ denote the smallest σ-algebra over which $X(s)$, $s \leq t$, is measurable. By the definition of a Markov process, $\mathscr{B}_t(x) \subset \mathscr{B}_t$. In many cases $\mathscr{B}_t$ will actually be $\mathscr{B}_t(x)$. Certainly, (3.1) must hold if $\mathscr{B}_t(x)$ replaces $\mathscr{B}_t$ there. For theoretical purposes, it is often desirable to choose a $\mathscr{B}_t$ which strictly includes $\mathscr{B}_t(x)$. A common case is where $\mathscr{B}_t$ measures variables $X(s)$, $Y(s)$, $s \leq t$, where $Y(\cdot)$ is some process which is independent of $X(\cdot)$, under each measure P_x. The inclusion of the $Y(\cdot)$ enables us to treat randomizations of strategies, for example, in various decision problems.

Our definition of a Markov process is not quite complete, but is adequate for our needs. In order to facilitate the mathematical exploitation of the Markov property, there is an additional requirement: for each $s > 0$ and $\omega \in \Omega$, there is an $\omega' \in \Omega$ such that $X(\omega, t+s) = X(\omega', t)$ for all $t \geq 0$. If all the other conditions of the definition are met, it is often not hard to either augment the space Ω or to find a new probability space on which all the previous conditions hold, but which also contain points corresponding to all "left shifts" as required by the last condition.

1.3.2 Nonhomogeneous Markov Processes

Almost everything remains the same as in the previous section except that we now have a *transition function* $P(\cdot, \cdot; \cdot, \cdot)$ which depends on four quantities rather than on three: it has values $P(x, t; \Gamma, t+s)$ for $t \geq 0$, $s \geq 0$, $x \in S$, and $\Gamma \in \mathscr{B}(S)$. Also $P(x, t; \Gamma, t) = I_\Gamma(x)$ and the nonhomogeneous Markov property (3.3) holds:

$$P_x\{X(t+s) \in \Gamma \mid \mathscr{B}_t\} = P(X(t), t; \Gamma, t+s) \qquad \text{w.p. } 1. \tag{3.3}$$

The quantity $P(x, t; \Gamma, t+s)$ is the probability that if $X(t) = x$, then $X(t+s) \in \Gamma$. So, in the nonhomogeneous case, the transition probabilities are not stationary. The *Chapman–Kolmogorov equation*, equivalent to (3.3), now takes the form

$$P(x, t; \Gamma, t+s) = \int_S P(x, t; dy, t+u)P(y, t+u, \Gamma, t+s) \tag{3.4}$$

for each $t \geq 0$, $x \in S$, $\Gamma \in \mathscr{B}(s)$ and $s \geq u \geq 0$.

Let $\mathscr{B}_t = \mathscr{B}_t(x)$. Then a property that is equivalent to (3.3), but is perhaps more intuitive, is: for each integer n, $t \geq 0$, $s \geq 0$, $\Gamma \in \mathscr{B}(S)$ and real $t_1, \ldots, t_n \leq t$, and each $x \in S$,

$$P_x\{X(t+s) \in \Gamma \mid X(t_i), i \leq n, X(t)\} = P_x\{X(t+s) \in \Gamma \mid X(t)\}$$
$$= P(X(t), t, \Gamma, t+s) \quad \text{w.p. 1.} \quad (3.5)$$

1.3.3 Strong Markov Processes

There is a subclass of Markov processes with some additional properties that are of great importance in applications. Suppose, first, that A is a set in S so that the entrance time τ defined by

$$\tau = \inf\{t : X(t) \in A\}$$

is a stopping time. Let $\tau < \infty$ w.p. 1 with respect to some P_x. We would usually expect that a relation like (3.6) would hold in physical applications.

$$P_x\{X(\tau+s) \in \Gamma \mid \mathscr{B}_\tau\} = P_x\{X(\tau+s) \in \Gamma \mid X(\tau)\}$$
$$= P(X(\tau), \tau; \Gamma, \tau+s) \qquad \text{w.p. 1.} \qquad (3.6)$$

Equation (3.6) says that if we know the values of $X(t)$ at the moment of hitting A, then the prior history is of no additional value in predicting the evolution of the path beyond time τ. As far as the author is aware, if a physical process can be modeled by a Markov process, then that Markov process has property (3.6).

A Markov process is said to be a *strong Markov process* if (3.6) holds for all random variables τ which are stopping times relative to $\{\mathscr{B}_t\}$. A Markov process is said to be a *Feller* (resp., strong *Feller*) process if for each $t > 0$ and each bounded real-valued continuous (resp., measurable) function $g(\cdot)$, the function of x, with values $E_x g(X(t))$, is continuous in x on S. Every *Feller process* whose paths are right continuous w.p. 1 is a strong Markov process (Dynkin [D4, p. 99]). Verification of the Feller property is a convenient way of verifying whether a process is a strong Markov process.

1.4 Stochastic Integrals

1.4.1 Integrals with Respect to a Martingale

In order to discuss Markov processes of the diffusion type and the relationship between them and elliptic and parabolic equations, it is necessary to define the integral of a function with respect to a continuous martingale. Let $\{Y(t), \mathscr{B}_t, \infty > t \geq 0\}$ be a continuous square integrable scalar-valued martingale. We need to define the object $\int_0^t f(\omega, s)\, dY(s)$ for a class of functions $f(\cdot, \cdot)$.

If $\langle Y \rangle_t \not\equiv 0$, then $Y(\cdot)$ is not of bounded variation (with some positive probability) and the Lebesgue–Stieltjes definition of the integral does not make sense. Nevertheless, by first defining the integral with respect to a class of step functions $f(\cdot, \cdot)$, we can then extend the definition to a class of nonstep functions via a suitable probabilistic limiting procedure. We will only state some definitions and results. For a fuller discussion see Doob [D2], Gikhman and Skorokhod [G2], Kunita and Watanabe [K14] or Wong [W2].

Let $\mathcal{N}$ denote the class of *real-valued measurable random functions* $f(\cdot, \cdot)$ on $\Omega \times [0, \infty)$ which are adapted to $\{\mathscr{B}_t\}$. The following functions and classes of functions are all elements of or subclasses of $\mathcal{N}$. *We say that* $f(\cdot) \in L_2(\langle Y \rangle)$ *or* $f_n(\cdot) \to f(\cdot)$ *in* $L_2(\langle Y \rangle)$, *resp., if*

$$E \int_0^T |f(t)|^2\, d\langle Y \rangle_t < \infty, \qquad \text{each} \quad T < \infty$$

or

$$f_n(\cdot), f(\cdot) \in L_2(\langle Y \rangle)$$

and

$$E \int_0^T |f_n(t) - f(t)|^2\, d\langle Y \rangle_t \to 0 \qquad \text{as} \quad n \to \infty, \quad \text{each} \quad T < \infty,$$

resp. Let $\bar{\mathcal{N}}_{2s}$ denote the class of random step functions in $L_2\langle Y \rangle$. If $f(\cdot) \in L_2(\langle Y \rangle)$ and there is a sequence $f_n(\cdot) \in \bar{\mathcal{N}}_{2s}$ such that $f_n(\cdot) \to f(\cdot)$ in $L_2(\langle Y \rangle)$, then we say that $f(\cdot) \in \bar{\mathcal{N}}_2$. If $f(\cdot) \in \bar{\mathcal{N}}_{2s}$, there is a sequence of stopping times $0 = t_1, \ldots,$ such that (by assumption, $f(\cdot) \in \bar{\mathcal{N}}_{2s}$ is right continuous)

$$f(t) = f(t_i), \qquad t \in [t_i, t_{i+1}).$$

We say that $f(\cdot) \in \mathcal{N}_2$ if there is a sequence of stopping times $\tau_1, \ldots,$ tending to ∞ w.p. 1 such that, for each n, the function $f_n(\cdot)$ defined by $f_n(t) = f(t)$, $t < \tau_n$, $f_n(t) = 0$, $t \geq \tau_n$ is in $\bar{\mathcal{N}}_2$.

Definition of the Stochastic Integral of a Step Function $f(\cdot)$

If $f(\cdot) \in \bar{\mathcal{N}}_{2s}$, then we define the stochastic integral $\psi(\cdot)$ of $f(\cdot)$ with respect to $Y(\cdot)$ by the formula (4.1). Let $f(\cdot)$ be constant on the intervals $[t_i, t_{i+1})$, where the $\{t_i\}$ are an increasing sequence of stopping times with respect to $\{\mathscr{B}_t\}$:

$$\psi(t) = \int_0^t f(s)\, dY(s) = \sum_{i=1}^{n-1} f(t_i)[Y(t_{i+1}) - Y(t_i)] + f(t_n)[Y(t) - Y(t_n)] \qquad \text{for} \quad t \in [t_n, t_{n+1}). \tag{4.1}$$

The pair $\{\psi(t),\ \infty > t \geq 0\}$ is a continuous square integrable martingale and

$$\langle \psi \rangle_t = \int_0^t |f(s)|^2 \, d\langle Y \rangle_s . \tag{4.2}$$

Let $u \geq t \geq \tau$ be finite stopping times with respect to $\{\mathscr{B}_t\}$, and let $f(\cdot)$ and $g(\cdot)$ be in $\bar{\mathscr{N}}_{2s}$. Then we can also verify (4.3)–(4.7) by simple calculations.

$$E_{\mathscr{B}_\tau} \int_0^t f(s)\, dY(s) = \int_0^\tau f(s)\, dY(s) \qquad \text{w.p. 1}, \tag{4.3}$$

$$E \int_0^t f(s)\, dY(s) = 0, \tag{4.4}$$

$$E \int_0^t f(s)\, dY(s) \int_0^u g(s)\, dY(s) = E \int_0^t f(s)g(s)\, d\langle Y \rangle_s, \tag{4.5}$$

$$\int_0^t f(s)\, dY(s) + \int_0^t g(s)\, dY(s) = \int_0^t (f(s) + g(s))\, dY(s), \tag{4.6}$$

$$E\left[\int_0^t f(s)\, dY(s) - \int_0^t g(s)\, dY(s)\right]^2 = E \int_0^t [f(s) - g(s)]^2\, d\langle Y \rangle_s . \tag{4.7}$$

Definition of the Stochastic Integral of $f(\cdot) \in \bar{\mathscr{N}}_2$

For any $f(\cdot) \in \bar{\mathscr{N}}_2$, there is a sequence $f_n(\cdot) \in \bar{\mathscr{N}}_{2s}$ for which

$$E \int_0^T |f_n(s) - f(s)|^2\, d\langle Y \rangle_s \to 0 \qquad \text{for each} \quad T < \infty \tag{4.8}$$

as $n \to \infty$. If $\langle Y \rangle$ is absolutely continuous with respect to Lebesgue measure, and has a uniformly bounded derivative (as with the case of a Wiener process), then for any $f(\cdot)$ satisfying $E \int_0^T |f(s)|^2\, ds < \infty$, each $T < \infty$, a sequence $f_n(\cdot) \in \bar{\mathscr{N}}_{2s}$ which satisfies (4.8) can easily be constructed. Thus such $f(\cdot)$ are in $\bar{\mathscr{N}}_2$. See Doob [D2, Chapter IX.5]; or Wong [W2, pp. 142–144]. Let $\{f_n(\cdot)\}$ be a sequence in $\bar{\mathscr{N}}_{2s}$ which satisfies (4.8), and define

$$\psi_n(t) = \int_0^t f_n(s)\, dY(s).$$

By (4.7),

$$E|\psi_n(t) - \psi_m(t)|^2 = E \int_0^t |f_n(s) - f_m(s)|^2\, d\langle Y \rangle_s , \qquad \text{each} \quad t < \infty.$$

In fact, by (2.2) and (2.3),

$$E \sup_{T \geq t \geq 0} |\psi_n(t) - \psi_m(t)|^2 \leq 4E \int_0^T |f_n(s) - f_m(s)|^2 \, d\langle Y \rangle_s, \tag{4.9}$$

$$P\left\{ \sup_{T \geq t \geq 0} |\psi_n(t) - \psi_m(t)| \geq \varepsilon \right\} \leq \left(E \int_0^T |f_n(s) - f_m(s)|^2 \, d\langle Y \rangle_s \right) \Big/ \varepsilon^2. \tag{4.10}$$

Equations (4.8) and (4.9) imply that $\{\psi_n(t)\}$ is a Cauchy sequence for each t. Hence, for each t, there is a random variable $\psi(t)$ which is the limit of $\psi_n(t)$ in mean square, and $\psi(t)$ is adapted to $\mathscr{B}_t$. The limit $\psi(t)$, corresponding to two different $\{f_n(\cdot)\}$ sequences in $\bar{\mathscr{N}}_{2s}$, each satisfying (4.8), are the same w.p. 1. All the processes are assumed to be separable. Hence, if we can find one particular limit function $\psi(\cdot)$ that is continuous w.p. 1, then any separable version must be continuous w.p. 1.

Let T be an arbitrary positive number. Select $f_n(\cdot) \in \bar{\mathscr{N}}_{2s}$ such that

$$E \int_0^T |f(s) - f_n(s)|^2 \, d\langle Y \rangle_s \leq 2^{-n},$$

and set $\varepsilon = 2^{-n/4}$ and $m > n$ in (4.10). Then the right-hand side of (4.10) is bounded above by $4 \cdot 2^{-n/2}$. The summability of this sequence and the Borel–Cantelli lemma imply that the series

$$\psi(t) = \sum_{n=1}^{\infty} [\psi_{n+1}(t) - \psi_n(t)] + \psi_1(t)$$

converges uniformly on $[0, T]$. Since each $\psi_n(\cdot)$ is continuous, so is $\psi(\cdot)$. We write $\psi(t) = \int_0^t f(s) \, dY(s)$.

It can also be verified that (4.3)–(4.7) hold for any $f(\cdot), g(\cdot) \in \bar{\mathscr{N}}_2$, and $u \geq t \geq \tau$, and that $\psi(\cdot)$ is a continuous square integrable martingale on $[0, \infty)$ with respect to $\{\mathscr{B}_t\}$.

Let $\{Y(t), \mathscr{B}_t, \infty > t \geq 0\}$ be an R^r-valued continuous square integrable martingale, $f(\cdot)$ a $q \times r$ matrix-valued function which is adapted to $\{\mathscr{B}_t\}$. Suppose that (4.11) holds for each $T < \infty$.

$$E \int_0^T \sum_{i,j} f_{ij}^2(s) \, d\langle Y_j \rangle_s < \infty. \tag{4.11}$$

Then we can define the stochastic integral $\psi(t) = \int_0^t f(s) \, dY(s)$ term by term. The pair $\{\psi(t), \mathscr{B}_t, \infty > t \geq 0\}$ is a continuous square integrable R^q-valued martingale, the ith component of $\psi(\cdot)$ is

$$\psi_i(t) = \int_0^t \sum_k f_{ik}(s) \, dY_k(s)$$

and the function $\langle\psi\rangle = \{\langle\psi_i, \psi_j\rangle, i, j = 1, \ldots, q\}$ is given by

$$\langle\psi_i, \psi_j\rangle_t = \int_0^t \sum_{k,l} f_{ik}(s) f_{jl}(s)\, d\langle Y_k, Y_l\rangle_s \tag{4.12}$$

or, more compactly, by

$$\langle\psi\rangle_t = \int_0^t [f(s) \cdot d\langle Y\rangle_s f'(s)]. \tag{4.13}$$

Let $\{Y(t), \mathscr{B}_t, \infty > t \geq 0\}$ be a scalar-valued martingale and let $f(\cdot) \in \bar{\mathscr{N}}_2$ (with respect to $\langle Y\rangle$). Define $\psi(t) = \int_0^t f(s)\, dY(s)$. Let $g(\cdot) \in \bar{\mathscr{N}}_2$ (with respect to $\langle\psi\rangle$). Then

$$E \int_0^T f^2(s) g^2(s)\, d\langle Y\rangle_s < \infty$$

for each $T < \infty$. We can now define the stochastic integral of $g(\cdot)$ with respect to $\psi(\cdot)$ by

$$\phi(t) = \int_0^t g(s)\, d\psi(s).$$

Furthermore,

$$\langle\phi\rangle_t = \int_0^t g^2(s)\, d\langle\psi\rangle_s = \int_0^t g^2(s) f^2(s)\, d\langle Y\rangle_s,$$

$$\phi(t) = \int_0^t g(s) f(s)\, dY(s) \qquad \text{w.p. } 1.$$

1.4.2 Local Martingales

There is a useful definition of the stochastic integral of functions $f(\cdot) \in \mathscr{N}_2$. Let $\{f_n(\cdot)\}$ and $\{\tau_n\}$ denote the functions and stopping times used in the definition of $\mathscr{N}_2$.

The stochastic integral

$$\psi_n(t) = \int_0^t f_n(s)\, dY(s)$$

is defined as in Section 1.4.1. We have that $P\{\tau_n \leq T\} \to 0$ as $n \to \infty$, for each $T \leq \infty$. Using this fact, we can show that there is a continuous function $\psi(\cdot)$ such that $\psi_n(t) \to \psi(t)$ in probability for each $t > 0$. By selecting a suitable subsequence, the convergence is uniform on each finite interval w.p. 1. We define the integral by

$$\psi(t) = \int_0^t f(s)\, dY(s).$$

The vector case is treated analogously. The integral $\psi(\cdot)$ is well defined since if $\{f_n^i(\cdot)\}$, $i = 1, 2$ are two approximating sequences, then the limits of the stochastic integrals $\psi_n^i(t)$ agree w.p. 1.

An adapted pair $\{X(t), \mathscr{B}_t, \infty > t \geq 0\}$ is said to be a *local martingale* if there is a sequence of increasing stopping times $\{\tau_n\}$ so that, for each t, $X(t) = \lim_{n\to\infty} X(t \cap \tau_n)$ and, for each n, the process $X_n(\cdot)$ with values $X_n(t) = X(t \cap \tau_n)$ is a martingale. If each $X_n(\cdot)$ is square integrable, then $X(\cdot)$ is said to be a *locally square integrable martingale*. Let $X_n(\cdot)$ be square integrable. Then, the increasing function $\langle X_n \rangle$ is well defined. Also, $\langle X_{n+1}\rangle_t \geq \langle X_n \rangle_t$ for each t, and we define $\langle X \rangle$ by $\langle X \rangle_t = \lim_n \langle X_n \rangle_t$. The limit is (w.p. 1) independent of the sequence $\{\tau_n\}$. If $f(\cdot) \in \mathscr{N}_2$, then its stochastic integral is a local martingale.

1.4.3 Itô's Lemma

In order to deal effectively with functions of stochastic integrals, we need a calculus for differentiating and integrating functions of the integrals. This need is answered by Itô's lemma (see Gikhman and Skorokhod [G2, Chapter 8.2], for the case where $Y(\cdot)$ is a Wiener process, and Kunita and Watanabe [K14, Theorem 2.2], otherwise).

Let r denote a given integer, T a positive number and let $g(\cdot, \cdot)$ be a real-valued function in† $C^{1,2}([0, T] \times R^r)$. Let $\{Y(t), \mathscr{B}_t, \infty > t \geq 0\}$ denote a continuous R^r-valued locally square integrable martingale, and let $f(\cdot)$ be an R^r-valued random function which is adapted to $\{\mathscr{B}_t\}$, and satisfies $\int_0^T |f(s)|\, ds < \infty$ w.p. 1. Define the process $X(\cdot)$ by

$$X(t) = \int_0^t f(s)\, ds + Y(t)$$

and suppose that $\langle Y \rangle$ is absolutely continuous with respect to Lebesgue measure. Define the operator D:

$$Dg(t, x) = \frac{\partial g(t, x)}{\partial t} + \sum_i f_i(t) \frac{\partial g(t, x)}{\partial x_i} + \frac{1}{2} \sum_{i,j} \frac{\partial^2 g(t, x)}{\partial x_i\, \partial x_j} \cdot \frac{d\langle Y_i, Y_j \rangle_t}{dt}. \tag{4.14}$$

Then Itô's Lemma states that

$$dg(t, X(t)) = Dg(t, X(t))\, dt + \sum_i \frac{\partial g(t, X(t))}{\partial x_i} \cdot dY_i(t). \tag{4.15}$$

† $C^{1,2}([0, T] \times R^r)$ denotes the space of real-valued continuous functions on $[0, T] \times R^r$, whose first t, x derivatives and second partial x derivatives are continuous.

The differential (4.15) is interpreted to mean

$$g(t, X(t)) = g(0, X(0)) + \int_0^t dg(s, X(s))$$

$$= g(0, X(0)) + \int_0^t Dg(s, X((s)) \, ds$$

$$+ \sum_i \int_0^t \frac{\partial g(s, X(s))}{\partial x_i} \cdot dY_i(s) \tag{4.16}$$

w.p. 1, where the last integral on the right-hand side is a stochastic integral (and also a continuous locally square integrable martingale). If, for each i, the function with values $[\partial g(s, X(s))/\partial x_i]$ is in $\bar{\mathcal{N}}_2$ (with respect to $\langle Y_i \rangle$), then the stochastic integral in (4.16) is a square integrable martingale with zero expectation.

If $\langle Y \rangle$ is not absolutely continuous with respect to Lebesgue measure, then $d\langle Y_i, Y_j \rangle_t$ replaces $(d\langle Y_i, Y_j \rangle_t / dt)\, dt$ in (4.15) and (4.16).

Suppose that $\{Y(t), \mathscr{B}_t, \infty > t \geq 0\}$ is a continuous vector-valued square integrable martingale with $\langle Y \rangle_t = tI$ and $Y(0) = EY(t) \equiv 0$. Then $Y(\cdot)$ is a standard Wiener process. We will only show that $Y(t)$ is normally distributed. Applying Itô's lemma to the function $g(x) = \exp i\lambda' x$ yields

$$\exp i\lambda' Y(t) = 1 + \int_0^t i\lambda' [\exp i\lambda' Y(s)] \, dY(s) - \tfrac{1}{2} \int_0^t |\lambda|^2 [\exp i\lambda' Y(s)] \, ds.$$

Denoting the characteristic function of $Y(t)$ by $Z_\lambda(t) \equiv E \exp i\lambda Y(t)$ and taking expectations yields

$$Z_\lambda(t) = 1 - \int_0^t (|\lambda|^2/2) Z_\lambda(s) \, ds,$$

which has the unique solution

$$Z_\lambda(t) = \exp - |\lambda|^2 t/2,$$

which is the characteristic function of a vector of independent normally distributed random variables with zero mean and covariance tI.

1.4.4 Representation of a Continuous Square Integrable Martingale $Y(\cdot)$ as a Stochastic Integral with Respect to a Wiener Process

Let $\{Y(t), \mathscr{B}_t, \infty > t \geq 0\}$ be a continuous vector-valued square integrable martingale on the probability space $(\Omega, \mathscr{B}, P)$. In all of the applications in this book, the increasing process $A(\cdot) = \langle Y \rangle$ will be absolutely continuous with respect to Lebesgue measure; i.e., there is some nonnegative matrix-valued Lebesgue measurable (w.p. 1) function $\Sigma(\cdot)$, adapted to $\{\mathscr{B}_t\}$ and such that $\langle Y\rangle_t = \int_0^t \Sigma(s)\,ds$. In such cases, $Y(\cdot)$ can be written as a stochastic integral with respect to Wiener process $w(\cdot)$, although we may have to enlarge the probability space to do so. Let us enlarge the probability space by adding a process $\Phi(\cdot)$, a R^r-valued standard Wiener process which is independent of $Y(\cdot)$. Suppose that $\Phi(\cdot)$ is defined on the Wiener process space $(\tilde{\Omega}, \tilde{\mathscr{B}}, \tilde{P})$, and let $\tilde{\mathscr{B}}_t$ denote the smallest σ-algebra which measures $\Phi(s)$, $s \le t$. Define the space $(\hat{\Omega}, \hat{\mathscr{B}}, \hat{P}) = (\Omega \times \tilde{\Omega}, \mathscr{B} \times \tilde{\mathscr{B}}, P \times \tilde{P})$, and let $\hat{\mathscr{B}}_t$ denote the smallest sub-σ-algebra of $\mathscr{B} \times \tilde{\mathscr{B}}$ which contains $\mathscr{B}_t \times \tilde{\Omega}$, and $\Omega \times \tilde{\mathscr{B}}_t$. Then $\{Y(t), \hat{\mathscr{B}}_t, \infty > t \ge 0\}$ is still a continuous square integrable martingale.

There are measurable $(r \times r)$ matrix-valued functions $P(\cdot, \cdot)$ and $D(\cdot, \cdot)$ which are adapted to $\{\hat{\mathscr{B}}_t\}$ and such that for each ω, t, $D(\omega, t)$ is diagonal with nonnegative elements $P(\omega, t)$ is orthonormal, and

$$\sum(\omega, t) = P'(\omega, t)D^2(\omega, t)P(\omega, t).$$

Denote the diagonal elements of $D(\omega, t)$ by $d_i(\omega, t)$. Define the $(r \times r)$ diagonal matrix $D^+(\omega, t)$ with elements $d_i^+(\omega, t)$ by

$$d_i^+(\omega, t) = \begin{cases} d_i^{-1}(\omega, t), & \text{if} \quad d_i(\omega, t) > 0, \\ 0, & \text{otherwise.} \end{cases}$$

Define

$$\sigma(\omega, t) = P'(\omega, t)D(\omega, t)$$

and define a process $w(\cdot)$ by

$$w(t) = \int_0^t D^+(\omega, s)P(\omega, s)\,dY(s) + \int_0^t [I - D^+(\omega, s)D(\omega, s)]\,d\Phi(s). \tag{4.17}$$

The stochastic integrals in (4.17) are well defined and they are orthogonal since $\langle Y, \Phi\rangle = 0$ (by the mutual independence of $Y(\cdot)$, $\Phi(\cdot)$). A direct calculation using $\langle Y, \Phi\rangle_{t'} = 0$ and (4.12) or (4.13) yields that

$$\langle w\rangle_t = tI,$$

which implies that $w(\cdot)$ is a Wiener process. Using the definitions of D, D^+, P, and of the differential of $w(\cdot)$ as given by (4.17), we get that (w.p. 1)

$$Y(t) = \int_0^t P'(\omega, s)D(\omega, s)\,dw(s) = \int_0^t \sigma(\omega, s)\,dw(s), \tag{4.18}$$

the desired representation.

If $\Sigma(\omega, t)$ is positive definite for almost all ω, t, then we can "recover" a Wiener process by "inverting" $Y(\cdot)$ via the definition $dw(t) = \Sigma^{-1/2}(\omega, t)\,dY(t)$. Otherwise, $\Sigma(\cdot, \cdot)$ is degenerate on a nonnull ω, t set, and even if $Y(\cdot)$ were defined by (4.18), we could not recover $w(\cdot)$ from $Y(\cdot)$ since, due to the degeneracy, the integration in (4.18) "loses" information on $w(\cdot)$.

1.5 Stochastic Differential Equations

Let $w(\cdot)$ denote a standard R^r-valued Wiener process and $\mathscr{B}_t = \mathscr{B}(w(s), s \le t)$. A large class of Markov processes are defined by the nonanticipative (with respect to $w(\cdot)$) solution of the equation

$$X(t) = x + \int_0^t f(X(s), s)\, ds + \int_0^t \sigma(X(s), s)\, dw(s). \tag{5.1}$$

Equation (5.1) is often written in the symbolic differential form

$$dx = f(x, s)\, ds + \sigma(x, s)\, dw(s). \tag{5.2}$$

Such processes are widely used in current stochastic control theory and in other applications in engineering, physics and economics, and encompass a wide variety of processes of practical usefulness. Functionals associated with these processes are intimately related to solutions of second-order partial differential elliptic and parabolic equations, and the study of the properties of these processes yields much valuable insight into the properties of those partial differential operators. In fact, the relationships between the process (5.1) and the partial differential equations will be heavily exploited in our study of approximations to the processes and to the partial differential equations.

We will give the classical Itô–Doob [D2] proof of existence and uniqueness of the solution to (5.1) under assumption A1.5.1.

A1.5.1 *$f(\cdot, \cdot)$ and $\sigma(\cdot, \cdot)$ are R^r-valued and $(r \times r)$ matrix-valued measurable functions on $R^r \times [0, \infty)$ and there is a real K such that*

$$|f(x, s) - f(y, s)| + |\sigma(x, s) - \sigma(y, s)| \le K|x - y|$$

$$|f(x, s)|^2 + |\sigma(x, s)|^2 \le K^2(1 + |x|^2)',$$

for each $s < \infty$, and $y, x \in R^r$.

Theorem 1.5.1 (see Doob [D2, Chapter VI]; Gikhman and Skorokhod [G2, Chapter 8]) *Under* A1.5.1 *for each $x \in R^r$, there is one and only one process $X(\cdot)$ which is adapted to $\{\mathscr{B}_t\}$ and satisfies* (5.1) *and (for each $T < \infty$)*

$$E \max_{T \ge s \ge 0} |X(s)|^2 < \infty. \tag{5.3}$$

The process is continuous w.p. 1.

PROOF The proof depends heavily on the use of the Lipschitz and growth conditions A1.5.1, Schwarz's inequality, and the martingale inequalities (2.2), (2.3), (4.9), and (4.10). The type of calculations that are used in the proof are used frequently in the analysis of stochastic differential equations and in the rest of this book. Note that if $X(\cdot)$ is nonanticipative (with respect to $w(\cdot)$), with $\int_0^T |X(s)|^2 \, ds < \infty$ w.p. 1 for each $T < \infty$, then $X(\cdot)$ must be continuous w.p. 1, since the stochastic integral can be constructed as a local martingale as in Section 1.4.2.

Let $Y(\cdot)$ denote an R^r-valued process which satisfies

(a) $Y(\cdot)$ is continuous w.p. 1.
(b) $Y(\cdot)$ is adapted to $\{\mathscr{B}_t\}$
(c) $E \max_{T \geq t \geq 0} |Y(s)|^2 < \infty$ for each $T < \infty$.

Define $\tilde{Y}(\cdot)$ by

$$\tilde{Y}(t) = x + \int_0^t f(Y(s), s) \, ds + \int_0^t \sigma(Y(s), s) \, dw(s). \tag{5.4}$$

The integrals are well defined and $\tilde{Y}(\cdot)$ satisfies (a) and (b) also. The inequalities (5.5), (5.6) imply that (c) holds for $\tilde{Y}(\cdot)$. The derivation of inequality (5.5) requires the use of the growth condition and Schwarz's inequality. The derivation of (5.6) uses the growth condition, the fact that the stochastic integral is a martingale and (2.3).

$$E \max_{T \geq t \geq 0} \left| \int_0^t f(Y(s), s) \, ds \right|^2 \leq E \max_{T \geq t \geq 0} t \int_0^t |f(Y(s), s)|^2 \, ds$$

$$\leq ET \int_0^T |f(Y(s), s)|^2 \leq ETK^2 \int_0^T (1 + |Y(s)|^2) \, ds < \infty. \tag{5.5}$$

$$E \max_{T \geq t \geq 0} \left| \int_0^t \sigma(Y(s), s) \, dw(s) \right|^2 \leq 4E \int_0^t |\sigma(Y(s), s)|^2 \, ds$$

$$\leq 4K^2 \int_0^T E(1 + |Y(s)|^2) \, ds. \tag{5.6}$$

Now, select a process $X_0(\cdot)$ which satisfies (a)–(c), and define $X_{n+1}(\cdot)$ $(n = 0, 1, \ldots)$ iteratively by setting (recursively) $X_n(\cdot) = Y(\cdot)$ and defining $X_{n+1}(\cdot) = \tilde{Y}(\cdot)$ in (5.4). Then all $X_n(\cdot)$ satisfy (a)–(c).

We now define

$$M_n(t) = E \max_{t \geq u \geq 0} |X_n(u)|^2$$

and sharpen the estimate (c), for $\{X_n(\cdot)\}$. We can write (again using the growth condition, Schwarz's inequality, and the martingale estimate)

$$
\begin{aligned}
M_{n+1}(t) &\le 4|x|^2 + 4E \max_{t\ge u\ge 0} \left| \int_0^u f(X_n(s), s)\, ds \right|^2 \\
&\quad + 4E \max_{t\ge u\ge 0} \left| \int_0^u \sigma(X_n(s), s)\, dw(s) \right|^2 \\
&\le 4|x|^2 + 4tK^2 \int_0^t E(1 + |X_n(s)|^2)\, ds \\
&\quad + 4K^2 \cdot 4E \int_0^t (1 + |X_n(s)|^2)\, ds,
\end{aligned}
\tag{5.7}
$$

which yields, for some real number K_0,

$$
M_{n+1}(t) \le K_0 + K_0 \int_0^t M_n(s)\, ds, \qquad t \le T. \tag{5.8}
$$

Iterating (5.8) yields that

$$
\varlimsup_{n\to\infty} E \max_{T\ge t\ge 0} |X_n(t)|^2 \le K_0 e^{K_0 T}. \tag{5.9}
$$

Define $\Delta_{n+1}(\cdot)$ by $\Delta_{n+1}(t) = X_{n+1}(t) - X_n(t)$. We will show that the sequence

$$
X(t) = \sum_{n=1}^{\infty} \Delta_n(t) + X_0(t) \tag{5.10}
$$

converges uniformly w.p. 1 to a process satisfying (a)–(c) and (5.1). Clearly, if the limit exists, it must satisfy (b) and, by Fatou's lemma and (5.9), (c). Also, any limit $X(\cdot)$ which satisfies (5.1) and has a bounded mean square value on each $[0, T]$, must be continuous.

First, we prove that the limit exists on an interval $[0, T]$. We have

$$
\begin{aligned}
\Delta_{n+1}(t) &= \int_0^t [f(X_{n+1}(s), s) - f(X_n(s), s)]\, ds \\
&\quad + \int_0^t [\sigma(X_{n+1}(s), s) - \sigma(X_n(s), s)]\, dw(s),
\end{aligned}
$$

and (again using the Lipschitz condition, Schwarz's inequality and the martingale estimate) there is a real number K_1 such that

$$
\begin{aligned}
R_{n+1}(t) \equiv E \max_{t\ge s\ge 0} |\Delta_{n+1}(s)|^2 &\le 2K^2 T \int_0^t E|\Delta_n(s)|^2\, ds \\
&\quad + 8K^2 \int_0^t E|\Delta_n(s)|^2\, ds \le K_1 \int_0^t R_n(s)\, ds
\end{aligned}
$$

for all $t < \infty$. Hence,

$$R_{n+1}(t) \leq K_2(K_1 t)^n/n!, \tag{5.11}$$

where

$$K_2 = K_1 \cdot E \max_{T \geq t \geq 0} |X_1(t) - X_0(t)|^2 \cdot T.$$

By Chebychev's inequality and (5.11),

$$P\left\{\sup_{T \geq t \geq 0} |\Delta_{n+1}(t)| \geq 2^{-n/2}\right\} \leq K_2 2^n (K_1 T)^n/n!. \tag{5.12}$$

By the summability of the right-hand side of (5.12) over n and the Borel–Cantelli lemma we can have

$$\sup_{T \geq t \geq 0} |\Delta_{n+1}(t)| \geq 2^{-n/2}$$

only finitely often w.p. 1. Hence (5.10) converges uniformly to a continuous function $X(\cdot)$ w.p. 1.

The convergence is also in mean square. To see this, we use the inequality [which is a consequence of $(a + b)^2 \leq 2a^2 + 2b^2$]

$$[\Delta_n + \Delta_{n+1} + \cdots + \Delta_{n+k-1}]^2 \leq 2\Delta_n^2 + 4\Delta_{n+1}^2 + \cdots + 2^k \Delta_{n+k-1}^2.$$

From this inequality and (5.11) we can readily compute that

$$E \sup_{T \geq t \geq 0} |X_{n+k}(t) - X_n(t)|^2 \to 0$$

as $n, k \to \infty$, which in turn implies the mean square convergence.

By the w.p. 1 convergence,

$$\int_0^t f(X_n(s), s)\, ds \to \int_0^t f(X(s), s)\, ds \tag{5.13a}$$

w.p. 1, as $n \to \infty$. By the mean square convergence,

$$E \max_{T \geq t \geq 0} \left| \int_0^t [\sigma(X_n(s), s) - \sigma(X(s), s)]\, dw(s) \right|^2$$

$$\leq 4K^2 E \int_0^T |X_n(s) - X(s)|^2\, ds \to 0$$

as $n \to \infty$. Hence

$$\int_0^t \sigma(X_n(s), s)\, dw(s) \to \int_0^t \sigma(X(s), s)\, dw(s) \quad \text{as} \quad n \to \infty, \tag{5.13b}$$

in mean square. Equations (5.13a) and (5.13b) and the fact that $X_n(t) \to X(t)$ w.p. 1 uniformly on finite intervals imply that $X(\cdot)$ satisfies (5.1).

Only the uniqueness remains to be shown. Let $X^i(\cdot)$, $i = 1, 2$, be two nonanticipative solutions to (5.1), each satisfying (5.3). Then

$$X^i(t) = x + \int_0^t f(X^i(s), s)\, ds + \int_0^t \sigma(X^i(s), s)\, dw(s), \qquad i = 1, 2.$$

Define $\delta(t) = |X^1(t) - X^2(t)|^2$. Then

$$\delta(t) \le 2\left|\int_0^t [f(X^1(s), s) - f(X^2(s), s)]\, ds\right|^2 + 2\left|\int_0^t [\sigma(X^1(s), s) - \sigma(X^2(s), s)]\, dw(s)\right|^2.$$

The use of the triple of the Lipschitz condition, Schwarz's inequality, and the martingale estimate for the stochastic integral on the last inequality yields

$$E\delta(t) \le 2tK^2 \int_0^t E\delta(s)\, ds + 2K^2 \int_0^t E\delta(s)\, ds,$$

which implies that $E\delta(t) = 0$ since $E\delta(t)$ is finite for all t. Q.E.D.

Properties of the Solution to (5.1) *under* A1.5.1

The process $X(\cdot)$ is a Feller process and a strong Markov process. The process is of the form of $X(\cdot)$ above (4.14) and the operator D of (4.14) takes the form (5.14). The operator is denoted by the symbol $\mathscr{L}_t$ in the particular case of a stochastic differential equation:

$$\mathscr{L}_t = \frac{\partial}{\partial t} + \mathscr{L} \equiv \frac{\partial}{\partial t} + \sum_i f_i(x, s)\frac{\partial}{\partial x_i} + \sum_{i,j} a_{ij}(x, t)\frac{\partial^2}{\partial x_i\, \partial x_j}, \tag{5.14}$$

where

$$a(\cdot, \cdot) = \sigma(\cdot, \cdot)\sigma'(\cdot, \cdot)/2.$$

For $g(\cdot, \cdot) \in C^{1,2}([0, T] \times R^r)$,

$$\mathscr{L}_t g(0, x) = \lim_{\Delta\to 0} [E_x g(\Delta, X_\Delta) - g(0, x)]/\Delta,$$

where E_x denotes the expectation, given that $X(0) = x$.

Proofs of the local properties (5.15), (5.16) are found in Doob [D2, Chapter VI.3]. For each finite interval $[0, T]$, there is a function $\theta(\cdot)$ such that $\theta(h) \to 0$ as $h \to 0$ and

$$\left| E[X(t + h) - X(t) | X(t)] - \int_t^{t+h} f(X(t), s)\, ds \right| \le (1 + |X(t)|)h\theta(h), \tag{5.15}$$

$$\left| E[(X(t+h) - X(t))(X(t+h) - X(t))' \,|\, X(t)] - 2\int_t^{t+h} a(X(t), s)\, ds \right|$$
$$\leq (1 + |X(t)|^2) h\theta(h), \tag{5.16}$$

where the $O(\cdot)$ functions are uniform in t and $X(t)$.

REMARKS Assumption A1.5.1 makes the study of the properties of the solution to (5.1) relatively simple. We will give two examples of the simple analysis that it allows.

Example 1 *Sensitivity to perturbations* Let $\varepsilon_n(\cdot)$ denote a nonanticipative process such that

$$E\int_0^T |\varepsilon_n(t)|^2\, dt \to 0, \quad \text{each} \quad T < \infty, \quad \text{as} \quad n \to \infty.$$

For each n, we can construct a solution to

$$X_n(t) = x + \int_0^t f(X_n(s), s)\, ds + \int_0^t \sigma(X_n(s), s)\, dw(s) + \int_0^t \varepsilon_n(s)\, ds \tag{5.17}$$

in the same manner as we constructed one to (5.1). In addition, the solution is unique. Define $\delta(t) = |X_n(t) - X(t)|^2$. Then

$$\delta(t) \leq 4\left|\int_0^t [f(X_n(s), s) - f(X(s), s)]\, ds\right|$$
$$+ 4\left|\int_0^t [\sigma(X_n(s), s) - \sigma(X(s), s)]\, dw(s)\right|^2 + 4\left|\int_0^t \varepsilon_n(s)\, ds\right|^2,$$

and there is a real K_1 such that

$$E\delta(t) \leq (1+t)K_1 \int_0^t E\delta(s)\, ds + 4t\int_0^t |\varepsilon_n(s)|^2\, ds. \tag{5.18}$$

Equation (5.18) implies that there is a real K_2 such that

$$E\delta(t) \leq K_2 e^{K_2 T}\int_0^t |\varepsilon_n(s)|^2\, ds.$$

Hence $X_n(t) \to X(t)$ w.p. 1 as $n \to \infty$.

Example 2 *Finite difference approximations* For each $\Delta > 0$, define the sequence $\{X_n^\Delta\}$ by $X_0^\Delta = x$ and, for $n \geq 0$,

$$X_{n+1}^\Delta = X_n^\Delta + f(X_n^\Delta, n\Delta)\Delta + \sigma(X_n^\Delta, n\Delta)[w(n\Delta + \Delta) - w(n\Delta)],$$

and suppose A1.5.1, and that $f(\cdot,\cdot)$, $\sigma(\cdot,\cdot)$ are continuous in both x and t.

Define the process $X^{\Delta}(\cdot)$ by

$$X^{\Delta}(t) = X_n^{\Delta} \quad \text{on} \quad [n\Delta, n\Delta + \Delta), \quad n \geq 0.$$

We will not go through the detailed calculations, but simply by use of some of the devices; namely, the Lipschitz condition, Schwarz's inequality and the martingale estimate, it can be proved that $X^{\Delta}(\cdot)$ tends to the solution of (5.1) in mean square, as $\Delta \to 0$. If there is a real $\beta \in (0, 1)$ such that the successive values of Δ are related by $\Delta(\text{new})/\Delta(\text{last}) \leq \beta$, then the convergence is uniform on any finite time interval, w.p. 1. See Kushner [K5, Chapter 10].

In much of Chapters 6–10, problems of perturbations (Example 1) or discrete time (Example 2) or other approximations will arise, and we will not usually assume that there is a Lipschitz condition. We will always assume that a type of uniqueness holds. The uniqueness and various techniques involving weak convergence theory (Chapter 2) will have to be used rather heavily, owing to the lack of a Lipschitz condition.

Solutions to (5.1) exist under very weak assumptions, and the powerful tools of weak convergence theory can be used to construct those solutions. The main problem is uniqueness. If the K in A1.5.1 depends on x, y but is bounded on finite sets, then there will be a unique solution up to a (random) moment of escape to infinity. Often, concepts in stability theory can be used to show that there is no (w.p. 1) finite escape time. In Theorem 1.5.1, $f(\cdot, \cdot)$ and $\sigma(\cdot, \cdot)$ can depend nonanticipatively on ω, provided that A1.5.1 still holds. Uniqueness also holds under a certain Hölder condition (Yamada and Watanabe [Y1]).

If $f(\cdot, \cdot)$ and $a(\cdot, \cdot)$ satisfy a uniform Hölder condition and are bounded and $\sigma(\cdot, \cdot)$ can depend nonanticipatively on ω, provided that A1.5.1 still to (5.1) and the family of solutions over $x = X(0)$ is a strong Markov, strong Feller process (Dynkin [D4, Chapter 11]). According to Strook and Varadhan [S3], if $a(\cdot, \cdot)$ is strictly positive definite and continuous then there is one and only one solution for each bounded and measurable $f(\cdot, \cdot)$, and the family $X(\cdot)$, over all $x = X(0)$, is also a strong Feller strong Markov process. It should be pointed out, however, that the references of this paragraph do not show that for each $w(\cdot)$ process there is a solution for each initial condition $X(0) = x$. In fact, the $w(\cdot)$ process is constructed rather indirectly from the putative solution $X(\cdot)$ corresponding to a specific $X(0) = x$ and the uniqueness is in the sense of the probability law.

CHAPTER 2

Weak Convergence of Probability Measures

This chapter introduces the basic ideas of the theory of weak convergence of a sequence of probability measures on complete and separable metric spaces, or equivalently, of convergence in distribution of a sequence of random values with values in such a space. The generality is not only of theoretical interest, for it brings powerful and useful tools to bear on problems in the convergence of a sequence of real-valued stochastic processes. In the cases in this book, this sequence will be an "approximating" sequence to a random process which we wish to study—or to estimate functions of. The development starts with a discussion of convergence in distribution for the simple case of a sequence of real-valued random variables, and then moves on to the more complex cases. No proofs are given, but the results that are needed in the sequel are stated and discussed.

2.1 Probability Measures on the Real Line. Real-Valued Random Variables

The ideas of weak convergence of probability measures find useful applications in many areas of probability and statistics, as well as in many areas of application of those subjects, particularly in operations research and control theory. They frequently provide fundamental tools in the study of approximations and they will be used in just this way in this book. Most of the book is devoted to approximations to diffusion processes, to elliptic and parabolic equations, and to optimal stochastic control problems. The approximations will involve simpler processes—particularly, interpolated Markov chains†—and the concepts of weak convergence theory will be used to prove that the probability measures of the approximating sequences converge to the probability measure of the desired limit process, and then to prove that various functionals of the simpler processes converge to the appropriate functionals of the limit. Part of this chapter consists of a survey of the ideas as presented in the comprehensive book of Billingsley [B3]. The book of Gikhman and Skorokhod [G2, Chapter 9] also contains much useful information. See also Iglehart [I1] for a discussion of other applications.

Suppose that $P_1, P_2, \ldots$ and P are probability measures on the real line R. The sequence $\{P_n\}$ is said to *converge weakly* to P (written $P_n \Rightarrow P$) if $P_n(-\infty, x] \to P(-\infty, x]$ at all points of continuity x of the distribution function $P(-\infty, \cdot]$. If X_n and X are random variables with distributions P_n and P, resp., then weak convergence is equivalent to convergence of $\{X_n\}$ to X in distribution.

Let $C(R)$ denote the class of bounded and continuous real-valued functions on R. *Then an equivalent definition of weak convergence $P_n \Rightarrow P$ is*

$$\int f(y)P_n\,(dy) \to \int f(y)P\,(dy) \tag{1.1}$$

for all $f(\cdot) \in C(R)$. We will take (1.1) as the definition, since it generalizes to abstract-valued random variables. Let $g(\cdot)$ be a bounded measurable real-valued function on R with discontinuity set D_g, and let $C_P(R)$ denote the set

† Let $\{X_n\}$ and $\{\Delta_n\}$ denote a discrete parameter process and a sequence of strictly positive random variables, resp. Define $T_0 = 0$, $T_n = \sum_{i=0}^{n-1} \Delta_i$. Then the process $X(\cdot)$ defined by $X(t) = X_n$ on $[T_n, T_{n+1})$ is said to be a (piecewise constant) interpolation of $\{X_n\}$. The process $X'(\cdot)$ defined by

$$X'(t) = X_n\left(\frac{T_{n+1} - t}{\Delta_n}\right) + X_{n+1}\left(\frac{t - T_n}{\Delta_n}\right)$$

is the linear interpolation.

of such $g(\cdot)$ that also satisfy $P\{D_g\} = 0$. Then, if (1.1) holds for all $f(\cdot)$ in $C(R)$, it holds for $f(\cdot)$ in $C_P(R)$ also. Weak convergence $P_n \Rightarrow P$ is equivalent to $P_n\{A\} \to P\{A\}$ for each Borel set A with† $P\{\partial A\} = 0$ and to $\overline{\lim}\, P_n\{A\} \leq P\{A\}$ for each closed A or to $\underline{\lim}\, P_n\{F\} \geq P\{F\}$ for each open set F (Gikhman and Skorokhod [G2, p. 447]; Billingsley [B3, Theorems 2.1 and 5.1]).

A sequence of measures $\{P_n\}$ is said to be tight if, for each $\varepsilon > 0$, there is a real number N_ε such that

$$P_n\{|X| \leq N_\varepsilon\} \geq 1 - \varepsilon, \qquad \text{all} \quad n. \tag{1.2}$$

If $\{P_n\}$ is tight, then for every subsequence of $\{P_n\}$, there is a further subsequence (say, denoted by $\{P'_n\}$) and a probability measure P so that $P'_n \Rightarrow P$. In the real variable case, this result is known as the Helly selection theorem (Loève [L3, Section 11.2, where the tightness is termed complete compactness]).

2.2 Probability Measures on Metric Spaces

The continuous parameter stochastic processes dealt with in this book will be defined either on some finite time interval $[0, T]$ or on the semi-infinite or infinite intervals $[0, \infty)$, $(-\infty, \infty)$, resp. Instead of viewing the processes as being a sequence of random variables whose parameter lies in a "time" set, it will sometimes be convenient to view them as abstract-valued random variables. We will not lose sight of the actual processes for very long, but the study of convergence will be greatly facilitated by the abstraction, since the powerful results concerning weak convergence of sequences of probability measures on abstract spaces can then be used.

Let $\mathscr{X}$ denote a complete separable metric space, and let P denote a probability measure on the Borel sets of $\mathscr{X}$. *Define $C(\mathscr{X})$ (resp., $C_P(\mathscr{X})$) to be the sets of real-valued bounded functions on $\mathscr{X}$ which are continuous (resp., continuous almost everywhere with respect to P).* Let $\{P_n\}$ denote a sequence of probability measures on the Borel subsets of $\mathscr{X}$. The sequence $\{P_n\}$ *converges weakly* to a probability measure P, if (1.1) holds for all $f(\cdot) \in C(\mathscr{X})$. If (1.1) holds on $C(\mathscr{X})$, it holds on $C_P(\mathscr{X})$. The statements below (1.1) all continue to hold if R is replaced by $\mathscr{X}$ and $\{|X| \leq N_\varepsilon\}$ is replaced by $\{X \in K_\varepsilon$, some compact set in $\mathscr{X}\}$ (Billingsley [B3, Chapter 1]; Gikhman and Skorokhod [G2, Chapter 9]). A sequence of probability measures $\{P_n\}$ on $\mathscr{X}$ is said to be relatively compact if each subsequence has a further subsequence which converges weakly to some probability measure on $\mathscr{X}$. (Strictly speaking, we should say on the Borel sets of $\mathscr{X}$, but omitting the qualifier should not cause confusion.)

† The boundary ∂A of a set A is defined to be the closure minus the interior of A.

Theorem 2.2.1 *If $\{P_n\}$ is a sequence of probability measures on the complete separable metric space $\mathscr{X}$, then tightness is a necessary and sufficient condition for $\{P_n\}$ to be relatively compact. Furthermore, every probability measure on $\mathscr{X}$ is tight* (Billingsley [B3, Chapter 1]).

Suppose that P_n is a probability measure on $\mathscr{X}$ induced by some vector-valued process $\xi_n(\cdot)$. The most frequently used tests for tightness of $\{P_n\}$ involve conditions on the processes $\{\xi_n(\cdot)\}$. These conditions will be readily verifiable in our cases. One of the main problems of the book is to prove that the "limiting" measures P correspond to some particular desired process (say, a controlled diffusion process).

Let $\mathscr{X} = \mathscr{X}_1 \times \mathscr{X}_2$, where $\mathscr{X}_i$ are complete separable metric spaces, and let $\{P_n\}$ denote a sequence of probability measures on $\mathscr{X}$, where P_n is the measure induced by a pair of random variables X_1^n, X_2^n, with X_i^n taking values in $\mathscr{X}_i$. Suppose that P_n^i is the measure induced by X_i^n on $\mathscr{X}_i$. Then the sequence $\{P_n\}$ is tight if each sequence $\{P_n^i\}$, $i = 1, 2$, is tight separately. This is a simple consequence of the fact that

$$P_n\{(X_1, X_2) \in A_1 \times A_2\} \le P_n^1\{X_1 \in A_1\} + P_n^2\{X_2 \in A_2\}$$

for any Borel $A_i \in \mathscr{X}_i$.

If P_n is the measure induced by an $\mathscr{X}$-valued random variable X_n, and $P_n \Rightarrow P$, where P corresponds to an $\mathscr{X}$-valued random variable X, we say that $X_n \to X$ in distribution.

It is usually more difficult to work with convergence in distribution, than with convergence w.p. 1. It turns out that by using a particular choice of the probability space, we can assume that all the random variables $\{X_n\}$ and X are defined on the same probability space, and that there is convergence w.p. 1, if there is convergence in distribution. First consider a special simple case. Let X_n denote a random variable satisfying†

$$P\{X_n = 0\} = 1 - 1/n, \qquad P\{X_n = 1\} = 1/n.$$

The sequence X_n converges to $X \equiv 0$ in distribution. If the $\{X_n\}$ are mutually independent, the convergence is *not* w.p. 1. Generally speaking, in our convergence studies, we are concerned mainly with properties of the "limit" X and with convergence of distributions of functionals of the $\{X_n\}$ to those of X. In these cases, we can alter the probability space in any way at all, as long as the distributions of each X_n and of X are not changed—it would not matter if we modified the joint distributions of the $(\{X_n\}, X)$. Suppose that $\tilde{\Omega} = [0, 1]$, $\tilde{\mathscr{B}}$ is the σ-algebra consisting of the Borel sets in $[0, 1]$, and $\tilde{P}$ is

† The subscript n is sometimes put on the random variable, and sometimes on the measure; notation is often abused by interchanging $P\{X_n \in A\}$ with $P_n\{X \in A\}$, but the interchange should cause no confusion even without an identification of the probability space.

the Lebesgue measure. Define random variables $\tilde{X}_n$ and $\tilde{X}$ on $(\tilde{\Omega}, \tilde{\mathscr{B}}, \tilde{P})$ by $\tilde{X} \equiv 0$, $\tilde{X}_n = 0$ on $[1/n, 1]$, $X_n = 1$ on $[0, 1/n)$. Then $\tilde{X}_n \to \tilde{X}$ w.p. 1 but for any Borel set A,

$$\tilde{P}\{\tilde{X}_n \in A\} = P\{X_n \in A\}, \qquad \tilde{P}\{\tilde{X} \in A\} = P\{X \in A\}. \tag{2.1}$$

There is a very useful theorem due to Skorokhod [S2, Theorem 3.1.1], which generalizes this result.

Theorem 2.2.2 *Let $\mathscr{X}$ be a complete separable metric space, let P_n and P be induced by $\mathscr{X}$-valued random variables X_n and X, resp., and let $P_n \Rightarrow P$. There is a probability space $(\tilde{\Omega}, \tilde{\mathscr{B}}, \tilde{P})$ with $\mathscr{X}$-valued random variables $\{\tilde{X}_n\}$ and $\tilde{X}$ defined on it, such that* (2.1) *holds for each Borel set $A \in \mathscr{X}$. Also $\tilde{X}_n \to \tilde{X}$* w.p. 1 *in the topology of $\mathscr{X}$.*

Theorem 2.1.2 has been extended in various ways. See Dudley [D3].

If $X_n \to X$ in distribution, we will often assume, without notice, that the convergence is also w.p. 1 in the topology of $\mathscr{X}$, and not use the tilde ($\tilde{}$) notation. The construction $(\tilde{\Omega}, \tilde{\mathscr{B}}, \tilde{P})$ is known as *Skorokhod imbedding.*

2.3 The Spaces $C^m[\alpha, \beta]$ of Continuous Functions

Let $C^m[\alpha, \beta]$ (resp., $C^m[0, \infty)$, $C^m(-\infty, \infty)$) denote the space of R^m-valued continuous functions on the finite interval $[\alpha, \beta]$ (resp. infinite intervals $[0, \infty)$, $(-\infty, \infty)$). If $m = 1$, we drop the superscript. The topology on all spaces is the topology of uniform convergence on finite intervals. In this topology all three spaces are complete, separable, and metric. These spaces appear in numerous applications of weak convergence theory, some of which will be seen in the rest of the book. The proofs of the following results are in Billingsley [B3, Section 8].

Theorem 2.3.1 *Let $\{X_n(\cdot)\}$ denote a sequence of processes with paths in $C^m[\alpha, \beta]$,* w.p. 1. *The sequence is tight*† *if and only if*

For each $\eta > 0$, there is an $N_\eta < \infty$ such that

$$P\{|X_n(\alpha)| \geq N_\eta\} \leq \eta, \qquad \text{all} \quad n, \tag{3.1a}$$

for each $\varepsilon > 0$, $\eta > 0$, there is a $\delta \in (0, 1)$ and an $n_0 < \infty$ such that

$$P\left\{\sup_{|t-s|<\delta;\, \alpha \leq t,\, s \leq \beta} |X_n(t) - X_n(s)| \geq \varepsilon\right\} \leq \eta$$

$$\text{for} \quad n \geq n_0. \tag{3.1b}$$

† By the statement "$\{X_n\}$ is tight," we mean that the corresponding sequence of probability measures induced by $\{X_n(\cdot)\}$ on $C^m[\alpha, \beta]$ is tight.

If for each $\varepsilon > 0$, $\eta > 0$, *there is a* $\delta \in (0, 1)$ *and* n_0 *such that*

$$P\left\{\sup_{s \le t \le s+\delta} |X_n(t) - X_n(s)| \ge \varepsilon\right\} \le \eta\delta, \quad \textit{for} \quad n \ge n_0 \quad \textit{and}$$

$$\textit{each} \quad s, \alpha \le s, s + \delta \le \beta, \tag{3.2}$$

then (3.1b) *holds.*

A sufficient condition for (3.2) *is the existence of a real* K *and an* $a > 0$, $b > 0$ *such that*

$$E|X_n(t) - X_n(s)|^a \le K|t - s|^{1+b} \qquad \textit{all} \quad n. \tag{3.3}$$

If the finite dimensional distributions of $\{X_n(\cdot)\}$ *converge to those of* $X(\cdot)$, *where* $X(\cdot)$ *has continuous paths* w.p. 1, *and if* $\{X_n(\cdot)\}$ *is tight, then* $\{X_n\}$ *converges to* $X(\cdot)$ *in distribution. To any measure* P *on* $C^m[\alpha, \beta]$, *there corresponds a continuous stochastic process* $X(\cdot)$ *whose finite dimensional distributions (say at times* $t_1, \ldots, t_n$) *are given by the expressions* $P\{x(\cdot): x(t_1) \in A_1, \ldots, x(t_n) \in A_n\}$, *where* $x(\cdot)$ *is the generic element of* $C^m[\alpha, \beta]$.

Criterion (3.2) is simpler to verify than (3.1b), since s is fixed inside the $P\{\ \}$. A supremum over a fixed interval must be estimated, rather than the more complicated supremum in (3.1b).

The theorem holds for $C^m[0, \infty)$ and $C^m(-\infty, \infty)$ if we set α equal to zero (or to any other fixed time in the appropriate interval) and if (3.1b), (3.2), (3.3) all hold on each interval $[0, T]$ or $[-T, T]$, where δ, K and n_0 can depend on T.

2.4 The Space $D^m[\alpha, \beta]$

Let $D^m[\alpha, \beta]$ denote the space of R^m-valued functions on the finite interval $[\alpha, \beta]$, which are right continuous on $[\alpha, \beta)$, have left-hand limits on $(\alpha, \beta]$, and are continuous at β. Let $D^m[0, \infty)$ and $D^m(-\infty, \infty)$ denote the spaces of R^m-valued functions on $[0, \infty)$ and $(-\infty, \infty)$, resp., which are right continuous and have left-hand limits. If $m = 1$, we drop the superscript.

Our basic approximating processes will be derived from a family of Markov chains $\{\xi_n^h\}$, where h indexes the family. The parameter h may be, for example, a finite difference interval. It will be necessary to interpolate the chains to obtain a family (also indexed by h) of continuous time processes $\{\xi^h(\cdot)\}$. These processes will be piecewise constant and continuous from the right, although the limits† (which will be diffusions) will have continuous

† The word limit is used loosely to mean the process which is the limit in distribution of $\{\xi_n(\cdot)\}$.

paths w.p. 1. It will be convenient to treat these processes as random variables with values in the spaces $D^m[0, T]$, $D^m[0, \infty)$ or $D^m(-\infty, \infty)$. First we consider $D^m[0, T]$, under a particular topology.

To see which type of topology we need, consider the following example. Let $f_n(\cdot)$ denote the member of $D[0, T]$, which is defined by $f_n(t) = 1$ for $t \in [T/2 + 1/n, T]$ and $f_n(t) = 0$ elsewhere, and let $f(\cdot)$ denote the function whose values equal 1 on $[T/2, T]$ and are zero elsewhere. Convergence in distribution of a sequence of probability measures $\{P_n\}$ on R to a probability measure P on R only requires convergence of the distribution functions $P_n(-\infty, \cdot]$ at the points of continuity of $P(-\infty, \cdot]$. Here, in order to get reasonably useful results, we need to put a topology on $D[0, T]$ under which $f_n(\cdot) \to f(\cdot)$ as $n \to \infty$. The space $D[0, T]$ will be endowed with the Skorokhod topology (Billingsley [B3, Section 14]) which is defined by the following metric: Let Λ denote the class of strictly increasing continuous maps from $[0, T]$ onto $[0, T]$. For arbitrary elements $x(\cdot)$, $y(\cdot)$ of $D[0, T]$, define the metric

$$d_T(x(\cdot), y(\cdot)) = \inf\left\{\varepsilon : \sup_{T \geq t \geq 0} |t - \lambda(t)| \leq \varepsilon, \sup_{T \geq t \geq 0} |x(t) - y(\lambda(t))| \leq \varepsilon, \text{ for some } \lambda(\cdot) \in \Lambda\right\}.$$

The time scale transformation $\lambda(\cdot)$ allows a distortion of the time scale of one of the functions $x(\cdot)$, $y(\cdot)$, so that it can better match the other of the two functions in a neighborhood of a discontinuity. The sequence $\{f_n(\cdot)\}$ defined previously converges to the above $f(\cdot)$ in this topology. If $\{x_n(\cdot)\}$ and $x(\cdot)$ are members of $D[0, T]$ and $x_n(\cdot)$ converges to $x(\cdot)$ in the metric d_T, then $x_n(t)$ converges to $x(t)$ at all continuity points t of $x(\cdot)$. If we delete an arbitrary neighborhood of the points of discontinuity of $x(\cdot)$, the convergence is uniform.

As shown in Billingsley [B3], $D[0, T]$ is separable but not complete under this metric. There is another metric d'_T (called d_0 in Billingsley [B3, Section 14]), which generates the same topology and under which the space is complete. For each $\lambda(\cdot) \in \Lambda$, define

$$\|\lambda\| = \sup_{s \neq t, T \geq s, t} \left| \log \frac{[\lambda(t) - \lambda(s)]}{(t - s)} \right|.$$

The metric d'_T is defined by

$$d'_T(x(\cdot), y(\cdot)) = \inf\left\{\varepsilon : \|\lambda\| \leq \varepsilon, \sup_{T \geq t \geq 0} |x(t) - y(\lambda(t))| \leq \varepsilon, \lambda \in \Lambda\right\}.$$

We will always assume that the metric is d'_T, which implies in particular that $D[0, T]$ is complete and separable. The product topology is used on the product spaces $D^m[0, T]$.

Tightness on $D^m[0, T]$

Let $\{X_n(\cdot), t \le T\}$ denote a sequence of random processes whose paths are elements of $D^m[0, T]$ w.p. 1 for each n, and which induce the sequence of probability measures $\{P_n\}$ on (the Borel sets of) $D^m[0, T]$. Generally, in applications, the properties of the measures (such as those required to prove tightness) are deduced from the properties of the processes—properties such as moment values, stochastic continuity, or path continuity. The development of criteria for the tightness of $\{P_n\}$ involves first a characterization of the compact sets in $D^m[0, T]$, then the selection of a particular sequence of compact sets $\{K_\varepsilon, \varepsilon \to 0\}$, and finally the development of conditions on the paths of the family $\{X_n(\cdot)\}$ of processes which guarantee that the paths lie in K_ε with a probability at least $1 - \varepsilon$ for all n.

The complete story is given in Billingsley [B3], and we will only state the results.

Theorem 2.4.1 ([B3, Theorem 15.3]) *The sequence $\{P_n\}$, (induced by the sequence of processes $X_n(\cdot)$), on $D^m[0, T]$ is tight if and only if* (4.1)–(4.4) *hold.*

For each real $\eta > 0$, there is a real $N_\eta < \infty$ such that

$$P\left\{\sup_{T \ge t \ge 0} |X_n(t)| \ge N_\eta\right\} \le \eta, \qquad \text{all} \quad n \ge 1. \tag{4.1}$$

For each real $\varepsilon > 0$, $\eta > 0$, there is a real $\delta \in (0, 1)$ and an $n_0 < \infty$ such that

$$P\left\{\sup_{T \ge t_2 \ge t \ge t_1 \ge 0,\ |t_2 - t_1| \le \delta} \min[|X_n(t) - X_n(t_1)|,\ |X_n(t_2) - X_n(t)|] \ge \varepsilon\right\} \le \eta, \qquad n \ge n_0, \tag{4.2}$$

$$P\left\{\sup_{\delta > t \ge s \ge 0} |X_n(t) - X_n(s)| \ge \varepsilon\right\} \le \eta, \quad n \ge n_0, \tag{4.3}$$

$$P\left\{\sup_{T \ge t \ge s \ge T - \delta} |X_n(t) - X_n(s)| \ge \varepsilon\right\} \le \eta, \quad n \ge n_0. \tag{4.4}$$

Equations (4.1), (4.3), and (4.4) are usually easy to verify. The moment condition (4.5) implies (4.2), and will often be used in the sequel (see Billingsley [B3, proof of Theorem 15.6]).

There is a real K for which

$$E|X_n(t_1) - X_n(t)|^2 |X_n(t_2) - X_n(t)|^2 \le K|t_2 - t_1|^2 \quad \text{for all} \quad T \ge t_2 \ge t \ge t_1 \ge 0. \tag{4.5}$$

Conditions (4.3) and (4.4) guarantee that the probability of a discontinuity on $[0, \delta)$ or on $(T - \delta, T]$ goes to zero, uniformly in n as $\delta \to 0$. They assure us that the "limits" are continuous (w.p. 1) at $t = 0$ and $t = T$. Equation (4.2) is a type of equicontinuity condition—basically neglecting a finite number of discontinuities: it requires a type of equicontinuity (with a probability arbitrarily close to unity, uniformly in n) on *either* one side or another of each $t \in (0, T)$.

Frequently, the processes corresponding to a limiting measure will have continuous paths w.p. 1. There is a criterion for this—in terms of properties of the $\{X_n(\cdot)\}$.

Theorem 2.4.2 (Billingsley [B3, Theorems 15.5 and 8.3]) *The sequence $\{P_n\}$ on $D^m[0, T]$ is tight if* (3.1a) *and either* (3.1b) *or* (3.2) *or* (3.3) *hold. If P is a weak limit of $\{P_n\}$ and $X(\cdot)$ is a process inducing P on $D^m[0, T]$, then $X(\cdot)$ has continuous paths* w.p. 1.

Let $X(\cdot)$ denote a process with paths in $D^m[0, T]$ w.p. 1 and define T_X as the set of $t \in (0, T]$ for which $P\{X(t^-) = X(t)\} = 1$.

In particular, $t = T$ is in T_X.

Theorem 2.4.3 (Billingsley [B3, Theorems 15.4 and 15.6]) *Let P_n be induced on $D^m[0, T]$ by a process $X_n(\cdot)$, whose paths are in $D^m[0, T]$* w.p. 1. *If* (4.2) *or* (4.5) *hold and if for each integer m and set $(t_1, \ldots, t_m) \in T_X$*

$$(X_n(t_1), \ldots, X_n(t_m)) \to (X(t_1), \ldots, X(t_m))$$

in distribution, then $X_n(\cdot) \to X(\cdot)$ in distribution.

We also have: If $X_n(\cdot) \to X(\cdot)$ in distribution and $(t_1, \ldots, t_m) \in T_X$, then $(X_n(t_1), \ldots, X_n(t_m)) \to (X(t_1), \ldots, X(t_m))$ as $n \to \infty$.

Define the finite dimensional projections as follows: For $t_1, \ldots, t_n \in [0, T]$ and $x(\cdot) \in D^m[0, T]$, define the projection (mapping $D^m[0, T]$ to R^{mn}) $\pi(t_1, \ldots, t_n; x) = ((x(t_1), \ldots, x(t_n))$.

Suppose that P is the weak limit of a tight sequence $\{P_n\}$ on $D^m[0, T]$. Since the finite dimensional projections are measurable [B3, p. 121], the measures induced by the projections can be used to construct a separable stochastic process $X(\cdot)$ on some sample space. By Billingsley [B3, Theorem 15.8], *the sample paths of the process $X(\cdot)$ are right continuous at $t = 0$, have left-hand limits at $t = T$ and are continuous on* $(0, T)$, *except for simple jumps.* If (3.1a,b) holds for $X(\cdot)$, then $X(\cdot)$ will be continuous w.p. 1, and will obviously have its paths in $D^m[0, T]$ w.p. 1.. Also, if $P\{X(t) = X(t^+)\} = 1$, $t \in (0, T)$, then we can define the paths to be right continuous, without altering the finite dimensional distributions. In both of these cases, we can assume that the process $X(\cdot)$ has paths in $D^m[0, T]$ w.p. 1, and, indeed, induces the measure $P\{\cdot\}$ on $D^m[0, T]$.

If $X_n(\cdot) \to X(\cdot)$ in distribution, then we can assume, via the Skorokhod imbedding, that the probability space is chosen so that $X_n(\cdot) \to X(\cdot)$ w.p. 1 in the topology of $D^m[0, T]$. In particular, there is (w.p. 1) pointwise convergence at all t where $X(\cdot)$ is continuous. If $X(\cdot)$ is continuous w.p. 1, then the convergence is uniform w.p. 1.

The space $D^m[\alpha, \beta]$ for arbitrary finite α, β, $\beta > \alpha$, is defined exactly as $D^m[0, T]$ was defined. We simply shift the time origin and set $T = \beta - \alpha$. In the special case where $\alpha = -\beta = -T$, denote the metric by d_T''.

The Space $D^m[0, \infty)$

There are several natural extensions of the metric d_T' to the space $D^m[0, \infty)$. The details of one of them appears in Lindvall [L1] and we will describe it roughly. Let $x(\cdot)$ denote a generic element of $D^m[0, \infty)$ and, for each integer j, define the function $g_j(\cdot)$ on $[0, \infty)$ by: $g_j(t) = 1$ for $t \le j$, $g_j(t) = 0$ for $t \ge j + 1$ and $g_j(t) = j + 1 - t$ on $[j, j + 1]$. Define $x^j(\cdot) = x(\cdot)g_j(\cdot)$. Considering $x^j(\cdot)$ as a function on $[0, j + 1]$, define the metric

$$d(x(\cdot), y(\cdot)) = \sum_{j=1}^{\infty} \frac{d_{j+1}'(x^j(\cdot), y^j(\cdot))}{1 + d_{j+1}'(x^j(\cdot), y^j(\cdot))} 2^{-j}. \tag{4.6}$$

Under this metric, $D^m[0, \infty)$ is a complete separable metric space. (Actually, Lindvall considers each $x^j(\cdot)$ as a function on $[0, \infty)$ and maps $[0, \infty)$ into $[0, 1]$ via the transformation $\phi(t) = -\log(1 - t)$.)

A sequence $\{X_n(\cdot)\}$ is tight on $D^m[0, \infty)$ if and only if $\{X_n^j(\cdot)\}$ is tight on $D^m[0, j + 1]$ for each integer j. Then, we need not verify (4.4).

The metric on the space $D^m(-\infty, \infty)$ is defined similarly to that on $D^m[0, \infty)$. Define $\bar{g}_j(\cdot)$ on $(-\infty, \infty)$ by $\bar{g}_j(t) = \bar{g}_j(-t) = g_j(t)$, and $x^j(\cdot) = x(\cdot)\bar{g}_j(\cdot)$, for an arbitrary $x(\cdot) \in D^m(-\infty, \infty)$. Considering $x^j(\cdot)$ to be an element of $D^m[-j - 1, j + 1]$, define the metric

$$d(x, y) = \sum_{j=1}^{\infty} \frac{d_{j+1}''(x^j(\cdot), y^j(\cdot))}{1 + d_{j+1}''(x^j(\cdot), y^j(\cdot))} 2^{-j}, \tag{4.7}$$

where d_j'' is the Skorokhod metric on $D^m[-j - 1, j + 1]$. All the statements concerning $D^m[0, \infty)$ are true here [except that neither (4.3) nor (4.4) plays a role anymore], provided that we replace all terms of the form $[0, a]$ $(a > 0)$ by $[-a, a]$. The $D^m(-\infty, \infty)$ case is only alluded to and not dealt with in detail in [L1], but the development would be almost identical to that given there for $D^m[0, \infty)$.

2.5 Weak Convergence on Other Spaces

We will sometimes use the spaces $R^+ = [0, \infty)$ with the usual Euclidean topology, and its one point compactification $\bar{R}^+ = [0, \infty]$.

The space L_2^m of R^m-valued square integrable functions on $[0, T]$ will also be used, but in a special way that does not require many of the concepts of weak convergence. Since $L_2^m[0, T]$ is complete and separable, each single probability measure on it is tight. As will be seen in Chapters 8 and 9, the L_p space is useful in problems arising in control theory, since the L_p spaces are usually more natural homes for the actual control than are the C or D spaces.

We just note the following. Let $w(\cdot, \cdot)$ (say R^r-valued) denote a Wiener process, $\mathscr{U}$ a compact convex set (in some Euclidean space R^m), and $u(\cdot)$ a $\mathscr{U}$-valued separable and measurable process on $[0, T]$ which is nonanticipative with respect to $w(\cdot, \cdot)$. Then $u(\omega, \cdot)$ lies in $L_2^m[0, T]$ for almost all ω since $u(\omega, \cdot)$ is Lebesgue measurable for almost all ω. The processes $u(\cdot, \cdot)$, $w(\cdot, \cdot)$ induce a measure on (the Borel sets of) $L_2^m[0, T] \times C^r[0, T] \equiv \mathscr{X}$. Let $v(\cdot)$, $x(\cdot)$ denote generic elements of $L_2^m[0, T]$ and $C^r[0, T]$, resp. Under the induced measure, any measurable functional of $v(\cdot)$ which depends only on $v(s)$, $s \le t$, is independent of any measurable function of $x(\cdot)$ which depends only on $x(\tau) - x(t)$, $\tau \ge t$ for each t.

Let $(\Omega', P', \mathscr{B}')$ denote a probability space with generic element ω' and let $U(\cdot)$, $W(\cdot)$ denote random variables with values in $L_2^m[0, T]$, $C^r[0, T]$, resp. That is, for each $\omega' \in \Omega'$, $U(\omega')$ is an equivalence class in $L_2^m[0. T]$ and $W(\omega')$ is an element of $C^r[0, T]$. Let $W_t(\omega')$ and $U_t(\omega')$ (for a fixed member of the equivalence class) denote the values at time t. Suppose that the marginal distribution induced on $C^r[0, T]$ is a Wiener measure and that for almost all ω', and any element of the equivalence class of $U(\omega')$, $U_t(\omega')$ is $\mathscr{U}$-valued for almost all t, where $\mathscr{U}$ is a compact convex set. Let $U(\cdot)$ be nonanticipative with respect to $W(\cdot)$ in the sense that any measureable function of $U(\cdot)$ that depends only on the part of $U(\cdot)$ on $[0, t]$ is independent of $\{W_\tau(\cdot) - W_t(\cdot), T \ge \tau \ge t\}$ for each $t \in [0, T]$. Then there is a probability space with $\mathscr{U}$- and R^r-valued separable and measurable processes $u(\cdot, \cdot)$ and $w(\cdot, \cdot)$, resp., defined on it, where $w(\cdot, \cdot)$ is a Wiener process, $u(\cdot, \cdot)$ is nonanticipative with respect to $w(\cdot, \cdot)$, and they induce the measure of $U(\cdot)$, $W(\cdot)$ on $\mathscr{X}$. In particular, for each ω' not in some null set, we can choose a member of the equivalence class of $U(\omega')$ such that the process with values $U_t(\omega')$ is separable, measurable, and $\mathscr{U}$-valued, where $U_t(\omega')$ is the value at t of the chosen member of the equivalence class of $U(\omega')$.

The purpose of these remarks is simply that it will occasionally be convenient to view a control as an abstract element of an L_2 space and sometimes as a $\mathscr{U}$-valued random process; thus it is desirable to know that the two views are consistent.

Sometimes it is convenient to set $T = \infty$. We then use $L^m_{2,l}$: *A function is in* $L^m_{2,l}$ *if its restriction to* $[0, T]$ *is in* $L^m_2[0, T]$, *for each* $T < \infty$. Let u, v be generic elements of $L^m_{2,l}$. If $|\ |_{L_2{}^m[T]}$ denotes the norm of $L^m_2[0, T]$, then $L^m_{2,l}$ is separable and complete under the metric

$$d(u, v) = \sum_{n=1}^{\infty} \frac{|u - v|_{L_2{}^m[n]}}{1 + |u - v|_{L_2{}^m[n]}} 2^{-n}$$

and when using $L^m_{2,l}$, we can and will suppose that the space is separable and complete.

CHAPTER 3

Markov Chains and Control Problems with Markov Chain Models

In this chapter we present a brief discussion of the equations that are satisfied by various functionals of Markov chains and by the cost functionals associated with optimal control problems on Markov chain models. Since the discretization methods of Chapters 5–10 yield equations that are satisfied by functionals of Markov chains and since such chains will be used to approximate diffusions, we present an outline of the relevant concepts in the theory of Markov chains in this chapter. More detail appears in Kushner [K5]. Our notation differs somewhat from the notation used in that reference. Since we are concerned with computational techniques, the state spaces of the Markov chains will be assumed to be finite. This finiteness will simplify the discussion in this chapter, but most of the results of the following chapters do not require finiteness of the approximating chains. Notation is introduced and the dynamic programming equations for the more standard types of control problems on Markov chains are given. There are some brief remarks on computational methods and the linear programming formulation of controlled Markov chains under side constraints is discussed. This will be useful when we treat control problems on diffusion models with side constraints.

3.1 Equations Satisfied by Functionals of Markov Chains

3.1.1 Notation

Let $\{X_n\}$ denote a *Markov chain* on the state space S. Let ∂S denote a *fixed subset* of S (a "boundary"), and define the *stopping time* $N = \min\{n : X_n \in \partial S\}$. If $X_n \in S - \partial S$ for all n, then following our usual convention with stopping times, we set $N = \infty$. If the chain is homogeneous, we denote the transition probability $P\{X_{n+m} = y | X_n = x\}$ by $p_{xy}^{(m)}$, for $x, y \in S$ and all integers $m, n \geq 0$, with $p_{xy}^{(1)}$ written as p_{xy}. If the chain is nonhomogeneous, we use the terminology $p_{xy}^{(m)}(n) = P\{X_{n+m} = y | X_n = x\}$, $x, y \in S$, for all integers $m, n \geq 0$, with $p_{xy}^{(1)}(n)$ written as $p_{xy}(n)$. If $f(\cdot)$ is a real-valued function on S, then we write the *expectation* as (assuming that it exists)

$$E_x f(X_m) = E[f(X_m) | X_0 = x] = \sum_y p_{xy}^{(m)} f(y) \qquad \text{(homogeneous case)},$$

$$E_{x,n} f(X_{n+m}) = E[f(X_{n+m}) | X_n = x] = \sum_y p_{xy}^{(m)}(n) f(y)$$

$$\text{(nonhomogeneous case)}.$$

Let $r(\cdot)$, $b(\cdot)$, $b_T(\cdot)$, $r_1(\cdot, \cdot)$, $b_1(\cdot, \cdot)$ denote *bounded real-valued functions* on $S - \partial S$, ∂S, $S - \partial S$, $(S - \partial S) \times [0, 1, \ldots)$ and $\partial S \times [0, 1, \ldots)$. The functions $r(\cdot)$ and $r_1(\cdot, \cdot)$ will be "running" costs or "cost rates" and $b(\cdot)$, $b_T(\cdot)$, $b_1(\cdot, \cdot)$ will be boundary costs.

Assume that the states are ordered in some definite way, so that quantities such as the transition matrix $P = \{p_{xy}, x, y \in S\}$ are well defined.

3.1.2 Cost Accumulated until ∂S is Reached. Homogeneous Case

Suppose that

$$E_x N < \infty \qquad \text{for all} \quad x \in S - \partial S \tag{1.1}$$

and define the quantity $R(x)$ by

$$R(x) = E_x\left[\sum_{i=0}^{N-1} r(X_i) + b(X_N)\right]. \tag{1.2}$$

The $R(x)$ is well defined by (1.1). Equation (1.2) can be rewritten in the form

$$\begin{aligned} R(x) &= r(x) + E_x E_x \left[\sum_{i=1}^{N-1} r(X_i) + b(X_N) \middle| X_1 \right] \\ &= r(x) + E_x R(X_1) \qquad \text{for} \quad x \notin \partial S, \\ &= b(x) \qquad \text{for} \quad x \in \partial S. \end{aligned} \tag{1.3}$$

Under (1.1), the right-hand side of (1.2) is the unique bounded solution to Eq. (1.3).

Suppose now that $r(\cdot) = 0$; drop (1.1) and consider the functional

$$R(x) = E_x b(X_N) I_{\{N < \infty\}} . \tag{1.4}$$

Then $R(\cdot)$ satisfies

$$R(x) = \begin{cases} E_x R(X_1), & x \in S - \partial S, \\ b(x), & x \in \partial S, \end{cases} \tag{1.5}$$

and the right-hand side of (1.4) is a solution to (1.5). It is not necessarily the unique solution since any function $\hat{R}(\cdot)$ that satisfies the boundary condition, and such that $\{\hat{R}(X_n)\}$ is a martingale, also satisfies (1.5). If $P_x\{N < \infty\} = 1$ each $x \in S - \partial S$, then the solution is unique (this is not necessarily true if S is not finite).

3.1.3 Cost Accumulated until a Fixed Terminal Time

Let M denote a fixed integer (a "terminal" time) and define the functional†

$$\begin{aligned} R(x, n) = E_{x, n} \Bigg[& \sum_{i=n}^{(M \cap N)-1} r_1(X_i, i) + b_T(X_M) I_{\{M < N\}} \\ & + b_1(X_N, N) I_{\{M \geq N\}} \Bigg], \qquad x \in S - \partial S, \quad n < M, \end{aligned} \tag{1.6}$$

$$\begin{aligned} R(x, M) &= b_T(x), \qquad x \in S - \partial S, \\ R(x, n) &= b_1(x, n), \qquad x \in \partial S, \quad n \leq M. \end{aligned} \tag{1.7}$$

The boundary conditions (1.7) are implicit in (1.6).

† The symbol I_A is the indicator function of the set A.

Equation (1.6) can be rewritten in the form

$$R(x, n) = r_1(x, n) + E_{x,n}E_{x,n}\left[\sum_{i=n+1}^{(M \cap N)-1} r_1(X_i, i) + b_T(X_M)I_{\{M<N\}} + b_1(X_N, N)I_{\{M \geq N\}} \,|\, X_{n+1}\right]$$

or

$$R(x, n) = r_1(x, n) + E_{x,n}R(X_{n+1}, n+1), \qquad n < M, \quad x \in S - \partial S, \quad (1.8)$$

with boundary conditions (1.7). The right-hand side of (1.6) is the unique solution to Eq. (1.8).

3.1.4 Discounted Cost

Let β denote a real number in (0, 1) and assume that $\{X_n\}$ is homogeneous. Define the function $R(\cdot)$ by

$$R(x) = E_x\left[\sum_{i=0}^{N-1} \beta^i r(X_i) + \beta^N b(X_N)\right]. \quad (1.9)$$

Equation (1.9) can be rewritten in the form

$$R(x) = r(x) + E_x E_x\left[\sum_{i=1}^{N-1} \beta^i r(X_i) + \beta^N b(X_N) \,|\, X_1\right].$$

Equivalently,

$$R(x) = \begin{cases} r(x) + \beta E_x R(X_1), & x \in S - \partial S \\ b(x), & x \in \partial S. \end{cases} \quad (1.10)$$

The right-hand side of (1.9) is the unique bounded solution to (1.10).

3.1.5 Average Cost per Unit Time

Suppose that the chain is homogeneous and the state space consists of a single positive recurrent, a-periodic class, with invariant measure (*which we always write as a row vector*) $\{\mu(x), x \in S\}$. Thus $\mu = \mu P$, where $P = \{p_{xy}, x, y \in S\}$. Define γ by

$$\gamma = E_\mu r(X) \equiv \sum_{x \in S} \mu(x) r(x)$$

or by

$$\gamma = \lim_{N \to \infty} (1/N) E_x \sum_{i=0}^{N-1} r(X_i).$$

Suppose that there is a bounded function $R(\cdot)$ on S and a real number $\tilde{\gamma}$, such that

$$R(x) + \tilde{\gamma} = r(x) + E_x R(X_1). \tag{1.11}$$

Then, by premultiplying each term of (1.11) by $\mu(x)$ and summing over x, and using $\mu = \mu P$, we have

$$\tilde{\gamma} = \gamma.$$

Conversely, there is a bounded function $R(\cdot)$ and a unique value $\tilde{\gamma}$ (which must then equal γ) which solves (1.11). The function $R(\cdot)$ in (1.11) is not unique, but $\tilde{\gamma}$ is.

The chains arising from the discretization methods used in Chapters 5–10 will often be periodic.† Let $\{X_n\}$ denote a homogeneous Markov chain which is a single positive recurrent class of period $v \geq 1$. The following remarks are taken from Chung [C1, Sections I.7, I.8, and I.15]. The limit $\mu(y)$ defined by

$$\mu(y) = \lim_{n\to\infty} (1/n) \sum_{i=0}^{n} p_{xy}^{(i)}, \qquad x, y \in S,$$

always exists, and does not depend on x. Let $f(\cdot)$, $g(\cdot)$ be real-valued functions on S. Define

$$\mathscr{F}'(f) = \sum_y \mu(y) f(y) \equiv E_\mu f(X).$$

Then, if $\mathscr{F}'(f)$ and $\mathscr{F}'(g)$ are bounded and $\mathscr{F}'(g) \neq 0$, then

$$\sum_{i=0}^{n} f(X_i) \Big/ \sum_{i=0}^{n} g(X_i) \to \mathscr{F}'(f)/\mathscr{F}'(g) \qquad \text{w.p. 1 as } n \to \infty.$$

The *row vector* $\mu = \{\mu(x), x \in S\}$ is the unique solution (up to a proportionality constant) to

$$\bar{\mu} = \bar{\mu} P.$$

Sometimes it is more convenient (faster) to compute μ as follows. The system with transition matrix P^v is a-periodic, and has an invariant measure which solves

$$\bar{\mu} = \bar{\mu} P^v.$$

It is readily verified that the μ given by

$$\mu = \frac{1}{v}[\bar{\mu} + \bar{\mu}P + \cdots + \bar{\mu}P^{v-1}]$$

satisfies $\mu = \mu P$. Of course, if μ is used to compute an expectation, it must be normalized so that $\sum_y \mu(y) = 1$, $\mu(y) \geq 0$ all y.

† Note, for example, that the random walk, which is the usual approximation to the Wiener process, has period 2. In particular, $X_{n+1} = X_n \pm 1$, each with probability $\frac{1}{2}$.

3.1.6 Methods of Solutions

Equations (1.3), (1.5), (1.8), and (1.10) are all linear. Under the stated assumptions, (1.3), (1.5), and (1.10) are of the vector form

$$\tilde{R} = Q\tilde{R} + A,$$

where $\tilde{R} = \{R(x), x \in S - \partial S\}$ is the quantity to be found. In particular, for (1.3), $A = \{r(x) + \sum_y b(y)p_{xy}, x \in S - \partial S\}$, and $Q = \{p_{xy}, x, y \in S - \partial S\}$. If $E_x N < \infty$, $x \in S - \partial S$, then the spectral radius of Q is strictly less than unity, and any of the methods for solving such linear equations can be used. See Varga [V1] for more details. See also the remarks at the end of section 3.4. The obvious way to solve (1.8) is via "backward iteration," starting with $n = M$. To get $\gamma = \tilde{\gamma}$ in (1.11), we need to calculate the invariant measure μ. This can be done by letting u^0 denote any probability measure on S and using the iteration $u^{n+1} = u^n P$ (a-periodic) or via any method that can calculate the eigenvector of P' corresponding to eigenvalue unity. (There is only one such eigenvalue unity under the ergodicity assumption.) If we wish to calculate $E_\mu f(X)$ (Section 3.1.5) when $\{X_n\}$ has period $v > 1$ and the chain is a single recurrent class, we can calculate $\bar{\mu}$ via the iteration $u^{n+1} = u^n P^v$, where u^0 is a probability on S and then use $\mu = (1/v)[\bar{u} + \cdots + \bar{u}P^{v-1}]$.

3.2 Optimal Stopping Problems

3.2.1 Introduction

One of the simplest of optimal control problems with a Markov chain model is the optimal stopping problem. Only one decision needs to be made: the process evolves until we decide to stop it. The cost is the sum of a cost associated with the evolution of the system until the stopping moment and a cost associated with the point at which we stop.

For a simple example, let X_n = accumulated number of heads minus tails in a coin tossing game with independent trials. Suppose that there is a cost c to be paid for each toss. If by stopping at time n when $X_n = y$ we get a terminal payoff of y, we may wish to find the stopping time m at which (x = initial state)

$$E_x[cm - X_m]$$

is minimized. The random stopping time m would usually depend only on the observed data; i.e., if $m \geq n$, then whether or not $m = n$ would depend only on the observed values of $X_0, \ldots, X_n$. Actually, we can enlarge the class of allowable stopping times without decreasing the minimal average return, provided that the larger class does not "secretly" give more information about $X_{n+1}, \ldots$ than is already contained in $X_0, \ldots, X_n$ for each n. For example, we can allow "randomized" stopping times. The extension will be formalized, and is quite useful for mathematical purposes.

The gambling example may seem frivolous, but it is a prototype of a large class of important problems in sequential hypothesis testing, control theory, or economics. There are numerous decision problems where only one decision (or action) is possible—whether it is termination or initiation of an investment, the firing of a weapon, or the timing of a single thrust of a motor, etc. An excellent reference is Shiryaev [S1], where the reader will find proofs of all the unproved assertions in this section. In the simple context of the finite Markov chain, the results are relatively simple to prove as special cases of Markov chain optimization problems (Kushner [K5]).

3.2.2 Problem Formulation

First, we need to introduce several classes of stopping times. We need to consider these several classes because, when we treat the stopping time problem for the diffusion, we will be able to show that the sequence of approximations (in Chapter 8) to the optimal (minimal) cost converges to the cost for a stopping problem for the diffusion, corresponding to a possibly randomized stopping time. In order to show that the particular limit is also the minimal cost, over the class of "path dependent nonrandomized" stopping times, we need to consider the larger randomized classes.

Let the *homogeneous* Markov chain $\{X_n\}$ be defined on the probability space $(\Omega, \mathscr{B}, \mathscr{B}_n, P_x)$, where $\{\mathscr{B}_n\}$ is a sequence of nondecreasing sub-σ-algebras of $\mathscr{B}$, X_n is $\mathscr{B}_n$ measurable and $P_x\{X_{n+m} = y \mid \mathscr{B}_n\} = P_{X_n}\{X_m = y\}$ w.p. 1. Equivalently,

$$P_x\{X_{n+m} = y \mid \mathscr{B}_n\} = \sum_z p_{zy}^{(m)} I_{\{X_n = z\}}. \tag{2.1}$$

Equation (2.1) is just the defining *Markov property* for the chain. However, $\mathscr{B}_n$ can measure other random variables besides $X_0, \ldots, X_n$; it may, for example, measure $\psi_0, \ldots, \psi_n$, where $\{\psi_n\}$ is a sequence which is independent of $\{X_n\}$ under each P_x. Let $\mathscr{B}_n(x)$ denote the minimal σ-algebra over which $X_0, \ldots, X_n$ is measurable.

Let $\mathscr{T}_{\mathrm{PM}}^0$ denote the class of *pure Markov stopping times m*, possibly taking infinite values. If $m \in \mathscr{T}_{\mathrm{PM}}^0$, then there is a set $A \subset S$ for which

$$\begin{aligned} m &= \min\{n : X_n \in A\} \\ &= \infty, \qquad \text{if } X_n \notin A, \text{ all } n < \infty. \end{aligned}$$

Conversely such sets A correspond to stopping times $m \in \mathscr{T}_{PM}^0$. For each $x \in S$, $\mathscr{T}_{PM}(x)$ and $\mathscr{T}_{PM}^+(x)$ *denote the subclasses of* $\mathscr{T}_{PM}^0(x)$ *for which* $P_x\{m < \infty\} = 1$ *and* $E_x m < \infty$, *resp.*

A stopping time m relative to $\{\mathscr{B}_n\}$ is in the *randomized class* $\mathscr{T}_R^0$ if $\{m = n\} \in \mathscr{B}_n$ for each n. For each $x \in S$, the subclasses $\mathscr{T}_R(x)$ and $\mathscr{T}_R^+(x)$ of $\mathscr{T}_R^0(x)$ denote the sets of stopping times for which $P_x\{m < \infty\} = 1$ and $E_x m < \infty$, resp. If $\mathscr{A}(v)$ is a class of stopping times, strategies, etc., for each v in some set [e.g., $\mathscr{A}(v) = \mathscr{T}_R(x)$, where $v = x$], we write $\mathscr{A} = \bigcap_v \mathscr{A}(v)$; i.e., we drop the argument if the defining property holds for all values of the argument. Whether or not $m = n$ for a stopping time $m \in \mathscr{T}_{PM}^0(x)$ depends only on the value of X_n, but whether or not $m = n$ for a stopping time $m \in \mathscr{T}_R^0(x)$ can depend on the values of all of $X_0, \ldots, X_n$ as well as, perhaps, additional variables (which, however, do not give more "information" on $X_{n+1}, \ldots$ than is already contained in $X_0, \ldots, X_n$).

Let $b(\cdot)$ and $r(\cdot)$ denote real-valued bounded functions on S. Suppose that there is a $r_0 > 0$ such that $\inf_x r(x) \geq r_0 > 0$. For each $m \in \mathscr{T}_R^0(x)$, define the function (which is infinite unless $m \in \mathscr{T}_R^+(x)$)

$$R(x, m) = E_x\left[b(X_m) + \sum_{i=0}^{m-1} r(X_i)\right], \qquad x \in S, \tag{2.2}$$

and define $V(\cdot)$ by

$$V(x) = \inf_{m \in \mathscr{T}_{PM}^+(x)} R(x, m). \tag{2.3}$$

At each instant, we must either stop or continue. Hence, the appropriate dynamic programming equation is

$$V(x) = \min[b(x), E_x V(X_1) + r(x)]. \tag{2.4}$$

Under our conditions on $r(\cdot)$ and $b(\cdot)$, an optimum m exists. Since we only need m such that $R(x, m) \leq b(x)$, the optimum m must satisfy†

$$E_x m \leq 2 \max_x |b(x)|/r_0 . \tag{2.5}$$

Equation (2.4) has a unique solution which is precisely the right-hand side of (2.3). It can be solved via the iteration

$$V^{n+1}(x) = \min[b(x), E_x V^n(X_1) + r(x)], \qquad V^0(x) \text{ arbitrary}, \tag{2.6}$$

for which it can be shown that $V^n(x) \to V(x)$ as $n \to \infty$. Actually, (2.2) is a special case of a cost equation for a more general control problem (Section 3.4) and there are algorithms for the computation of $V(\cdot)$ which converge

† (2.5) follows from $-\max_y |b(y)| + (E_x m) r_0 \leq V(x) \leq b(x)$.

much faster than does (2.6) (see Kushner and Kleinman [K9] and Kushner [K5], and the remarks at the end of Section 3.4).

A sufficient condition for the stopping problem to have an optimal solution is that there be a solution to (2.4), and, in that case, an optimal stopping rule is given by the minimizing rule in (2.4). Then, our optimal policy consists in stopping at the first n for which $V(X_n) = b(X_n)$; i.e., on first hitting the stopping set

$$B = \{x : V(x) = b(x)\}.$$

For each $\varepsilon > 0$, define the set

$$B_\varepsilon = \{x : V(x) \geq b(x) - \varepsilon\}.$$

If a stopping time m is such that $X_m \in B_\varepsilon$, then

$$R(x, m) \leq V(x) + \varepsilon.$$

Also (Shiryaev [S1, Chapter II.7])

$$V(x) = \inf_{m \in \mathscr{T}_{\mathrm{R}}^{+}(x)} R(x, m). \tag{2.7}$$

The proof of (2.7) is not hard but will be omitted. It depends on the fact that the existence of a solution to (2.4) guarantees that there is an optimal solution m^* in the class $\mathscr{T}_{\mathrm{PM}}^{+}$, with value $V(x)$, and then uses the minimization property in (2.4) and the fact that $R(x, m^*) = V(x)$, to compare m^* to any $m \in \mathscr{T}_{\mathrm{R}}^{+}$.

3.3 Controlled Markov Chains: Families of Controlled Strategies

For the simple stopping problem of Section 3.2, there was no need to formally introduce the notion of a controlled Markov chain (of which the stopping problem is actually a special case). In a controlled Markov chain, we have some choice over the transition probabilities, and can select them to minimize some cost function or to satisfy some constraint. The controlled model has numerous applications—in many different areas. See Derman [D1], Howard, [H2], Kushner [K5], Wagner [W1], and Hillier and Lieberman [H1]. Most of the results in Sections 3.3–3.8 can be found in Kushner [K5].

Let $\mathscr{U}(x)$ denote the action or control set associated with the state x; $\mathscr{U}(x)$ will either be a finite set of discrete points or a compact set in some Euclidean space. The one-step transition probabilities $p_{xy}(\cdot)$ are functions of the control. If $\mathscr{U}(x)$ is not discrete, then we will assume that it is independent of x and that the $p_{xy}(\cdot)$ are continuous functions on $\mathscr{U} = \mathscr{U}(x)$. For any $\alpha \in \mathscr{U}(x)$, $\sum_y p_{xy}(\alpha) = 1$ and $p_{xy}(\alpha) \geq 0$. The initial time is $n = 0$ unless otherwise specified. A quantity

$$\pi = (u_0, u_1, \ldots)$$

is a pure *Markov control law or policy* if each u_i is a function on S and if the value $u_i(x)$ at x lies in $\mathscr{U}(x)$. With the use of the law π, the (possibly nonstationary) transition probabilities are defined by

$$p_{xy}(u_n(x)) = P\{X_{n+1} = y \mid X_n = x, \text{ law } \pi \text{ used}\}.$$

The terms E_x^π and P_x^π denote expectation and probability, given initial condition x and that law π is used. If $u_i(\cdot) = u(\cdot)$ for some function $u(\cdot)$ and all i, then the law is said to be stationary and we may write E_x^u, P_x^u for E_x^π, P_x^π, resp. We define $u_n(X_n) \equiv U_n$, the *actual sample control* action used at time n. The class of *pure Markov laws* is denoted by $\mathscr{C}_{\text{PM}}$.

We need to define a *randomized class* $\mathscr{C}_{\text{R}}$ of control laws and will do it in a way that is suited to our later needs.

Let $(\Omega, P, \mathscr{B})$ denote a probability space. Let the sequences of random variables $\{\psi_n, X_n\}$ be defined on the space and let $\mathscr{B}_n$ denote the minimal σ-algebra over which $\{\psi_i, X_i, i \leq n\}$ are measurable. Each ψ_n takes values in a set A, and X_n in S. Let $u_n(\cdot)$ be a measurable function of $\{\psi_i, X_i, i \leq n\}$ with values in $\mathscr{U}(x)$ if $X_n = x$. Let $U_n = u_n(\psi_0, \ldots, \psi_n, X_0, \ldots, X_n)$ denote the actual sample value. Suppose that the distribution function of ψ_n, given $X_0, \ldots, X_n; \psi_0, \ldots, \psi_{n-1}$, is given. *The sequence*

$$\pi = (u_0, u_1, \ldots)$$

is said to be in $\mathscr{C}_{\text{R}}$ *if the law of evolution of* $\{X_n\}$ *is given by* (3.1) (where there is an obvious abuse of notation) for each initial condition $z \in S$.

According to our terminology, the sample space may depend on the policy π since $\{\psi_n\}$ depends on π.

$$\begin{aligned} P_z^\pi\{X_0 = z\} &= 1, \\ P_z^\pi\{X_1 = y \mid \psi_0, X_0\} &= p_{zy}(u_0(\psi_0, z)), \\ P_z^\pi\{X_{n+1} = y \mid \psi_0, \ldots, \psi_n; X_0, \ldots, X_n = x\} & \\ = p_{xy}(u_n(\psi_0, \ldots, \psi_n; z, X_1, \ldots, X_{n-1}, x)). \end{aligned} \tag{3.1}$$

In applications in the later chapters, the random variables $\{\psi_n\}$ will either be explicitly given or will be easily constructable from the given randomized control law. For an example, suppose that we desire a law satisfying

$$P\{U_n = \pm a \mid X_0, \ldots, X_n\} = \tfrac{1}{2}.$$

Then we only need let $\{\psi_n\}$ be independent and have ψ_n take values ± 1, each with probability $\frac{1}{2}$; to generate U_n, we need only toss a fair coin and let $U_n = \pm a$ if $\psi_n = \pm 1$.

The class $\mathscr{C}_R$ and the classes of controls and stopping times defined in the following are introduced owing to the nature of proofs and approximations in Chapter 9. In Chapter 9, the approximation to the optimal control problem for the diffusion will be an optimal control problem for a Markov chain and the sequence of minimum cost functions corresponding to the sequence of approximations will converge to a cost functional for a controlled diffusion with a possibly randomized control. In order to show that the limiting function is indeed minimal, we will actually need to approximate arbitrary controls so that they can be applied to the chain. The various classes of controls appear in this context.

Let us introduce some additional terminology. Let G_1 denote an arbitary set and define $N = \min\{n : X_n \notin G_1\}$. With the subscript c standing for either PM or R, let $\mathscr{C}_c(x, G_1)$ and $\mathscr{C}_c^+(x, G_1)$, resp., denote the sets of policies π in $\mathscr{C}_c$ for which $P_x^\pi\{N < \infty\} = 1$ and $E_x^\pi N < \infty = 1$, resp. As usual, $\mathscr{C}_c(G_1) = \bigcap_x \mathscr{C}_c(x, G_1)$, etc. Let $\mathscr{T}_c^0(\pi)$, $\mathscr{T}_c(x, \pi)$, and $\mathscr{T}_c^+(x, \pi)$, resp. (and the intersections $\mathscr{T}_c(\pi) = \bigcap_x \mathscr{T}_c(x, \pi)$, etc.) denote the classes of stopping times m for the controlled problem when the control policy is fixed at $\pi \in \mathscr{C}_c$ and either $P_x^\pi\{m < \infty\} \le 1$, $P_x^\pi\{m < \infty\} = 1$ or $E_x^\pi m < \infty$, resp. (and where the properties hold for all x, etc.).

3.4 Optimal Control until a Boundary Is Reached

We return to the cost structure of Section 3.1.2. Suppose that we wish to control the chain until the first time that the set ∂S is reached. Let $\tilde{r}(\cdot, \cdot)$ denote a *bounded real-valued function* on $\{x, \alpha : x \in S - \partial S, \alpha \in \mathscr{U}(x)\}$. *If $\mathscr{U}(x)$ is not discrete and finite, then suppose that $\tilde{r}(x, \cdot)$ is continuous on $\mathscr{U}(x)$.* Assume that either (a) or (b) holds. (Recall that $N = \min\{n : X_n \notin S - \partial S \equiv G_1\}$.)

Case (a) $\min_{x, \alpha} \tilde{r}(x, \alpha) > 0$, and there is at least one† $\pi \in \mathscr{C}_{PM}^+(S - \partial S)$.

Case (b) $\sup_{\substack{\pi \in \mathscr{C}_R \\ x}} E_x^\pi N < \infty$.

† $\mathscr{C}_{PM}^+(S - \partial S) = \mathscr{C}_{PM}^+(G_1) = \bigcap_x \mathscr{C}_{PM}^+(x, G_1)$, where $G_1 = S - \partial S$; i.e., $\pi \in \mathscr{C}_{PM}^+(S - \partial S)$ implies that $E_x^\pi N < \infty$, all x.

Case (a) or (b) can often be verified from the data of the problem, as we shall see in some of the examples in this volume. For each $\pi \in \mathscr{C}_{\mathrm{R}}$, define the cost

$$R(x, \pi) = E_x^\pi \left[\sum_{i=0}^{N-1} \tilde{r}(X_n, U_n) + b(X_N) \right] \tag{4.1}$$

and define

$$V(x) = \inf_{\pi \in \mathscr{C}_{\mathrm{PM}}} R(x, \pi). \tag{4.2}$$

Under (*a*), there is at least one $\pi \in \mathscr{C}_{\mathrm{PM}}$ under which (4.1) is finite for each x and under (b), (4.1) is finite for all $\pi \in \mathscr{C}_{\mathrm{R}}$, $x \in S - \partial S$. Hence under (a) or (b), $V(x)$ is finite for each x. By a formal dynamic programming argument, $V(\cdot)$ satisfies (4.3)

$$\begin{aligned} V(x) &= \inf_{\alpha \in \mathscr{U}(x)} [E_x^\alpha V(X_1) + \tilde{r}(x, \alpha)], \qquad x \in S - \partial S, \\ &= b(x), \qquad x \in \partial S, \end{aligned} \tag{4.3}$$

where for a function $f(\cdot)$, $E_x^\alpha f(X_1)$ is defined by $\sum_y p_{xy}(\alpha) f(y)$. Under either (a) or (b), (4.3) has one and only one solution, which is given by the right-hand side of (4.2) and there is an optimal policy in $\mathscr{C}_{\mathrm{PM}}$ which is stationary. Also, under both (a) and (b),

$$\inf_{\pi \in \mathscr{C}_{\mathrm{R}}} R(x, \pi) = V(x). \tag{4.4}$$

Computational techniques for solving (4.3) are given in Kushner [K5] and Kushner and Kleinman [K9].

Remark on Computational Methods

The methods for solving equations such as (4.3) are very similar to those for the purely linear problem, where there is no infimum operation. A comparison of various techniques appears in Kushner and Kleinman [K9] and we will merely list some alternatives. Let $V^0(\cdot)$ denote an arbitrary bounded vector and assume either Case (a) or (b) (the methods also work for the discounted problem).

Suppose $V^n(\cdot)$ is given. In the *Jacobi method*, $V^{n+1}(\cdot)$ is given by

$$V^{n+1}(x) = \inf_{\alpha \in \mathscr{U}(x)} \left[\sum_y p_{xy}(\alpha) V^n(y) + \tilde{r}(x, \alpha) \right], \qquad x \notin \partial S. \tag{4.5}$$

The boundary conditions are always

$$V^n(x) \equiv b(x), \qquad x \in \partial S. \tag{4.6}$$

Suppose that there are K points in $S - \partial S$ and order them in some way. The *Gauss–Seidel iteration* is

$$V^{n+1}(x) = \inf_{\alpha \in \mathscr{U}(x)} \left[\sum_{y=1}^{x-1} p_{xy}(\alpha) V^{n+1}(y) + \sum_{y=x}^{K} p_{xy}(\alpha) V^n(y) + \tilde{r}(x, \alpha) \right],$$

$$x = 1, \ldots, K. \quad (4.7)$$

Choose a parameter $\omega \geq 1$. The *accelerated Gauss–Seidel iteration* is

$$\tilde{V}^{n+1}(x) = \min_{\alpha \in \mathscr{U}(x)} \left[\sum_{y=1}^{x-1} p_{xy}(\alpha) \tilde{V}^{n+1}(y) + \sum_{y=x}^{K} p_{xy}(\alpha) V^n(y) + \tilde{r}(x, \alpha) \right] \quad x = 1, \ldots, K, \quad (4.8)$$

$$V^{n+1}(x) = \omega \tilde{V}^{n+1}(x) + (1 - \omega) V^n(x)$$

Iterations (4.5) and (4.7) always converge to the correct solution, while (4.8) does so only for ω in an interval $[1, \alpha)$, $\alpha > 1$. Method (4.7) never converges more slowly than method (4.5), and (4.8) is preferable to (4.7) for suitably chosen ω. Usually, the convergence is considerably faster for (4.8) for suitable ω. The ordering of the points of $S - \partial S$ plays an important role in determining the convergence rates of both (4.7) and (4.8). The use of probabilistic intuition can be quite helpful in selecting a good ordering; a clever choice can accelerate the convergence considerably. See [K5, Chapter 4]. Some further remarks appear in Chapter 6.

3.5 Optimal Discounted Cost

Define, for $\pi \in \mathscr{C}_R$ and $\beta \in (0, 1)$,

$$R(x, \pi) = E_x^\pi \left[\sum_{i=0}^{N-1} \beta^i \tilde{r}(X_n, U_n) + \beta^N b(X_n) \right] \quad (5.1)$$

and define

$$V(x) = \inf_{\pi \in \mathscr{C}_{PM}} R(x, \pi). \quad (5.2)$$

The costs are well defined and finite without Case (a) or (b). A straightforward, formal dynamic programming calculation yields that $V(x)$ satisfies

$$\begin{aligned} V(x) &= \inf_{\alpha \in \mathscr{U}(x)} [\beta E_x^\alpha V(X_1) + \tilde{r}(x, \alpha)], && x \in S - \partial S, \\ &= b(x), && x \in \partial S. \end{aligned} \quad (5.3)$$

Equation (5.3) has a unique bounded solution, which is given by the right-hand side of (5.2). There is an optimal policy in $\mathscr{C}_{PM}$ which is stationary and (4.4) holds here also.

3.6 Optimal Stopping and Control

The optimal stopping problem of Section 3.2 is actually a special case of the problem of Section 3.4. To see the correspondence, add a fictitious state $\{\mathbf{0}\}$ to S and let $\partial S = \{\mathbf{0}\}$. Let $\mathscr{U}(x) = \{0, 1\}$ and set the *terminal cost* $b(\mathbf{0})$ in (4.1) equal to zero. For $x \neq \mathbf{0}$ define the controlled transition probabilities $p_{xy}(\cdot)$ and costs $\tilde{r}(x, \alpha)$ by $p_{xy}(0) = p_{xy}$, $\tilde{r}(x, 0) = r(x)$, $p_{x\mathbf{0}}(1) = 1$, and $\tilde{r}(x, 1) = b(x)$, *where this* $b(\cdot)$ *is the stopping cost term of the stopping problem.* With $r(x) \geq r_0 > 0$, we have Case (a) of Section 3.4, where $m + 1 = N = \min\{n: X_n = \mathbf{0}\}$.

In many examples, we are able to control the process up until the stopping time. This is, obviously, still a problem of the type dealt with in Section 3.4, and we will now formulate it in those terms. The set $\mathscr{U}(x)$ will contain the action 1 (stopping) and the various possible control actions other than the stopping action. Set $\mathscr{U}_0(x) = \mathscr{U}(x) - \{1\}$. Again, $\partial S = \{\mathbf{0}\}$ and $m + 1 = N = \min\{n : X_n = \mathbf{0}\}$.

For each $\pi \in \mathscr{C}_R$ and $m \in \mathscr{T}_R^+(x, \pi)$, define the cost

$$R(x, m, \pi) = E_x^\pi\left[\sum_{i=0}^{m-1} \tilde{r}(X_i, U_i) + b(X_m)\right], \qquad x \neq \mathbf{0}, \tag{6.1}$$

and define

$$V(x) = \inf_{\pi \in \mathscr{C}_{PM},\, m \in \mathscr{T}_{PM}^+(x, \pi)} R(x, m, \pi). \tag{6.2}$$

It follows from Section 3.4 that $V(\cdot)$ is the unique solution to the dynamic programming equation

$$V(x) = \min\left\{b(x), \min_{\alpha \in \mathscr{U}_0(x)} [E_x^\alpha V(X_1) + \tilde{r}(x, \alpha)]\right\}, \qquad x \neq \mathbf{0}, \tag{6.3}$$
$$V(\mathbf{0}) = 0.$$

There are optimal π and m in the classes of pure Markov stationary policies and stopping times, resp., and also

$$V(x) = \inf_{\pi \in \mathscr{C}_R,\, m \in \mathscr{T}_R^+(x, \pi)} R(x, m, \pi). \tag{6.4}$$

The optimal stopping set is

$$B = \{x : V(x) = b(x)\}. \tag{6.5}$$

If $V(x) < b(x)$, then we do not stop and an optimal action is that which attains the minimum over $\mathscr{U}_0(x)$ in (6.3).

Define the ε-stopping set

$$B_\varepsilon = \{x : V(x) \geq b(x) - \varepsilon\}$$

and

$$m_\varepsilon = \min\{n : X_n \in B_\varepsilon\}.$$

If π denotes an optimal control policy, then

$$R(x, m_\varepsilon, \pi) \le V(x) + \varepsilon. \tag{6.6}$$

We can easily alter the problem to add the requirement of *compulsory* stopping on first reaching a set A, with associated stopping cost $b(\cdot)$. Equation (6.3) then holds for $x \notin A \cup \{\mathbf{0}\} \equiv \partial S$ and we need to impose the additional boundary conditions $V(x) = b(x)$ for $x \in A$.

3.7 Impulsive Control Systems

There is a generalization of the optimal stopping problem which is of some importance in applications where a sequence of decisions must be made and where one must also determine the optimal timing and "magnitude" of these decisions or actions as they will be called. The Markov chain version of this problem is just a special case of the discounted problem of Section 3.5. However, since the special case will arise when we treat the continuous time version of the problem (which is not a special case of the continuous time version of the discounted problem) we will describe it briefly here.

Let $\{0\}$ represent the decision "*no action.*" Let $Q(x)$ denote the set of *actual control actions*, which is assumed finite for each x. Define $\bar{Q}(x) = Q(x) + \{0\}$. Let $x + \gamma \in S$ for each $x \in S$, $\gamma \in Q(x)$. With $Q(x)$, q_i, and Q_i replacing $\mathscr{U}(x)$, u_i, and U_i, resp., we define the control law $\pi = \{q_0, q_1, \ldots\}$; Q_i denotes the *actual sample value* of the control action at time i. Let $\beta \in (0, 1)$ and let $p(\cdot, \cdot)$ be a bounded function on $\{x, \gamma\colon x \in S, \gamma \in \bar{Q}(x)\}$, with $p(\cdot, 0) = 0$. Define the transition probabilities

$$\begin{aligned} p_{xy}(\gamma) &= 1 \qquad \text{if} \quad y = x + \gamma, \quad \gamma \in Q(x) \\ p_{xy}(0) &= p_{xy}. \end{aligned} \tag{7.1}$$

Again, let $\mathscr{C}_{\mathrm{PM}}$ and $\mathscr{C}_{\mathrm{R}}$ denote the sets of pure Markov and randomized control laws, resp. The control law is called impulsive owing to the "sudden" and deterministic change in the state which is implied by (7.1) when there is an action with value in $Q(x)$.

For each $\pi \in \mathscr{C}_{\mathrm{R}}$, define the cost functional

$$R(x, \pi) = E_x^\pi \sum_{i=0}^{\infty} \beta^i [p(X_i, Q_i) + r(X_i)] \tag{7.2}$$

and

$$V(x) = \inf_{\pi \in \mathscr{C}_{\mathrm{PM}}} R(x, \pi). \tag{7.3}$$

The function defined by the right-hand side of (7.3) is the unique solution to the dynamic programming equation

$$V(x) = \min\left\{\beta E_x^0 V(X_1) + r(x), \inf_{\gamma \in Q(x)} [\beta E_x^\gamma V(X_1) + p(x, \gamma) + r(x)]\right\}. \quad (7.4)$$

The E_x^0 implies that the transition probabilities are p_{xy}. There is a pure Markov stationary optimal control, and we can also replace $\mathscr{C}_{\text{PM}}$ by $\mathscr{C}_{\text{R}}$ in (7.3) without changing the value.

An alternative notation, which is useful in the continuous time version of the model, involves viewing the problem as one in which there is a sequence of action times $m_1, m_2, \ldots$ and a sequence of corresponding actions $\bar{Q}_1, \bar{Q}_2, \ldots$, where $\bar{Q}_i = Q_{m_i}$ and $\bar{Q}_i \in Q(X_{m_i})$. $\bar{Q}_i$ is the ith actual (nonzero) action. We then rewrite $R(x, \pi)$ in the form $R(x, \{m_i, \bar{Q}_i\})$. The $\{m_i\}$ is a sequence of nondecreasing stopping times with respect to an appropriate sequence of nondecreasing σ-algebras. In this sense the model is an extension of the stopping problem; instead of one action or transaction taking place, there is a sequence of actions or transactions and we must determine their timing and value. The functions $p(\cdot, \cdot)$ and $r(\cdot)$ represent the transactions cost and the "running" rate, resp. The action set is defined by

$$B = \left\{x : V(x) = \inf_{\gamma \in Q(x)} (\beta E_x^\gamma V(X_1) + p(x, \gamma) + r(x))\right\} \quad (7.5)$$

and any minimizing function $\gamma(x)$ yields an optimal policy. It is not necessary that $P_x\{m_i < \infty\} = 1$; for, indeed, we may possibly never have an action.

The problem can be generalized by allowing a continuously acting control. Reintroduce the control set $\mathscr{U}(x)$ and suppose that at each state x and time n we have a choice of selecting a control from $Q(x)$ or from $\mathscr{U}(x)$: If we select a control value γ from $Q(x)$, then the transition probabilities are of the impulsive form (7.1); if we select a control value α from $\mathscr{U}(x)$, then the transition probabilities take the value $p_{xy}(\alpha)$ as discussed in Section 3.3. Then we assume that $\mathscr{U}(x)$ is either discrete and finite or compact and in the latter case, $p_{ij}(\cdot)$ is assumed continuous. The control policy takes the form $\pi = (u_1, q_1, u_2, q_2, \ldots)$ and the cost functional is (7.2) with $\tilde{r}(X_i, U_i)$ replacing $r(X_i)$ there. The dynamic programming equation is still (7.4) but with $\beta E_x^0 V(X_1) + r(x)$ replaced by

$$\inf_{\alpha \in \mathscr{U}(x)} [\beta E_x^\alpha V(X_1) + \tilde{r}(x, \alpha)] \quad (7.6)$$

and all the previous conclusions concerning the optimal policy continue to hold. In order to emphasize the impulsive control aspect of the problem, we may write $R(x, \pi)$ as $R(x, \{m_i, U_i, \bar{Q}_i\})$, where $\bar{Q}_i$ is the ith actual impulsive action as previously defined.

A stopping boundary ∂S and boundary cost $b(\cdot)$ can clearly be added to the formulation.

3.8 Control over a Fixed Time Interval

There is a natural "control" analogy for the chain with cost structure (1.6) and the reader is referred to Sections 3.1.3 and 3.3 for the terminology. Let $\tilde{r}_1(\cdot, \cdot, \cdot)$ be a real-valued function, continuous in its third argument and defined on

$$\{x, n, \alpha : x \in S - \partial S,\ n = 1, \ldots, M,\ \alpha \in \mathscr{U}(x)\}.$$

For $\pi \in \mathscr{C}_{\mathrm{R}}$, define the cost

$$R(x, n, \pi) = E_{x,n}^{\pi}\left[\sum_{i=n}^{(M \cap N)-1} \tilde{r}_1(X_i, i, U_i) + b_T(X_M) I_{\{M<N\}} + b_1(X_N, N) I_{\{M \ge N\}}\right], \quad x \notin \partial S, \quad n < M, \tag{8.1}$$

with boundary condition

$$\begin{aligned} R(x, M, \pi) &= b_T(x), && x \in S - \partial S, \\ R(x, n, \pi) &= b_1(x, n), && x \in \partial S, \quad n \le M. \end{aligned} \tag{8.2}$$

Define

$$V(x, n) = \inf_{\pi \in \mathscr{C}_{\mathrm{PM}}} R(x, n, \pi). \tag{8.3}$$

Then the right-hand side of (8.3) is the unique solution to the dynamic programming equation (8.4) with boundary conditions (8.2):

$$V(x, n) = \inf_{\alpha \in \mathscr{U}(x)} [E_{x,n}^{\alpha} V(X_{n+1}, n+1) + \tilde{r}_1(x, n, \alpha)], \quad n < M, \quad x \notin \partial S. \tag{8.4}$$

There is an optimal pure Markov control law and (8.3) continues to hold if $\mathscr{C}_{\mathrm{PM}}$ is replaced by $\mathscr{C}_{\mathrm{R}}$.

3.9 Linear Programming Formulation of the Markov Chain Control Problems

The control problems on Markov chain models with which we previously dealt can be formulated as linear programming problems (see Derman [D1], Kushner and Kleinman [K10], Kushner and Chen [K8]. Most of the following results can be found in the second reference). There does not seem to be any numerical advantage to doing so if we stick to the problem formulations of Sections 3.2–3.8. However, in many cases, there are additional constraints on the state occupancies or on the control usage, which cannot easily be accounted for in the usual dynamic programming formulation but where a linear programming formulation will still be valid. For convenience, we will assume that $\mathscr{U}(x)$ contains exactly L elements, although we can also treat the case of convex $\mathscr{U}(x)$ via a generalized programming approach. Let K be the number of states in S and set $S = \{1, \ldots, K\}$. There is no loss of generality in supposing that $\mathscr{U}(x) = \{1, 2, \ldots, L\}$, which we shall do.

If there are constraints of the type that we will be concerned with, then there may not be an optimal control for the constrained problem in the class of pure Markov controls and it may be necessary to consider randomized controls. Generally, we will need to consider only the class of stationary random Markov controls $\mathscr{C}_{SRM}$. Each $\pi = (u, u, \ldots)$ in $\mathscr{C}_{SRM}$ is characterized by a collection of real numbers $\{\gamma_{ij}, i = 1, \ldots, M; j = 1, \ldots, L\}$, where

$$\gamma_{ij} \geq 0, \quad \gamma_i = \{\gamma_{i1}, \ldots, \gamma_{iL}\}, \qquad \sum_j \gamma_{ij} = 1.$$

Thus, for each $i \in S$, γ_i is a distribution over the control actions $\mathscr{U}(i)$. For such a control law π, we have the transition probabilities

$$\begin{aligned} P^{\pi}\{X_{n+1} = j \mid X_0, \ldots, X_n = i\} &= P^{\pi}\{X_{n+1} = j \mid X_n = i\} \\ &= \sum_k p_{ij}(k)\gamma_{ik} = p_{ij}(\gamma_i). \end{aligned}$$

Thus the chain $\{X_n\}$ is a Markov chain under the law π even though π is not pure Markov.

The usefulness of the introduction of randomized Markov laws can be seen by a simple example. Suppose that there is one constraint and only two possible pure strategies, π_1 and π_2. Suppose that π_1 satisfies the constraint but gives a large cost, while π_2 does not satisfy the constraint but gives a small cost. It is not hard to see that randomizing "fills in the gaps" and yields a strategy which satisfies the constraint and at a smaller cost than given by π_1.

Control until the Boundary Is Reached

We consider the problem of Section 3.4. The sets S and ∂S are defined as previously *but we modify the problem by adding a fictitious rest or absorbing state* $\{\mathbf{0}\}$, and extend $p_{ij}(\cdot)$, and $\tilde{r}(i, \cdot)$ by defining

$$\begin{aligned} &p_{i\mathbf{0}}(\alpha) = 1 \quad \text{for } i \in \partial S, \text{ all } \alpha \in \mathscr{U}(i); \qquad p_{\mathbf{00}}(\alpha) \equiv 1, \\ &\tilde{r}(i, \alpha) = b(i) \quad \text{for } i \in \partial S, \text{ all } \alpha; \qquad \tilde{r}(\mathbf{0}, \alpha) \equiv 0. \end{aligned} \tag{9.1}$$

Let M_{ij} denote the average number of times that $X_n = i$ and control action $U_n = j \in \mathscr{U}(i)$ occur simultaneously and let $(\mu_1, \ldots, \mu_K)$ denote the initial probability distribution; namely, $P\{X_0 = i\} = \mu_i$, $i = 1, \ldots, K$. Of course, $\{M_{ij}\}$ depends on the control law. We shall see how to determine the optimum $\{M_{ij}\}$ over all laws in $\mathscr{C}_{\text{SRM}}$. The optimum $\{M_{ij}\}$ will determine the optimum control law. In the Markov dynamic programming formulation, it is not necessary to specify μ; the solution is optimal for all μ. In the linear programming formulation, if $\mu_i > 0$ for all i, then the resulting law is optimal for all other initial μ if there are no additional constraints of the type (9.4).

Under any $\pi \in \mathscr{C}_{\text{SRM}}$, the linear system (9.2) of relations among the M_{ij}

$$\begin{aligned} &\sum_l M_{il} \equiv M_i = \mu_i + \sum_{k,j} p_{ji}(k) M_{jk}, \qquad \text{all} \quad i, j \neq \mathbf{0}, \\ &M_{ij} \geq 0 \end{aligned} \tag{9.2}$$

is a direct consequence of the Chapman–Kolmogorov "conservation of probability" equation for the steady state probabilities under the law π. If $\{M_{ij}\}$ solves (9.2), then $\gamma_{ij} = M_{ij}/M_i$ can be interpreted to be the probability that the control action $j \in \mathscr{U}(i)$ is used when the state is i. If $E_i^\pi N < \infty$, then (4.1) is well defined for $x = i$, and the cost equation (4.1) takes the form [recall the modifications (9.1)]

$$R(x, \pi) = E_x^\pi \sum_{n=0}^{\infty} \tilde{r}(X_n, U_n) = \sum_{i,j} M_{ij} \tilde{r}(i, j) \equiv z, \qquad \mu_x = 1, x \neq \mathbf{0}. \tag{9.3}$$

The dynamic programming problem is equivalent to minimizing z under the constraints (9.2) and conversely. The simplex procedure will yield a pure Markov control (where $\gamma_{ij} = 1$ or 0) in this case. Under (a) or (b) of Section 3.4, z has a finite minimum under (9.2). In fact, the dual system to (9.2) and (9.3) is precisely the dynamic programming equation (4.3).

State Space Constraints

For any set of real numbers $\{\alpha_{ij}^s, i = 1, \ldots, K; j = 1, \ldots, L; s = 1, \ldots\}$, $\{b_s, s = 1, \ldots\}$, the linear constraints (9.4) can be added to (9.2)

$$\sum_{i,j} \alpha_{ij}^s M_{ij} \leq b_s, \qquad s = 1, \ldots. \tag{9.4}$$

If l such constraints are added, then the simplex procedure will yield an optimal policy (if one exists) with which the controls for as many as l states are randomized. Under Case (a) or (b) of Section 3.4, there is an optimal policy in $\mathscr{C}_{\mathrm{SRM}}$ with finite cost if there is one policy in $\mathscr{C}_{\mathrm{SRM}}$ which guarantees satisfaction of (9.4) with a corresponding finite cost (9.3).

Examples Suppose that we want to constrain the use of an action α_0, so that α_0 is used no more than b_1 times on the average, until the boundary ∂S is reached. Then $l = 1$ in (9.4), and (9.4) has the form

$$\sum_i M_{i\alpha_0} \leq b_1.$$

For another example, suppose that we want to constrain the average number of times that a set $S_1 \subset S - \partial S$ is hit before ∂S is hit to be no more than b_1. Then (9.4) takes the form

$$\sum_{i \in S_1,\, j} M_{i1} \leq b_1.$$

The Optimal Stopping Problems

To put the optimal stopping into the framework of (9.2)–(9.3), set $\mathscr{U}(x) = \{1, 0\}$, $1 \equiv$ stop, $0 \equiv$ continue, $\partial S =$ empty. We will retain the fictitious absorbing or rest state $\{\mathbf{0}\}$ and use (9.5) in lieu of (9.1).

$$\begin{aligned} &\tilde{r}(i, 0) = r(x), \qquad \tilde{r}(i, 1) = b(x), \qquad b(\mathbf{0}) = r(\mathbf{0}) = 0, \\ &p_{ij}(0) = p_{ij}, \qquad p_{i\mathbf{0}}(1) = 1; \qquad i, j \neq \mathbf{0}. \end{aligned} \tag{9.5}$$

The following types of constraints may be of interest. If we wish (E_μ^π denotes the expectation under π, μ denotes the probability measure on X_0 and m the stopping time)

(a) $E_\mu^\pi m \leq b_2$ (bound on the average stopping time), then set

$$\sum_i M_{i0} \leq b_2.$$

If we wish

(b) $P_\mu^\pi\{X_m \in S_1\} \leq b_3$ (bound on probability of stopping in the set S_1), then set

$$\sum_{i \in S_1} M_{i1} \leq b_3.$$

Discounted Control until a Boundary Is Reached

If state space constraints are added to the problem of Section 3.5, we again may not have an optimum policy in the class of pure Markov controls but there is one in $\mathscr{C}_{\mathrm{SRM}}$.

Define the discounted occupancy vector

$$\tilde{M}_{ij} = E_{\mu}^{\pi} \sum_{n=0}^{\infty} I_{ij}^{n} \beta^{n},$$

where $I_{ij}^{n} = 1$ if $X_n = i$ and $U_n = j$. The $\{\tilde{M}_{ij}\}$ satisfy constraints [K10]

$$\sum_{l} \tilde{M}_{il} \equiv \tilde{M}_{i} = \mu_i + \beta \sum_{k,j}^{\infty} p_{ji}(k) \tilde{M}_{jk}, \qquad \text{all } i \neq \mathbf{0},$$
$$\tilde{M}_{ij} \geq 0. \tag{9.6}$$

State space and control usage constraints of the discounted form

$$\sum_{i,j} \alpha_{ij}^{s} \tilde{M}_{ij} \leq b_s, \ldots, s = 1, \ldots \tag{9.7}$$

can be considered and cost (5.1) can be written as [notation (9.1) for $\tilde{r}(\cdot, \cdot)$]

$$R(\mu, \pi) = E_{\mu}^{\pi} \sum_{i=0}^{\infty} \beta^{n} \tilde{r}(X_n, U_n) = \sum_{i,j} \tilde{M}_{ij} \tilde{r}(i, j) \equiv z. \tag{9.8}$$

If the constraints (9.6) and (9.7) are feasible, then there is a feasible set $\{\tilde{M}_{ij}\}$ which minimizes z and the minimizing $\{\tilde{M}_{ij}\}$ yields the optimal policy via $\gamma_{ij} = \tilde{M}_{ij}/\tilde{M}_{i}$, where feasibility is simply a question of whether there is a $\pi \in \mathscr{C}_{\text{SRM}}$ which guarantees (9.7).

There are also linear programming versions of the optimal impulsive control and finite time optimal control problems.

CHAPTER 4

Elliptic and Parabolic Equations and Functionals of Diffusions

One aim of this chapter is to describe some further properties of the solution to stochastic differential equations when there is either a control or when the Lipschitz condition is replaced by a weaker condition. We shall be concerned with properties such as causality and uniqueness; also, several classes of control functions will be defined. These results and definitions will be used in later chapters.

The techniques introduced from Chapter 5 onward will deal with methods for approximating values of a large class of functionals of diffusions and of optimal costs and controls in optimal control problems with diffusion models. Since these functionals are often weak solutions of degenerate or nondegenerate elliptic or parabolic partial differential equations, we will also be able to calculate approximate values to the solutions to these equations. We will state some of the functionals of interest and briefly describe some of their relations to partial differential equations. Heavy use will be made of Strook and Varadhan [S5] and Dynkin [D4, Chapter 13]. See also

Fleming and Rishel [F1], Friedman [F3], and Karoui [K1], where a number of results concerning the partial differential equations satisfied by functionals of diffusions are discussed.

The relations between diffusions and linear elliptic and parabolic equations and between optimally controlled diffusions and nonlinear elliptic and parabolic equations will be treated in a partly formal manner. We aim to cite some of the main relationships and to set up part of the structure of equations, processes, and terminology for later use.

4.1 Assumptions and Uniqueness Results: No Control

Assume

A4.1.1 *In the homogeneous case, let $f(\cdot)$ and $\sigma(\cdot)$ denote bounded continuous R^r and $r \times r$ matrix-valued functions, resp., on R^r. In the nonhomogeneous case, let $f(\cdot, \cdot)$, $\sigma(\cdot, \cdot)$ be bounded and continuous on $R^r \times [0, \infty)$. Suppose that, for each $X(0) = x$, (1.5.1) has a unique (nonanticipative) solution in the sense of probability distributions; i.e., any two nonanticipative solutions (perhaps corresponding to different Wiener processes) induce the same distributions on $C^r[0, \infty)$. Assumption* A4.1.1 *(homogeneous or nonhomogeneous case) will be used throughout the book whenever there is no control function explicitly introduced.*

The uniqueness property has some important implications. For simplicity of notation, only the homogeneous case will be treated. In Theorem 4.1.1, define $(\Omega', \mathscr{B}', \mathscr{B}'_t, P'_x)$ as follows: $\Omega' = C^r[0, \infty)$, $\mathscr{B}'$ = Borel algebra on $C^r[0, \infty)$, $\mathscr{B}'_t$ is the smallest sub-σ-algebra of $\mathscr{B}'$ containing the "*projections*" $\{x(\cdot): x(s) < \alpha\}$ for all $s \le t$ and all $\alpha \in R^r$, and $P'_{x,w}$ is the regular conditional probability of $X(\cdot)$ given $w(\cdot)$ when $X(0) = x$. Let P denote Wiener measure on Ω'. Define $\Omega = \Omega' \times \Omega'$, $P_x = P'_{x,w} \times P$, and let $\mathscr{B}$ and $\mathscr{B}_t$ denote the smallest σ-algebras on Ω containing $\mathscr{B}' \times \mathscr{B}'$ and $\mathscr{B}'_t \times \mathscr{B}'_t$, resp. The first Ω' will "carry" $X(\cdot)$ and the second $w(\cdot)$ and we can assume that $X(\cdot)$, $w(\cdot)$, $X(0) = x$, are defined on $(\Omega, \mathscr{B}, \mathscr{B}_t, P_x)$ for each x. Let P'_x denote the measure that $X(\cdot)$ induces on $C^r[0, \infty)$ when $X(0) = x$.

Theorem 4.1.1 *Assume* A4.1.1. *Then $X(\cdot)$ is a strong Markov and a Feller process on $(\Omega, \mathscr{B}, \mathscr{B}_t, P_x, x \in R^r)$.*

Proof For any bounded set A, the family $\{P'_x, x \in A\}$ is tight on $C^r[0, \infty)$. Let $\{x_n\}$ denote a bounded sequence which tends to x, as $n \to \infty$, and let P'_{x_n} be weakly convergent with limit $\bar{P}$. The limit process solves (1.5.1) with $X(0) = x$ and some Wiener process $w(\cdot)$ (although we may have to augment the probability space by adding an independent Wiener process). By the uniqueness, $\bar{P} = P'_x$. Thus, the subsequence is irrelevant. Now, let $g(\cdot)$ denote a bounded and continuous real-valued function on R^r. This convergence implies that $E_{x_n} g(X(t)) \to E_x g(X(t))$ as $n \to \infty$; i.e., the process is Feller. Since the Feller property together with the Markov property imply the strong Markov property, we need only show the Markov property.

Let A be an open set in R^r, and $g_n(\cdot)$ a sequence of continuous functions which converges to $I_A(\cdot)$. Since $E_x g_n(X(t))$ is continuous in x, and $E_x g_n(X(t)) \to P(x, t, A)$, where $P(x, t, A) = P_x\{X(t) \in A\}$, the function $P(\cdot, t, A)$ is measurable for each open A and each $t \geq 0$. This implies that it is measurable for each t and Borel set A. We have

$$X(t+s) = X(t) + \int_t^{t+s} f(X(u))\, du + \int_t^{t+s} \sigma(X(u))\, dw(u),$$

where $w(u) - w(t)$, $u \geq t$, is independent of $X(t)$ for each $t \geq 0$. By uniqueness, the distribution of $X(t+s)$, given $X(u)$, $u \leq t$ (or, given $X(u)$, $w(u)$, $u \leq t$) with $X(t) = y$, must be $P(y, s, \cdot)$. Thus the uniqueness and the measurability of each $P(\cdot, t, A)$ imply that

$$P_x\{X(t+s) \in A \mid \mathscr{B}_t\} = P(X(t), s, A)$$

w.p. 1 for each $s \geq 0$, $t \geq 0$, and Borel set A. Q.E.D.

4.2 Functionals of Uncontrolled Diffusions

Let G be a *bounded open set with compact closure* $\bar{G}$ *and boundary* ∂G. The following process $X(\cdot)$ is the diffusion (1.5.1). For all functionals except (2.4), we suppose that $f(\cdot)$, $\sigma(\cdot)$ do not depend on t. Define

$$\tau = \min\{t : X(t) \notin G\}, \qquad \tau' = \inf\{t : X(t) \notin \bar{G}\}.$$

As usual, if $X(t) \in G$ (resp. $\bar{G}$), all $t < \infty$, then $\tau = \infty$ (resp., $\tau' = \infty$) for that path. Then (Dynkin [D4, Chapter IV.1]), τ and τ' are stopping times (perhaps taking infinite values). Let $k(\cdot)$ and $b(\cdot)$ denote bounded real-valued continuous functions on $\bar{G}$ and ∂G, resp.

Next, several functionals of $X(\cdot)$ will be defined. In subsequent sections, we will discuss equations which these functionals satisfy.

For each $x \in G$ for which $E_x \tau < \infty$, define

$$R(x) = E_x \left[\int_0^\tau k(X(s))\, ds + b(X(\tau)) \right]. \tag{2.1}$$

Let λ_0 be a positive number and $\lambda(\cdot)$ be a real-valued bounded continuous function on $\bar{G}$ satisfying

$$\min_{x \in \bar{G}} \lambda(x) \geq \lambda_0 > 0.$$

Define

$$R(x) = E_x \int_0^\tau A(t)k(X(t))\, dt + E_x A(\tau)b(X(\tau)), \tag{2.2}$$

where

$$A(0, t) \equiv A(t), \qquad A(u, t) = \exp - \int_u^t \lambda(X(s))\, ds. \tag{2.3}$$

Let $T > 0$ denote a given real number, and $b_1(\cdot, \cdot)$, $b_T(\cdot)$ real-valued bounded continuous functions on $\partial G \times [0, T]$ and $\bar{G}$, resp., and define the functional [here $k(\cdot)$, $f(\cdot)$, $\sigma(\cdot)$ can depend on t if we wish]

$$\begin{aligned} R(x, t) = E_{x, t} \int_t^{T \cap \tau} k(X(s))\, ds + E_{x, t} b_1(X(T \cap \tau), T \cap \tau) I_{\{\tau \leq T\}} \\ + E_{x, t} b_T(X(T)) I_{\{\tau > T\}} . \end{aligned} \tag{2.4a}$$

The discounted version of (2.4a) is

$$\begin{aligned} R(x, t) = E_{x, t} \int_t^{T \cap \tau} [A(t, s)] k(X(s))\, ds \\ + E_{x, t}[A(t, T \cap \tau)] b_1(X(T \cap \tau), T \cap \tau) I_{\{\tau \leq T\}} \\ + E_{x, t}[A(t, T)] b_T(X(T)) I_{\{\tau > T\}}. \end{aligned} \tag{2.4b}$$

Finally, let $\mu(\cdot)$ be an invariant measure for the process $X(\cdot)$. Let $P(x, t, \cdot)$ define the measure with values $P_x\{X(t) \in A\} = P(x, t, A)$. Suppose that $P(x, t, \cdot)$ converges weakly to $\mu(\cdot)$, as $t \to \infty$, and define γ by

$$\gamma = \lim_{t \to \infty} \frac{1}{t} \int_0^t E_x k(X(s))\, ds = \int k(y)\mu(dy). \tag{2.5}$$

If in (2.1), (2.2), and (2.4), τ' replaces τ, we denote the corresponding functional by R'.

4.3 Partial Differential Equations Associated with Functionals of Diffusions. $a(\cdot)$ Uniformly Positive Definite

If $a(\cdot)$ is uniformly positive definite [i.e., there is a real $K > 0$ such that $x'a(y)x \geq K|x|^2$ for all x, y] then there exists a unique solution to (1.5.1) under weaker conditions on $f(\cdot)$, $\sigma(\cdot)$ than the Lipschitz condition.

Theorem 4.3.1 (Dynkin [D4, Theorem 5.11]) *Let $a_{ij}(\cdot), f_i(\cdot)$ be bounded and satisfy a uniform Hölder condition and let $a(\cdot)$ be uniformly positive definite. Then there exists a homogeneous diffusion process with differential generator $\mathscr{L}$ (see* 1.5.14). *Its transition density $p(x, t, y)$ [the density of the Markov transition function $P(x, t, dy)$] is the fundamental solution of the differential equation (in variables x, t, where x = initial state)*

$$(\partial u/\partial t)(x, t) = \mathscr{L}u(x, t),$$

with initial condition (3.1)

$$u(x, t) \to \delta(x - y) \qquad \textit{as } t \to 0.$$

The process is strong Feller and Markov, and is a unique solution to (1.5.4).

Actually, Dynkin's definition of a diffusion process is not quite the same as merely satisfying (1.5.1) and he does not actually prove that (1.5.1) is satisfied by what he calls a diffusion process, in the sense that there is a solution to (1.5.1) for each $x = X(0)$, for the same $w(\cdot)$ process. However, the function $P(\cdot, \cdot, \cdot)$ can be used to define a family of measures $\{P_x\}$ (on, say, $C^r[0, \infty)$) which generate a homogeneous strong Markov strong Feller process $X(\cdot)$ (on the sample space $C^r[0, \infty)$, with the appropriate σ-algebras) and it can be shown that

$$E_x[X(t + \Delta) - X(t) \mid X(s), s \le t] = f(X(t))\Delta + o(\Delta),$$

$$\mathrm{cov}_x[X(t + \Delta) - X(t) \mid X(s), s \le t] = 2a(X(t))\Delta + o(\Delta),$$

$$P_x\{|X(\Delta) - x| > \varepsilon\} = o(\Delta),$$

where $o(\Delta)$ is uniform in (ω, t). These last estimates imply† that there is a Wiener process $w(\cdot)$, such that $X(\cdot)$ is nonanticipative with respect to $w(\cdot)$ and (w.p. 1) Eq. (1.5.1) is satisfied. The process $w(\cdot)$ may depend on x.

A similar construction exists for the nonhomogeneous case. A proof of the fact that (1.5.1) has a unique solution (which is a strong Markov, strong Feller process) under weaker conditions than those of Theorem 4.3.1 [still with $a(\cdot)$ uniformly positive definite] appears in Strook and Varadhan [S3]. The proof of uniqueness is much more difficult than that given in Theorem 1.5.1 since we cannot use the easy and explicit calculations associated with the Lipschitz condition.

Equation (3.1) is known as Kolmogorov's backward equation [since it is an equation in the *initial condition* x of the transition density $p(x, t, y)$].

† The implication is not entirely obvious. The equations imply that $X(t) - x - \int_0^t f(X(s))\, ds$ is a continuous martingale with quadratic variation $\int_0^t 2a(X(s))\, ds$. The result then follows from Section 1.4.4 (see also Doob [D2, pp. 286–291]), provided that we augment the probability space by adding an independent Wiener process.

Suppose that the derivatives $\partial f_i(x)/\partial x_i$, $\partial a_{ij}(x)/\partial x_i$, $\partial^2 a_{ij}(x)/\partial x_i\,\partial x_j$ are bounded and satisfy a Hölder condition on R^r. Then (Dynkin [D4, Vol. I, p. 168]) for each x, $p(x, \cdot, \cdot)$ satisfies the adjoint equation to (3.1) (in the *current* variables t, y):

$$\frac{\partial p}{\partial t} = \mathscr{L}^* p = \sum_{i,j=1}^{r} \frac{\partial^2 [a_{ij}(y)p]}{\partial y_i\,\partial y_j} - \sum_{i=1}^{r} \frac{\partial}{\partial y_i}[f_i(y)p]. \tag{3.2}$$

In Chapter 7, we shall study numerical approximations to the (weak sense) transition density function $p(\cdot, \cdot, \cdot)$, without the positivity assumption on $a(\cdot)$, provided only that A4.1.1 holds.

A boundary point x is said to be regular (τ') *if*

$$\lim_{y \to x,\, y \in G} P_y\{\tau' \geq \varepsilon\} = 0 \qquad \textit{for all } \varepsilon > 0 \quad \textit{and all such sequences } \{y\}. \tag{3.3}$$

The boundary ∂G is regular (τ') *if each $x \in \partial G$ is regular* (τ'). *There are equivalent definitions for regular* (τ). *Clearly, regular* (τ') *implies regular* (τ) *since $\tau \leq \tau'$.*

Lemma 4.3.1 *The condition* (3.4) *implies* (3.3)

$$P_x\{\tau' > 0\} = 0 \qquad \textit{at} \quad x \in \partial G. \tag{3.4}$$

If $x \in \partial G$ is not an absorbing point of the process, then (3.3) *implies* (3.4).

PROOF We shall give only an outline of the first statement and not prove the second one. For each bounded set A, the family of measures $\{P_y, y \in A\}$ induced by Eq. (1.5.1) is tight on $C^r[0, \infty)$. If, say, $y_n \to x$ and $\{P_{y_n}\}$ converges weakly to a measure P, then there is a Wiener process $w(\cdot)$ and a corresponding solution $X(\cdot)$ to (1.5.1) with $X(0) = x$, such that $X(\cdot)$ induces the probability P on $C^r[0, \infty)$. By uniqueness of the measures induced by (1.5.1) for each fixed initial condition (Assumption A4.1.1), $P_x = P$.

Now let $y_n \to x$, $y_n \in G$, $x \in \partial G$ and assume (3.4). The set $\{\tau' \geq \varepsilon\}$ is a closed set in $C^r[0, \infty)$, and by the result of the last paragraph $P_{y_n} \to P_x$ weakly. Hence (Theorem 2.1)

$$\overline{\lim_{y_n \to x}} P_{y_n}\{\tau' \geq \varepsilon\} \leq P_x\{\tau' \geq \varepsilon\} = 0,$$

which proves the assertion.

Another useful fact is that $P_x\{\tau' > 0\} = 0$ or 1 by the zero–one law for Markov processes (see Dynkin [D4, Corollary 1, p. 84 and Theorem 3.3, p. 87], and note that $\{\tau' > 0\} \in \mathscr{N}_{0+}$ and $\mathscr{N}_{0+} \subset \mathscr{M}_{0+}$, in Dynkin's terminology).

Now let us continue with the case where $a(\cdot)$ is uniformly positive definite. We say that *a point $y \in \partial G$ can be touched by an open cone from the outside if* *there is an open cone* C *and an* $\varepsilon > 0$ *such that* (ϕ = empty set)

$$\{x : x - y \in C, \ |x - y| < \varepsilon\} \cap G = \phi.$$

If $y \in \partial G$ and there is a neighborhood N_y of y and a function $\phi(\cdot)$ on N_y, which is differentiable at y and such that $\phi_x(y) \neq 0$, and if

$$G \cap N_y = \{x : \phi(x) > 0\} \cap N_y,$$

then ∂G is said to be *differentiable* at y. If ∂G is differentiable at y, then it can be touched by an open cone from the outside at y (Dynkin [D4, Lemma 13.4]). All points on ∂G which can be touched by an open cone from the outside are regular (τ') (Dynkin [D4, Theorem 13.8]). Actually Dynkin uses (3.4) as the definition of regularity (τ'). By Lemma 4.3.1, that definition is equivalent to the one used here, provided that the point is not absorbing (which is true if $a(\cdot)$ is uniformly positive definite).

The next theorem gives (strong) conditions under which (2.2) actually satisfies a partial differential equation.

Theorem 4.3.2 (Dynkin [D4, Theorem 13.16]) *Let* $\lambda(\cdot)$, $k(\cdot)$ *satisfy a Hölder condition and* $\lambda(x) \geq 0$ *(rather than* > 0*) and let* ∂G *be regular* (τ). *Then, under the additional conditions of Theorem* 4.3.1, *the functional* (2.2) *is twice continuously differentiable and satisfies*

$$\begin{aligned} &\mathscr{L}V(x) - \lambda(x)V(x) + k(x) = 0, \qquad x \in G, \\ &V(y) \to b(x), \qquad \textit{as} \quad y \to x \in \partial G. \end{aligned} \tag{3.5}$$

The positive definiteness condition on $a(\cdot)$ seems to be satisfied only rarely in applications to stochastic control theory. In many cases, the $R(\cdot)$ are not even continuous. There are important cases where $a(\cdot)$ is not strictly positive definite, but where the solution to (1.5.1) is still a strong Markov and strong Feller process; in particular, where a continuous (in x, y) transition density $p(x, t, y)$ exists: For an example, consider the case

$$dX_1 = X_2 \, dt, \qquad dX_2 = g(X)\, dt + dw, \tag{3.6}$$

where $g(\cdot)$ is a suitably smooth function. A number of results concerning smoothness of functionals of processes of the type (3.6) and related partial differential equations appear in Fleming and Rishel [F1]. We will now go directly to the general degenerate case.

4.4 $a(\cdot)$ Degenerate

By $R(\cdot)$ (resp., $R'(\cdot)$), we mean the functional given by (2.2) (resp., (2.2) with τ' replacing τ). In the degenerate case, neither $R(\cdot)$ nor $R'(\cdot)$ may be continuous at all $x \in \bar{G}$. If $R'(\cdot)$ is to be continuous at $x \in \partial G$, we usually need that x be a regular (τ') point and similarly for $R(\cdot)$ (with τ replacing τ' in the definition of regularity). We will next cite some results of Strook and Varadhan [S5]. *Assume that* $f(\cdot) \in C^1(R^r)$, $a(\cdot) \in C^2(R^r)$, *and that there is a twice continuously differentiable real-valued function* $\phi(\cdot)$ *defined on some neighborhood* N of ∂G *such that*

$$\partial G = \{x : \phi(x) = 0\} \cap N,$$
$$G \cap N = \{x : \phi(x) > 0\} \cap N,$$
$$|\phi_x(x)| > 0 \qquad \text{on} \quad \partial G.$$

The boundary function $\phi(\cdot)$ can be defined "locally," but we will not do so.

Define the sets

$$\Sigma_3 = \{x : x \in \partial G,\ \phi_x'(x)a(x)\phi_x(x) > 0\},$$
$$\Sigma_2 = \{x : x \in \partial G - \Sigma_3,\ (f(x) - \tilde{a}(x))'\phi_x(x) < 0\},$$

where the ith component of the vector $\tilde{a}(x)$ is defined by

$$\sum_j (\partial/\partial x_j)a_{ij}(x),$$

and define

$$\Sigma = \Sigma_2 \cup \Sigma_3,$$
$$\Sigma_2^* = \{x : x \in \partial G - \Sigma_3,\ \mathscr{L}\phi(x) < 0\},$$
$$\Sigma^* = \Sigma_2^* \cup \Sigma_3.$$

[Note that the $a(\cdot)$ defined in Strook and Varadhan [S5] is twice our $a(\cdot)$.] Define

$$\Gamma = \text{set of regular } (\tau') \text{ points on } \partial G.$$

Then [S5, Section 7]

$$\Sigma \cup \Sigma^* \subset \Gamma \subset \bar{\Sigma} = \bar{\Sigma}^*, \tag{4.1}$$

$$P_x\{\tau = \tau'\} = 1, \qquad \text{a.e. } x \text{ in } G, \tag{4.2}$$

$$P_x\{X(\tau) \in \Sigma \cap \Sigma^* \mid \tau < \infty\} = 1, \qquad \text{a.e. } x \text{ in } G, \tag{4.3}$$

$$P_x\{X(\tau') \in \Sigma \cap \Sigma^* \mid \tau' < \infty\} = 1, \qquad \text{a.e. } x \text{ in } G, \tag{4.4}$$

$$P_x\{X(\tau') \in \Gamma \mid \tau' < \infty\} = 1. \tag{4.5}$$

Equation (4.2) implies that $R(x) = R'(x)$ for a.e. x in G and that both functions $R(\cdot)$ and $R'(\cdot)$ are well defined if

$$\sup_{x \in G} E_x \tau' < \infty \qquad \text{or} \qquad \inf_{x \in G} \lambda(x) > 0. \tag{4.6}$$

Equations (4.3) and (4.4) imply that it is enough to specify the boundary function $b(\cdot)$ on $\Sigma \cap \Sigma^*$ if we are only concerned with the solutions $R(\cdot)$ or $R'(\cdot)$ for almost all $x \in G$.

Define a weak solution $u(\cdot)$ to

$$\mathscr{L}u = \lambda(x)u - k(x), \tag{4.7}$$

with boundary condition $b(\cdot)$, to be a bounded measurable function $u(\cdot)$, which satisfies

$$\int u(x)\mathscr{L}^*\psi(x)\,dx = \int [\lambda(x)u(x) - k(x)]\psi(x)\,dx$$

for all† $\psi(\cdot) \in C_0^\infty(G)$ and for which

$$\lim_{y\to x,\ y\in G} u(y) = b(x) \qquad \text{for} \quad x \in \Sigma_2 \cup \Sigma_3 .$$

Then [S5, Corollary 8.2] under (4.6), $R'(\cdot)$ is a weak solution to (4.7), and any other weak solution to (4.7) equals $R'(\cdot)$ at almost all $x \in G$.

In general in the sequel we will not use all the conditions on $a(\cdot), f(\cdot)$ and ∂G of this section, but will seek ways of approximating the functionals (2.1)–(2.5) directly. Of course, since these functionals are intimately associated with certain elliptic or parabolic equations, there will still be a relationship between our conditions and some of those required to treat the equations.

Example It is not always easy to determine the set Γ of regular (τ') boundary points when $a(\cdot)$ is not strictly positive definite. However, often the dynamics of the problem yield this information without much trouble. Let us consider the two-dimensional process

$$dX_1 = X_2\,dt, \qquad dX_2 = f_2(X)\,dt + \sigma\,dw,$$

in the set $\bar{G} = \{x : |x_1| \le \alpha,\ |x_2| \le \alpha\}$. See Figure 4.1. The sets (c, d) and (f, a) are inaccessible since $\dot{X}_1$ points strictly inward there. Similarly, $[b, c)$ and $[e, f)$ are in Γ since $\dot{X}_1$ points strictly outward there. The sets $[a, b]$ and $[d, e]$ are also in Γ via a law of the iterated logarithm and the 0–1 law for Markov processes. The basic idea is the following. The function $X_2(\cdot)$ satisfies

$$X_2(t) = X_2(0) + \int_0^t f_2(X(s))\,ds + \sigma w(t).$$

† $C_0^\infty(G)$ is the class of R^r-valued infinitely differentiable functions on $\bar{G}$, which take the value zero on ∂G.

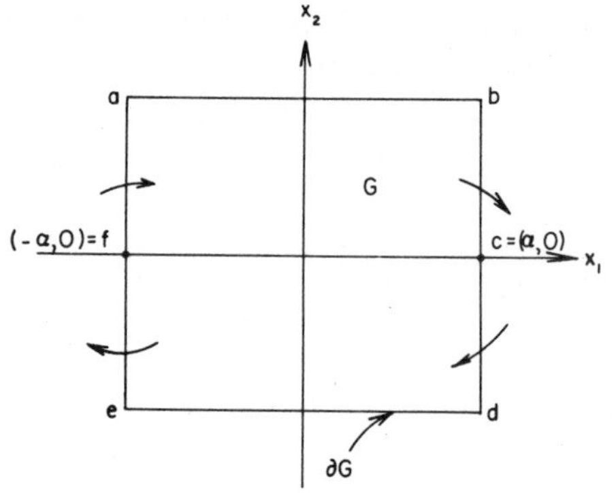

FIG. 4.1.

Since $w(\cdot)$ obeys the law of the iterated logarithm, so does $X_2(\cdot)$. Thus if $X_2(t) = +\alpha$ (or $-\alpha$), then $X_2(t+s)$ crosses the line $+\alpha$ (or $-\alpha$) infinitely often in any interval $[t, t+s_0)$, $s_0 > 0$. Hence, the upper and lower boundaries are regular. The same conclusion would hold if the constant σ were replaced by a function $\sigma(x)$, provided that $\inf_x \sigma^2(x) > 0$. In this case, a simple time change transforms $\int_0^t \sigma(X(s))\, dw(s)$ into a Wiener process.

The only remaining points are f and c. There are several ways of showing that these points are in Γ. The simplest is to use a result of Strook and Varadhan [S5, Section 6], where a number of criteria for regularity are given. Consider c, where the local boundary function is $\phi(x) = (\alpha - x_1)$. The following result is Theorem 6.1 of Strook and Varadhan [S5].

Let $\sigma(\cdot)$ be infinitely differentiable and suppose that $f(\cdot)$, $\phi(\cdot)$ are sufficiently smooth so that the derivatives below exist. Suppose that there is an integer s such that $\mathcal{L}^s\phi^2(x) \neq 0$ at $x = y \in \partial G$ and let n be the first such s. If $\mathcal{L}^s\phi(x) = 0$ at $x = y$, for $s \leq n/2$, then $P_y\{\tau' > 0\} = 0$.

In our case, $n = 3$ and the criterion holds at both points $y = (\alpha, 0)$ and $(-\alpha, 0)$.

4.5 Partial Differential Equations Formally Satisfied by Path Functionals

If we arbitrarily assume that the $R(\cdot)$ (resp., $R(\cdot, \cdot)$) are in $C^2(G)$ ($C^{1,2}([0, T] \times G)$, resp.), then Itô's lemma can be applied to obtain the relevant differential equations. We will only do some formal calculations, under the previous smoothness assumption. See Kushner [K5, Chapter 11] for still formal but more detailed derivations. Similar calculations appear in Fleming and Rishel [F1]. Let $'$ denote transpose.

Consider the functional (2.1) and let ρ be a stopping time such that $E_x \rho < \infty$. Then, assuming that $R(\cdot)$ is smooth, Itô's lemma yields

$$R(X(\rho \cap \tau)) - R(x) = \int_0^{\rho \cap \tau} \mathscr{L}R(X(s))\, ds + \int_0^{\rho \cap \tau} R'_x(X(s))\sigma(X(s))\, dw(s).$$

Taking expectations and using the definition of $R(\cdot)$, we get

$$E_x R(X(\rho \cap \tau)) - R(x) = E_x \int_0^{\rho \cap \tau} \mathscr{L}R(X(s))\, ds = -E_x \int_0^{\rho \cap \tau} k(X(s))\, ds.$$

Now let $N_\varepsilon(x)$ denote the set $\{y : |x - y| < \varepsilon\}$ and assume that $x \in G$ and $N_\varepsilon(x) \in G$. Define $\rho = \inf\{t : X(t) \notin N_\varepsilon(x)\}$. Then $\rho \le \tau$. If we divide the two right-hand sides of the last equation by $E_x \rho$ and let $\varepsilon \to 0$, we formally get

$$\mathscr{L}R(x) + k(x) = 0, \qquad x \in G. \tag{5.1a}$$

In order for the functional $R(\cdot)$ in (2.1) and for the formal derivation to make sense, we require that $\sup_{x \in G} E_x \tau < \infty$ if $k(\cdot) \not\equiv 0$. If $k(\cdot) \equiv 0$, we require that $P_x\{\tau < \infty\} = 1$, $x \in G$. These conditions imply that

$$\lim_{x \to y,\, x \in G} R(x) \to b(y), \qquad y \text{ regular } (\tau) \text{ on } \partial G. \tag{5.1b}$$

A similar formal derivation yields that (2.2) formally satisfies

$$\begin{aligned} &\mathscr{L}R(x) - \lambda(x)R(x) + k(x) = 0, \qquad x \in G, \\ &\lim_{x \to y} R(x) = b(y), \qquad y \text{ regular } (\tau) \text{ on } \partial G. \end{aligned} \tag{5.2}$$

It can also be shown formally that if $R(\cdot)$ satisfies (5.1), (5.2), or (5.4), then it is the desired cost functional.† We will check the calculation for (5.2). Let $F(t) = A(t)R(X(t))$, $t \le \tau$, and assume that (5.2) holds. Then

$$\begin{aligned} dF(t) = {} & [-\lambda(X(t))F(t) + A(t)\mathscr{L}R(X(t))]\, dt \\ & + A(t)R'_x(X(t))\sigma(X(t))\, dw(t), \qquad t < \tau. \end{aligned}$$

Substituting $\lambda(x)R(x) - k(x)$ for $\mathscr{L}R(x)$ yields

$$dF(t) = -[A(t)k(X(t))\, dt] + A(t)R'_x(X(t))\sigma(X(t))\, dw(t), \qquad t < \tau.$$

Integrating the above differential, using the fact that $A(0) = 1$, and taking expectations, yields us

$$E_x F(t \cap \tau) = R(x) - E_x \int_0^{t \cap \tau} A(s)k(X(s))\, ds. \tag{5.3}$$

† Since the calculation is formal, to avoid problems with the way that the boundary condition is realized, assume that the set of points $X(\tau)$ is regular (τ) w.p. 1, under P_x, $x \in G$. This holds if $P_x\{\tau = \tau'\} = 1$, $x \in G$.

Since $R(X(t)) \to b(X(\tau))$ as $t \to \tau$ [by the boundary condition in (5.2)], we have that $E_x F(t \cap \tau) \to E_x A(\tau) b(X(\tau))$ as $t \to \infty$. This together with (5.3) implies that $R(x)$ equals the right-hand side of (2.2).

By a formal derivation similar to that which leads to (5.1) from (2.1), we can show that (2.4a) satisfies

$$
\begin{aligned}
(\partial/\partial t + \mathscr{L})R(x, t) + k(x) = 0, &\qquad x \in G, \quad t < T, \\
R(x, T) = b_T(x), &\qquad x \in G, \\
\lim_{s \to t,\, x \to y,\, x \in G} R(x, s) \to b_1(y, t), &\qquad y \text{ regular } (\tau) \text{ on } \partial G,\ t \le T.
\end{aligned}
\tag{5.4}
$$

The discounted functional (2.4b) also formally satisfies (5.4), but with $k(x)$ replaced by $k(x) - \lambda(x)R(x, t)$.

The γ of (2.5) is formally associated with the equation

$$\mathscr{L}V(x) - \gamma + k(x) = 0, \tag{5.5}$$

in the sense that if $E_x V(X(t))/t \to 0$ as $t \to \infty$, then $\gamma = E_\mu k(X)$. (See also [K5, p. 332].) To see this note that, by Itô's lemma, (5.5) implies

$$[E_x V(X(t)) - V(x)]/t = E_x \int_0^t (\gamma - k(X(s)))\, ds/t.$$

4.6 The Characteristic Operator of the Diffusion

The Weak Infinitesimal Operator

Let $\mathscr{H}$ denote the space of bounded measurable real-valued functions on R^r. If for $f(\cdot) \in \mathscr{H}$,

$$[E_x f(X(t)) - f(x)]/t, \qquad t \to 0,$$

converges boundedly and pointwise to a function $g(\cdot) \in \mathscr{H}$ and if $E_x g(X(t)) \to g(x)$ as $t \to 0$, then we write $f(\cdot) \in \mathscr{D}(\tilde{\mathscr{A}})$, $\tilde{\mathscr{A}} f = g$, where $\tilde{\mathscr{A}}$ is the *weak infinitesimal operator of the Markov process* $X(\cdot)$. On $C^2(R^r) \cap \mathscr{H}$, $\tilde{\mathscr{A}} = \mathscr{L}$.

The *characteristic operator* $\mathscr{U}$ is, in a sense, an extension of $\tilde{\mathscr{A}}$. We will discuss it briefly because in a way we will be approximating $\mathscr{U}$ rather than $\mathscr{L}$. Equations (5.1), (5.2), and (5.4) frequently do not hold as stated in the strong sense (or even in a weak sense). But even in those cases, they often do hold if we replace $\mathscr{L}$ by the operator $\mathscr{U}$.

Let x denote a point in R^r and let U_n denote a sequence of neighborhoods of x such that $U_n \downarrow x$ and $E_x \tau_n < \infty$, where $\tau_n = \min\{t : X(t) \notin U_n\}$. If the limit in (6.1) exists

$$\lim_{n \to \infty} \frac{E_x f(X(\tau_n)) - f(x)}{E_x \tau_n} = g(x), \tag{6.1}$$

we write $\mathscr{U}f(x) = g(x)$. If $\mathscr{U}f(x) = g(x)$ for all x in a set B and the ratio on the left-hand side of (6.1) is bounded uniformly in n and $x \in B$, then we say that $f(\cdot) \in \mathscr{D}(\mathscr{U})$ in B. If $E_x \tau_n = \infty$ for all n, set $\mathscr{U}f(x) = 0$. For a more precise definition, see Dynkin [D4, Section 5.3]. In particular, the function $f(\cdot)$ defined by $f(x) \equiv E_x b(X(\tau))$ is in $\mathscr{D}(\mathscr{U})$ in G and $\mathscr{U}f(x) = 0$. If $\sup_{x \in G} E_x \tau < \infty$, then the function $f(\cdot)$ defined by $f(x) \equiv E_x \int_0^\tau k(X(s))\, ds$ is in $\mathscr{D}(\mathscr{U})$ in G and $\mathscr{U}f(x) + k(x) = 0$. The operators $\mathscr{U}$ and $\mathscr{L}$ are local. But the action of $\mathscr{U}$ is more reflective of the path properties than is $\mathscr{L}$, which is a differential operator and its properties are related to path properties only under some "smoothing" assumption. Some further comments on the relationship between $\mathscr{U}$ and our numerical procedures appear in Section 5.2.

4.7 Optimal Control Problems and Nonlinear Partial Differential Equations

In this section we will give the nonlinear partial differential equations that are formally satisfied by certain minimal cost functionals, define several classes of control strategies, and state some assumptions which will be used in Chapters 8 and 9. A more detailed treatment (under some additional assumptions) of the relationships between partial differential equations and optimal stochastic control theory appears in Fleming and Rishel [F1].

Several of the classes of control laws and stopping times are introduced owing to the nature of the approximation procedures used in Chapters 8 and 9. Our sequence of approximations will converge to a cost functional for a controlled or stopped diffusion. The introduction of the various classes of strategies is required in order to prove the optimality (or minimality) of the limit of the sequence of approximations to the optimal cost functional.

4.7.1 The Optimal Stopping Problem

Let $(\Omega, \mathscr{B}, \mathscr{B}_t, P_x, X(\cdot), x \in R^r)$ be a strong Markov diffusion process, where for each x, there is a $w(\cdot)$ with the properties: $X(\cdot)$, $w(\cdot)$ solves (1.5.1) under P_x; $w(s)$, $s \le t$, is $\mathscr{B}_t$ measurable; $w(s) - w(t)$, $s \ge t$, is independent of any $\mathscr{B}_t$ measurable function under P_x.

Assume A4.1.1 and

A4.7.1 *$b(\cdot)$ and $k(\cdot)$ are continuous and bounded real-valued functions on R^r and there is a real $k_0 > 0$ such that $\inf_{x \in R^r} k(x) \geq k_0$.*

Now we define several classes of stopping times analogous to those used in Section 3.2.2. A *stopping time ρ is said to be pure Markov if there is a Borel set A such that*

$$\rho = \inf\{t : X(t) \in A\}.$$

Let $\bar{\mathcal{T}}^0_{\text{PM}}$ denote the class of pure Markov stopping times ρ.

In general (unless otherwise defined) a bar over a symbol (such as $\mathcal{T}$ or $\mathcal{C}$) indicates that the underlying process is a *continuous parameter process*. The symbol without the bar denotes that the underlying process has a *discrete parameter. Thus, $\bar{\mathcal{T}}_{\text{PM}}(x)$, $\bar{\mathcal{T}}_{\text{PM}}$, $\bar{\mathcal{T}}^+_{\text{PM}}(x)$, and $\bar{\mathcal{T}}^+_{\text{PM}}$ are all defined analogously to the symbol without the bar in the Markov chain case. A stopping time ρ is said to be randomized if it is adapted to $\{\mathcal{B}_t\}$; i.e., if $\{\rho \leq t\} \in \mathcal{B}_t$ for all $t \geq 0$. Let $\bar{\mathcal{T}}^0_{\text{R}}$ denote the class of randomized stopping times, and define $\bar{\mathcal{T}}_{\text{R}}(x)$, $\bar{\mathcal{T}}_{\text{R}}$, $\bar{\mathcal{T}}^+_{\text{R}}(x)$ and $\bar{\mathcal{T}}^+_{\text{R}}$ in the obvious way, analogously to the definitions in the discrete parameter case of Chapter 3.*

For each x and each $\rho \in \bar{\mathcal{T}}^+_{\text{R}}(x)$, define the cost

$$R(x, \rho) = E_x\left[\int_0^\rho k(X(s))\, ds + b(X(\rho))\right] \tag{7.1}$$

and define

$$V(x) = \inf_{\rho \in \bar{\mathcal{T}}_{\text{PM}}{}^+(x)} R(x, \rho). \tag{7.2}$$

It is also true that (Shiryaev [S1])

$$V(x) = \inf_{\rho \in \bar{\mathcal{T}}_{\text{R}}{}^+} R(x, \rho). \tag{7.3}$$

In fact, Shiryaev [S1] proves that it is enough to minimize (7.2) over ρ which are defined by first entry into closed sets. The proof of (7.3) involves showing that we can approximate the cost arbitrarily closely by restricting attention to stopping times with values $\{k2^{-n}, k = 0, 1, \ldots\}$ for some n, and then showing that for the discrete parameter process $\{X_k\}$, defined by $X_k = X(k2^{-n})$, we cannot reduce the cost by using randomized rather than pure Markov strategies. Owing to the lower bound k_0, it is sufficient to consider stopping times ρ satisfying

$$E_x \rho \leq 2 \sup_x |b(x)|/k_0 . \tag{7.4}$$

The arguments given in connection with the stopping problem for the Markov chain suggest that for small Δ, $V(\cdot)$ should approximately satisfy

$$V(x) = \min[b(x), E_x V(X(\Delta)) + k(x)\Delta]. \tag{7.5}$$

Let B denote the stopping set $\{x : V(x) = b(x)\}$, and assume that $V(\cdot)$ is sufficiently smooth in the interior of $R^r - B$. Then, in $R^r - B$, where $V(x) < b(x)$, (7.5) yields

$$0 = \lim_{\Delta \to 0} \frac{E_x V(X(\Delta)) - V(x)}{\Delta} + k(x)$$

hence (by Itô's lemma)

$$\begin{aligned} \mathscr{L}V(x) + k(x) &= 0, \qquad x \in R^r - B, \quad V(x) \le b(x), \\ V(x) &= b(x), \qquad x \in B. \end{aligned} \tag{7.6}$$

Of course (7.6) is formal and we will not seek to justify it. Our approximate solutions to (7.6) will converge to the right-hand side of (7.2).

Define $B_\varepsilon = \{x : V(x) \ge b(x) - \varepsilon\}$; and let $\rho = \inf\{t : X(t) \in B_\varepsilon\}$. Then $R(x, \rho) \le V(x) + \varepsilon$ and B_ε is the ε-optimal stopping set.

The main reason for introducing the various classes of stopping times as well as the various classes of controls to be introduced next is that they appear naturally in the computational problem, and we must relate the infima of the costs over the various classes.

4.7.2 Optimal Control until a Boundary Is Reached

Let G denote a bounded open set in R^r with boundary ∂G and let $\mathscr{U}$ be a compact set in some Euclidean space, say R^m.

Assume

A4.7.2 *$k(\cdot, \cdot), f(\cdot, \cdot)$, and $\sigma(\cdot)$ are R, R^r, and $(r \times r)$ matrix-valued continuous and bounded functions on $R^r \times \mathscr{U}$, $R^r \times \mathscr{U}$, and R^r, resp.*

We shall discuss functionals associated with the controlled stochastic process

$$X(t) = x + \int_0^t f(X(s), u(s))\, ds + \int_0^t \sigma(X(s))\, dw(s). \tag{7.7}$$

First we introduce several classes of control functions. Whether the control function is of a form with values $u(x(t))$, $u(\omega, t)$, $u(w(\cdot), t)$, or $u(x(t), t)$, we *usually* denote the value simply by $u(t)$. In what follows, we denote by $(\Omega, \mathscr{B}, \mathscr{B}_t, P_x^u)$ a probability space on which there are defined random processes $X(t)$, $u(t)$, $w(t)$, $t < \infty$, which satisfy (7.7) for† $t < \infty$ and $X(0) = x \in G$. The sequence of σ-algebras $\{\mathscr{B}_t\}$ is nondecreasing and $\mathscr{B}_t \subset \mathscr{B}$. The function $w(\cdot)$

† It is possible to continue, if the processes were defined only up to $\tau = \min\{t : X(t) \notin G\}$. It is (notationally) somewhat more convenient to define them for all t.

is a Wiener process under P_x^u, and the stochastic integral in (7.7) is constructed using the measure P_x^u. The $X(\cdot)$, $u(\cdot)$ are assumed to be nonanticipative with respect to $w(\cdot)$, and $w(s)$, $X(s)$, $s \leq t$, are $\mathscr{B}_t$ measurable, and $w(s) - w(t)$, $s \geq t$, is independent of $\mathscr{B}_t$ for each t under P_x^u. The objects Ω, $\mathscr{B}$ and $\mathscr{B}_t$ can also depend on u and x. We are concerned mainly with the measures, and with expectations of functionals of $X(\cdot)$ and $u(\cdot)$, and not with the probability space *per se*. The use of a possibly different sample space for each x, u is not too common in stochastic control theory. But it suits the purposes of Chapter 9 quite well.

Suppose that there is a probability space of the previously described type, where $u(\cdot, \cdot)$ is a $\mathscr{U}$-valued measurable ω, t function, $u(\cdot, t)$ is $\mathscr{B}_t$ measurable and (7.7) has a nonanticipative solution. If the solution is unique (in the sense defined below), then we say that $u \in \overline{\mathscr{C}}_{\mathrm{R}}(x)$. We say that the solution is unique if the probability law† *of $\{X(s), u(s), s < \infty\}$ is uniquely determined by that of $\{u(s), w(s), s < \infty\}$, irrespective of the probability space. The subclasses of $\overline{\mathscr{C}}_{\mathrm{R}}(x)$ for which $P_x^u\{\tau < \infty\}$ and $E_x^u \tau < \infty$ are denoted by $\overline{\mathscr{C}}_{\mathrm{R}}(x, G)$, $\overline{\mathscr{C}}_{\mathrm{R}}^+(x, G)$,* resp.

Let $\overline{\mathscr{C}}_{\mathrm{W}}(x)$ denote the subclass of $\overline{\mathscr{C}}_{\mathrm{R}}(x)$ which consists of measurable $(w(\cdot), t)$ functions and for which $u(w_1(\cdot), t) = u(w_2(\cdot), t)$ if $w_i(\cdot)$ are any continuous R^n-valued functions on $[0, \infty)$ that are equal on $[0, t]$. Similarly, define $\overline{\mathscr{C}}_{\mathrm{W}}(x, G)$ and $\overline{\mathscr{C}}_{\mathrm{W}}^+(x, G)$.

Let $\overline{\mathscr{C}}_{\mathrm{PM}}$ denote the class of $\mathscr{U}$-valued Borel measurable "feedback" functions with values $u(x, t)$, for which (7.7) has a unique (in the sense of A4.1.1) nonanticipative solution which is a strong Markov and a Feller process. Define the classes $\overline{\mathscr{C}}_{\mathrm{PM}}(x, G)$, $\overline{\mathscr{C}}_{\mathrm{PM}}(G)$, $\overline{\mathscr{C}}_{\mathrm{PM}}^+(x, G)$ and $\overline{\mathscr{C}}_{\mathrm{PM}}^+(G)$ in the obvious way. In particular, $u \in \overline{\mathscr{C}}_{\mathrm{PM}}^+(G)$ if $u \in \overline{\mathscr{C}}_{\mathrm{PM}}$ and $E_x^u \tau < \infty$, $x \in G$.

If $f(\cdot, \cdot)$ and $\sigma(\cdot)$ satisfy a uniform Lipschitz condition in x, then for any measurable nonanticipative and $\mathscr{U}$-valued (ω, t) function $u(\cdot)$, there is a unique solution to (7.7). Uniqueness seems to be a critical property. With it we can approximate or perturb the dynamics and show that the distributions converge to those for the unperturbed problem as the perturbations go to zero. Such a property is rather essential in applications.

REMARK ON NOTATION Sometimes the notation $u \in \overline{\mathscr{C}}_{\mathrm{R}}(x)$ (or with W replacing R or with a superscript) is used *loosely* in the sense that it implies the following. There is a probability space with $u(\cdot)$, $w(\cdot)$ defined on it, with $u(\cdot)$ being nonanticipative, another probability space with $\tilde{X}(\cdot)$, $\tilde{u}(\cdot)$, $\tilde{w}(\cdot)$ (solving 7.7) defined on it, where $\tilde{X}(\cdot)$, $\tilde{u}(\cdot)$ are nonanticipative and $(\tilde{u}(\cdot), \tilde{w}(\cdot))$ has the same law as does $(u(\cdot), w(\cdot))$, and there is uniqueness of the

† By "probability law," we mean either the multivariate distributions (for almost all s) or the measure induced by $X(\cdot)$, $u(\cdot)$ on $C^r[0, \infty) \times L_{2,l}^m$.

solution to (7.7) in the sense described above. The solution $X(\cdot)$ or $\tilde{X}(\cdot)$ need not be defined on the original probability space. This looseness should cause no problems in Chapter 9. In the stopping and control problem, stopping times $\tilde{\tau}$ may be added to the above formulation.

Functionals of the Controlled Diffusion

For $u(\cdot)$ in $\overline{\mathscr{C}}_{\mathrm{R}}^{+}(x, G)$ or in $\overline{\mathscr{C}}_{\mathrm{PM}}^{+}(x, G)$, define the cost

$$R(x, u) = E_x^u \int_0^\tau k(X(s), u(s))\, ds + E_x^u b(X(\tau)) \tag{7.8}$$

and define†

$$V(x) = \inf_{u \in \overline{\mathscr{C}}_{\mathrm{R}}^{+}(x, G)} R(x, u). \tag{7.9}$$

Suppose that an optimal control exists in $\overline{\mathscr{C}}_{\mathrm{PM}}^{+}(G)$, assume that $V(\cdot)$ is sufficiently smooth, and let us apply a formal dynamic programming argument. Let $u(\cdot)$ denote the optimal pure Markov control and $v(\cdot)$ any other control in $\overline{\mathscr{C}}_{\mathrm{PM}}^{+}(G)$. Let $\tilde{X}(\cdot)$, $\tilde{\tau}$ denote the solution and escape time corresponding to the use of $u(\cdot)$ on $[t \cap \tilde{\tau}, \tilde{\tau})$ and $v(\cdot)$ before. Then, writing

$$u(s) = u(X(s)), \qquad \tilde{u}(s) = u(\tilde{X}(s)), \qquad v(s) = v(\tilde{X}(s)),$$

we have

$$V(x) = E_x^u \left[\int_0^\tau k(X(s), u(s))\, ds + b(X(\tau))\right], \tag{7.10}$$

where $X(\cdot)$ is the solution corresponding to $u(\cdot)$ and (by optimality of $u(\cdot)$)

$$V(x) \le E_x^v \int_0^{t \cap \tilde{\tau}} k(\tilde{X}(s), v(s))\, ds$$

$$+ E_x^v E^u_{\tilde{X}(t \cap \tilde{\tau}),\, t \cap \tilde{\tau}} \left[\int_{t \cap \tilde{\tau}}^{\tilde{\tau}} k(\tilde{X}(s), \tilde{u}(s))\, ds + b(\tilde{X}(\tilde{\tau}))\right].$$

Hence

$$V(x) \le E_x^v \int_0^{t \cap \tilde{\tau}} k(\tilde{X}(s), v(s))\, ds + E_x^v V(\tilde{X}(t \cap \tilde{\tau})), \tag{7.11}$$

or equivalently

$$\frac{E_x^v V(\tilde{X}(t \cap \tilde{\tau})) - V(x)}{E_x^v(t \cap \tilde{\tau})} + \frac{E_x^v \int_0^{t \cap \tilde{\tau}} k(\tilde{X}(s), v(s))}{E_x^v(t \cap \tilde{\tau})} \ge 0.$$

† If $k(\cdot) \equiv 0$, we only need that $u \in \overline{\mathscr{C}}(x, G)$.

A *formal* limit yields

$$\mathscr{L}^v V(x) + k(x, v) \geq 0 \qquad \text{for any value} \quad v \in \mathscr{U}, \tag{7.12}$$

where we define

$$\mathscr{L}^v = \sum_{i,j} a_{ij}(x)(\partial^2/\partial x_i\, \partial x_j) + \sum_i f_i(x, v)(\partial/\partial x_i).$$

Equations (7.11) and (7.12) yield the formal dynamic programming equation

$$\begin{aligned} \inf_{v \in \mathscr{U}} [\mathscr{L}^v V(x) + k(x, v)] &= 0, \qquad x \in G, \\ V(x) = b(x) &\qquad \text{on} \quad \partial G. \end{aligned} \tag{7.13}$$

Again, we will not rigorously justify (7.13), but use it to approximate the infima of the costs given by the right-hand side of (7.9). The approximations will converge under broad conditions, whether or not we know the smoothness of $V(\cdot)$. We define the solution of (7.13) to be the right-hand side of (7.9)

It will be useful in the sequel to be able to define $u(\cdot)$ for $t \geq \tau$ in such a way that τ' can be defined. This cannot always be done. *If* $u \in \overline{\mathscr{C}}_R(x)$ *and* $P_x^u\{\tau = \tau' \mid \tau' < \infty\} = 1$, *then we say that u is in* $\mathscr{C}_R^0(x, G)$. *Similarly, define the subclasses* $\hat{\mathscr{C}}_R(x, G)$, $\hat{\mathscr{C}}_R^+(x, G)$ *to be the subclasses of* $\hat{\mathscr{C}}_R^0(x, G)$ *for which* $P_x^u\{\tau' < \infty\} = 1$ *and* $E_x^u \tau' < \infty$. *The subclasses* $\mathscr{C}_W^0(x, G)$, $\hat{\mathscr{C}}_W(x, G)$, *and* $\hat{\mathscr{C}}_W^+(x, G)$ *are defined in the obvious way.*

If we introduce the continuous bounded discount factor $\lambda(\cdot)$ such that there is a real number λ_0 satisfying $\inf_x \lambda(x) \geq \lambda_0 > 0$, and replace (7.8) by

$$R(x, u) = E_x^u\left[\int_0^\tau A(s)k(X(s), u(s))\, ds + A(\tau)b(X(\tau))\right], \tag{7.14}$$

where

$$A(t) = \exp - \int_0^t \lambda(X(s))\, ds,$$

then we only need that $u \in \overline{\mathscr{C}}_R(x)$. In this case, a formal dynamic programming argument yields that (7.15) replaces (7.13)

$$\begin{aligned} \inf_{v \in \mathscr{U}} [\mathscr{L}^v V(x) + k(x, v) - \lambda(x)V(x)] &= 0, \qquad x \in G, \\ V(x) = b(x) &\qquad \text{on } \partial G. \end{aligned} \tag{7.15}$$

4.7.3 Optimal Stopping and Control until a Boundary Is Reached

In some problems we may desire to stop the process before ∂G is reached, with an associated penalty. From a formal point of view, this problem is a simple combination of the problems of Sections 4.7.1 and 4.7.2, except that the classes of stopping times depend on the control u. We will introduce the relevant notation.

If $u \in \bar{\mathscr{C}}_{\mathrm{R}}(x)$, define the sets of stopping times (all with respect to $\{\mathscr{B}_t\}$), $\bar{\mathscr{T}}^0_{\mathrm{R}}(u, x)$, $\bar{\mathscr{T}}_{\mathrm{R}}(u, x)$, and $\bar{\mathscr{T}}^+_{\mathrm{R}}(u, x)$ [and similarly $\bar{\mathscr{T}}^0_{\mathrm{W}}(u, x)$, $\bar{\mathscr{T}}^0_{\mathrm{PM}}(u)$, etc.] in the obvious way: the measure P^u_x simply replaces P_x in the definitions of the same terms when u is deleted.

For each pair (ρ, u) in (7.18), define the cost

$$R(x, \rho, u) = E^u_x \int_0^{\tau \cap \rho} k(X(s), u(s))\, ds + E^u_x b(X(\tau \cap \rho)), \tag{7.16}$$

and define the infimum over (ρ, u) in (7.18) by

$$V(x) = \inf_{\rho, u} R(x, \rho, u). \tag{7.17}$$

$$\bar{\mathscr{T}}^+_{\mathrm{R}}(u, x) \times \bar{\mathscr{C}}_{\mathrm{R}}(x) \quad \text{or} \quad \bar{\mathscr{T}}^0_{\mathrm{R}}(u, x) \times \bar{\mathscr{C}}^+_{\mathrm{R}}(x, G). \tag{7.18}$$

Let B denote the stopping set $\{x : V(x) = b(x)\}$, and note that ∂G is in B. Then under appropriate smoothness assumptions on $V(\cdot)$, a formal dynamic programming argument shows that

$$\begin{aligned} \min_{v \in \mathscr{U}} [\mathscr{L}^v V(x) + k(x, v)] &= 0, \qquad x \in \bar{G} - B, \\ V(x) &= b(x) \qquad \text{in } B, \\ V(x) &\le b(x) \qquad \text{in } G. \end{aligned} \tag{7.19}$$

4.7.4 Fixed Time of Control

Let $k(\cdot, \cdot, \cdot)$ be a bounded continuous function on $R^r \times \mathscr{U} \times [0, T]$. For $u(\cdot) \in \bar{\mathscr{C}}_{\mathrm{R}}(x)$, define the cost functional

$$\begin{aligned} R(x, t, u) = E^u_{x,t}\Bigg[\int_t^{T \cap \tau} k(X(s), u(s), s)\, ds &+ b_T(X(t)) I_{\{\tau > T\}} \\ &+ b_1(X(\tau), \tau) I_{\{\tau \le T\}}\Bigg], \end{aligned} \tag{7.20}$$

and define

$$V(x, t) = \inf_{u \in \bar{\mathscr{C}}_{\mathrm{R}}(x)} R(x, t, u). \tag{7.21}$$

The formal dynamic programming equation is

$$\inf_{v \in \mathscr{U}} [(\partial/\partial t + \mathscr{L}^v)V(x, t) + k(x, v, t)] = 0, \qquad t < T, \quad x \in G.$$
$$V(x, t) = \begin{cases} b_1(x, t), & x \in \partial G, \quad t \leq T \\ b_T(x), & x \in G. \end{cases} \tag{7.22}$$

A more rigorous approach for several classes of problems appears in Fleming and Rishel [F1].

4.7.5 The Impulsive Control Problem

Assume A4.1.1 (homogeneous case). *Let λ and p_0 denote positive real numbers, $\mathscr{V}$ a compact set in R^r, and $p(\cdot, \cdot)$ a continuous real-valued bounded function on $R^r \times \mathscr{V}$ satisfying $p(x, \alpha) \geq p_0$, $x \in R^r$, $\alpha \in \mathscr{V}$.*

We continue to use the terminology of Sections 4.7.1 and 4.7.2 but the continuously acting control will be replaced by a sequence of actions $\{u_i\}$ and the single stopping time will be replaced by a nondecreasing sequence $\{\tau_i\}$ of stopping times with respect to $\{\mathscr{B}_t\}$. Some or all of the τ_i may take infinite values, the controlled diffusion is replaced by the jump process (7.23), and u_i is $\mathscr{B}_{\tau_i}$ measurable and $\mathscr{V}$-valued.

We require that $\mathscr{B}_t$ measure $w(s)$, $s \leq t$, and $X(s)$, $s < t$, where $X(\cdot)$ is defined by (7.23), and that $w(s) - w(t)$, $s \geq t$, be independent of any $\mathscr{B}_t$ measurable function for each $t \geq 0$. If there is jump at t, then $X(t) = X(t^-) +$ (a $\mathscr{B}_t$ measurable quantity). Thus, $\mathscr{B}_t$ also measures $X(t)$. The right continuity of $X(\cdot)$ is purely a convention. There is no *a priori* reason why it could not be defined to be left continuous.

$$X(0) = x,$$
$$X(t) = X(\tau_i) + \int_{\tau_i}^{t} f(X(s))\, ds + \int_{\tau_i}^{t} \sigma(X(s))\, dw(s), \quad t < \tau_{i+1}, \tag{7.23}$$
$$X(\tau_i) = X(\tau_i^-) + u_i\,.$$

Let $\bar{\mathscr{A}}_{\mathrm{R}}(x)$, $\bar{\mathscr{A}}_{\mathrm{W}}(x)$, and $\bar{\mathscr{A}}_{\mathrm{PM}}$ denote the sets of such sequences $\{\tau_i, u_i\}$ according to the case. In the pure Markov case, there are Borel sets $\{A_i\}$ such that $\tau_i = \inf\{t : t \geq \tau_{i-1}, X(t) \in A_i\}$ and each $u_i(\cdot)$ is a measurable function of x.

By definition, if $\{\tau_i, u_i\}$ is in one of the $\bar{\mathscr{A}}$ or $\bar{\mathscr{A}}(x)$ classes, then there is a probability space $(\Omega, \mathscr{B}, P)$, a sequence of nondecreasing sub-σ-algebras $\mathscr{B}_t$ of $\mathscr{B}$, and a set $\{\tau_i, u_i\}$, $X(\cdot)$, $w(\cdot)$ satisfying the above properties, and the solution of (7.23) is unique in the sense that the law of $(\{\tau_i, u_i\}, w(\cdot))$ implies the law of $(\{\tau_i, u_i\}, X(\cdot))$.

Define the costs†

$$R(x, \{\tau_i, u_i\}) = E_x\left[\sum_{i=1}^{\infty} (\exp - \lambda\tau_i) p(X(\tau_i^-), \bar{u}_i) + \int_0^{\infty} (\exp - \lambda s) k(X(s))\, ds\right]. \tag{7.24}$$

$$V(x) = \inf_{\{\tau_i, u_i\} \in \bar{\mathscr{A}}_{PM}} R(x, \{\tau_i, u_i\}). \tag{7.25}$$

The problem is a generalization of the stopping problem—there is a sequence of discrete decision times rather than just one single decision time. A formal dynamic programming argument gives the following result.

Define $B = \{x : V(x) = \min_{\alpha \in \mathscr{V}} [p(x, \alpha) + V(x + \alpha)]\}$; B is the action set. If $x \in B$, we take the action which yields the minimum. Otherwise we do not act. For $x \notin B$,

$$\mathscr{L}V(x) + k(x) - \lambda V(x) = 0. \tag{7.26}$$

A continuously acting control can also be added. Under certain additional assumptions, the impulsive control problem has been discussed by Bensoussan and Lions [B1, B2], from a point of view that is rather different than the one which we will take in Chapter 8.

† If there are several simultaneous actions, then there is an ambiguity in the definitions (7.23) and (7.24). We will follow a procedure illustrated by the example: Let $\tau_1 = \tau_2$. Define $X(\tau_1) = X(\tau_1^-) + u_1 = X(\tau_2^-)$, $X(\tau_2) = X(\tau_2^-) + u_2$.

CHAPTER 5

A Simple Application of the Invariance Theorems

The purpose of this chapter is simply to illustrate on a simple example the approximation ideas (as well as some of the needed assumptions) which we will develop further in the subsequent chapters. With the exception of one result at the end of the chapter, no proofs are given.

The chapter begins with an example showing how the concepts of weak convergence of measures can give us very useful generalizations of the classical limit theorems for sequences of vector-valued random variables. We then apply this result to a problem arising in the numerical analysis of a simple two-point boundary value problem for a second-order ordinary differential equation. We take a finite difference approximation to the equation. If the approximation is chosen with a little care, then the resulting finite difference equation has a probabilistic interpretation, which we can exploit via weak convergence methods to get the convergence of the finite difference solution to the true solution.

5.1 A Functional Limit Theorem

We will try to illustrate the power of the ideas of weak convergence of measures in applications to problems in the numerical analysis of elliptic equations. First, we consider a special and simple convergence problem for a sequence of random variables. For each n, let $\{\psi_k^n, k = 0, 1, \ldots\}$ denote a sequence of independent, identically distributed random variables which satisfy, for some $A \in (0, 1)$,

$$P\{\psi_k^n = 1/n\} = P\{\psi_k^n = -1/n\} = A/2 \leq \tfrac{1}{2},$$
$$P\{\psi_k^n = 0\} = 1 - A,$$

and define the random walk $\{\xi_k^n\}$ by

$$\xi_{k+1}^n = \xi_k^n + \psi_k^n, \qquad \xi_0^n = 0.$$

Let $n \to \infty$, but keep k/n fixed at a constant t. Then the classical central limit theorem immediately yields that $n^{1/2}\xi_k^n$ converges in distribution to a normally distributed random variable, with zero mean and variance At. As a consequence of this convergence, we know that $Ef(n^{1/2}\xi_k^n) \to Ef(X)$ as $n \to \infty$, $k/n = t$, where X is normally distributed with mean 0 and variance At and $f(\cdot)$ is any bounded almost everywhere continuous function. The value of $Ef(X)$ can be approximated by $Ef(n^{1/2}\xi_k^n)$ if desired, although in this case, the limit will often be easier to calculate.

Now let us consider a strengthening of the convergence result. Define $t_k^n = k/n^2$ and $\Delta t_k^n = 1/n^2$, and define the piecewise constant process (with paths in $D[0, \infty)$) $\xi^n(\cdot)$ by

$$\xi^n(t) = \xi_k^n \qquad \text{on} \quad [t_k^n, t_{k+1}^n).$$

By a straightforward calculation, it can be shown that there is a real number K for which

$$E|\xi^n(t + \Delta) - \xi^n(t)|^4 \leq E\left|\sum_{k:t<t_{k+1}^n \leq t+\Delta} \psi_k^n\right|^4 \leq K[\Delta + O(1/n^2)]^2. \quad (1.1)$$

By Theorem 2.4.2, $\{\xi^n(\cdot)\}$ is tight† on $D[0, \infty)$ and if $\xi(\cdot)$ is a separable process to which a subsequence converges in distribution, then $\xi(\cdot)$ is continuous w.p. 1.

The limit is actually a Wiener process with covariance At. There are several ways to show this. For the first method, we start by calculating the limit of the characteristic functions of a weakly convergent subsequence.

† It is only necessary to check the conditions of the theorem for $\Delta \geq 1/2n^2$.

For any integer m and any real $0 \le t_1 < t_2 \cdots < t_m$ and real $\lambda_1, \ldots, \lambda_m$, we can get (via the weak convergence)

$$E \exp i\left(\sum_{j=1}^{n} \lambda_j[\xi^n(t_j) - \xi^n(t_{j-1})]\right) \to \exp - \sum_{j=1}^{n} A\lambda_j^2(t_j - t_{j-1})/2.$$

The limit does not depend on the subsequence. Thus the multivariate distributions converge to those of said Wiener process. This and tightness implies the asserted convergence (Theorem 2.4.3). Often it is difficult to show *à priori* that the finite dimensional distributions coverge to those of a specified process, and other methods must be used.

A second method does not require the fact that the multidimensional distributions converge, but uses some results in martingale theory. Note that (1.1) implies that $\{\xi^n(t)\}$ and $\{|\xi^n(t)|^2\}$ are uniformly integrable for any $t \ge 0$ since $E|\xi^n(t)|^4$ is bounded uniformly in n for each $t \ge 0$. Fix a particular weakly convergent subsequence and index it by n. Let $\xi(\cdot)$ denote the limit process; it is continuous w.p. 1. We will argue that the limit is a continuous martingale with quadratic variation At, hence a Wiener process with covariance At by Section 1.4.4. To see this, let q denote an arbitrary integer, $t_1, \ldots, t_q \le t$ be arbitrary real numbers, and $g(\cdot)$ be a bounded real-valued continuous function on R^q.

By a direct evaluation, we see that

$$Q_n \equiv Eg(\xi^n(t_1), \ldots, \xi^n(t_q))(\xi^n(t + \Delta) - \xi^n(t)) = 0. \tag{1.2}$$

The function on $D[0, \infty)$ with values $g(x(t_1), \ldots, x(t_q)) \cdot (x(t + \Delta) - x(t))^\alpha$ at $x(\cdot) \in D[0, \infty)$, $\alpha = 1, 2$, is not bounded, but owing to the uniform integrability of $\{|\xi^n(s)|^\alpha\}$, $\alpha = 1, 2$, each $s \ge 0$, and the weak convergence, we have

$$Q_n \to Eg(\xi(t_1), \ldots, \xi(t_q))(\xi(t + \Delta) - \xi(t)), \tag{1.3}$$

which equals zero by (1.2). Equation (1.3) and the arbitrariness of $q, t_1, \ldots, t_q$, and $g(\cdot)$ imply that

$$E[\xi(t + \Delta) - \xi(t) \mid \xi(s), s \le t] = 0,$$

and hence that $\xi(\cdot)$ is a continuous martingale with respect to some familly of σ-algebras. Similarly, by the uniform integrability and weak convergence,

$$R_n \equiv Eg(\xi^n(t_1), \ldots, \xi^n(t_q))(\xi^n(t + \Delta) - \xi^n(t))^2$$
$$\to Eg(\xi(t_1), \ldots, \xi(t_q))A\Delta; \tag{1.4}$$

$$R_n \to Eg(\xi(t_1), \ldots, \xi(t_q))[\xi(t + \Delta) - \xi(t)]^2. \tag{1.5}$$

Equations (1.4) and (1.5) and the arbitrariness of $q, t_1, \ldots, t_q$, and of $g(\cdot)$ imply that

$$E[(\xi(t + \Delta) - \xi(t))^2 \mid \xi(s), s \le t] = A\Delta.$$

Hence the quadratic variation is At and $\xi(\cdot)$ is a Wiener process. Since the limit of any convergent subsequence is a Wiener process, the entire original sequence $\{\xi^n(\cdot)\}$ tends to a Wiener process in distribution.

A third method depends on the fact that there is a Wiener process $w(\cdot)$ and for each n an increasing sequence of random variables $\{\tau_k^n\}$ so that ψ_k^n has the distribution of $w(\tau_{k+1}^n) - w(\tau_k^n)$ (see Breiman [B4]).

The convergence of the process $\xi^n(\cdot)$ to $\xi(\cdot)$ in distribution is considerably stronger and more useful than the convergence of $\xi^n(t)$ to $\xi(t)$ in distribution for each $t \geq 0$. In particular, if $g(\cdot) : D[0, \infty) \to R$ is bounded and almost everywhere continuous with respect to the measure that $\xi(\cdot)$ induces on $D[0, \infty)$, then $Eg(\xi^n(\cdot)) \to Eg(\xi(\cdot))$. For example, let $g(\cdot)$ be the minimum of T and the first hitting time of a point $\bar{x}$. The next section deals with a useful application to numerical analysis.

5.2 An Application to Numerical Analysis

To provide a simple introduction to some of the ideas and techniques of succeeding chapters, we consider a relatively simple problem of numerical analysis; namely, solving the differential equation

$$\begin{gathered} a(x)V_{xx}(x) + f(x)V_x(x) + k(x) = 0, \qquad x \in (0, 1) \\ V(0) = \alpha, \qquad V(1) = \beta, \end{gathered} \tag{2.1}$$

$$a(x) \geq 0, \qquad a(\cdot), b(\cdot), k(\cdot) \qquad \text{bounded, continuous}$$

$$\inf_{0 \leq x \leq 1} (|a(x)| + |f(x)|) > 0.$$

The last line is clearly necessary (but not sufficient) if (2.1) is to have a bounded solution for each continuous $k(\cdot)$. For any real-valued function $g(\cdot)$ define

$$g^+(x) = \max[0, g(x)], \qquad g^-(x) = \max[0, -g(x)].$$

Let h be a positive real number such that h^{-1} is an integer, and suppose that the stochastic differential equation (2.2) has a unique solution in the sense of probability distributions.

$$dX = f(X)\,dt + (2a(X))^{1/2}\,dw \tag{2.2}$$

Let $\tau = \inf\{t : X(t) \notin (0, 1)\}$. By Section 4.5, (2.1) is the equation formally satisfied by (if $\sup_{x \in (0,1)} E_x \tau < \infty$)

$$\begin{aligned} R(x) = E_x \int_0^\tau k(X(s))\,ds &+ \alpha P_x\{X(\tau) = 0\} \\ &+ \beta P_x\{X(\tau) = 1\}, \qquad x \in (0, 1). \end{aligned} \tag{2.3}$$

We will try to approximate the solution to (2.1) by a special finite difference method (finite difference interval h), which will have a very nice probabilistic interpretation. The solution is *defined* to be (2.3), so we will want to show that the approximate solutions tend to (2.3) as $h \to 0$. Use the approximations

$$V_{xx}(x) \to [V(x+h) - 2V(x) + V(x-h)]/h^2 \tag{2.4}$$

$$\begin{aligned} V_x(x) &\to [V(x+h) - V(x)]/h \qquad \text{if} \quad f(x) \geq 0 \\ &\to [V(x) - V(x-h)]/h \qquad \text{if} \quad f(x) < 0. \end{aligned} \tag{2.5}$$

The choice in (2.5) is very important for reasons which will soon be clear. Let $V^h(\cdot)$ denote the solution to the finite difference equations and define

$$Q_h(x) = 2a(x) + h|f(x)|,$$

$$p^h(x, x \pm h) = [a(x) + hf^{\pm}(x)]/Q_h(x),$$

$$\Delta t^h(x) = h^2/Q_h(x), \qquad x = 0, \pm h, \dots.$$

Define $p^h(x, y) = 0$ for $y \neq x \pm h$, where y, x are multiples of h. Next, substitute (2.4) and (2.5) into (2.1), collect terms, divide by $Q_h(x)$, the negative of the coefficient of $V^h(x)$, and get

$$V^h(x) = V^h(x+h)p^h(x, x+h) + V^h(x-h)p^h(x, x-h) + k(x)\,\Delta t^h(x),$$

$$x = h, \dots, 1-h, \tag{2.6}$$

$$V^h(0) = \alpha, \qquad V^h(1) = \beta.$$

Equation (2.6) has an interesting probabilistic interpretation. The $p^h(x, y)$ are ≥ 0 and sum over y to unity for each x. Thus $\{p^h(x, y)\}$ is the transition probability for a Markov chain on the state space $0, \pm h, \dots$. Let $\{\xi_n^h, n = 0, 1, \dots\}$ denote the random variables of the chain, and define N_h, the escape time of the chain from $(0, 1)$, by $N_h = \min\{n : \xi_n^h \notin (0, 1)\}$. Equation (2.6) can be rewritten in the form

$$\begin{aligned} V^h(x) &= E_x V^h(\xi_1^h) + k(x)\,\Delta t^h(x) \\ V^h(0) &= \alpha, \qquad V^h(1) = \beta. \end{aligned} \tag{2.7}$$

Let $b(\cdot)$ be any continuous function such that $b(1) = \beta$, $b(0) = \alpha$. If $E_x N_h < \infty$, then by Section 3.1.2, (2.7) has the unique solution

$$V^h(x) = E_x\left[\sum_{i=0}^{N_h - 1} k(\xi_i^h)\,\Delta t_i^h + b(\xi_{N_h}^h)\right], \tag{2.8}$$

where we use $\Delta t_i^h = \Delta t^h(\xi_i^h)$. If $k(x) \equiv 0$ and $P_x\{N_h < \infty\} = 1$, then (2.7) has the unique solution

$$V^h(x) = E_x b(\xi_{N_h}^h) I_{\{N_h < \infty\}}.$$

Equation (2.8) has a superficial similarly to (2.3); if we interpret Δt_i^h as a time interval, then the sum in (2.8) resembles a Riemann sum approximation to the integral in (2.3). In fact, the resemblence is very close as we shall see and under certain conditions (which are not stringent in applications), $V^h(x) \to R(x)$ as $h \to 0$.

First, let us note that

$$\begin{aligned} E[\xi_{n+1}^h - \xi_n^h \mid \xi_n^h = x] &= f(x)\,\Delta t^h(x) \\ \operatorname{cov}[\xi_{n+1}^h - \xi_n^h \mid \xi_n^h = x] &= 2a(x)\,\Delta t^h(x) + (h\,|f(x)| \\ &\quad - f^2(x)\,\Delta t^h(x))\,\Delta t^h(x) \\ &= 2a(x)\,\Delta t^h(x) + o(\Delta t^h(x)). \end{aligned} \tag{2.9}$$

The terms in (2.10) are just what we would get [see Eqs. (1.5.15), and (1.5.16)] for the conditional increments of the solution to (2.2) over an interval† Δ. These calculations suggest that there are limit theorems connecting $\{\xi^h(\cdot)\}$ and the diffusion $X(\cdot)$. This is, indeed, just where the ideas of weak convergence enter. But first we must interpolate $\{\xi_n^h\}$ into a continuous parameter process. The $\Delta t^h(x)$ factors in (2.8) and (2.9) suggest the following interpolation. Define t_i^h and $\xi^h(\cdot)$ by

$$t_n^h = \sum_{i=0}^{n-1} \Delta t_i^h, \qquad t_0^h = 0,$$

$$\xi^h(t) = \xi_n^h \qquad \text{on} \quad [t_n^h, t_{n+1}^h).$$

So $\xi^h(\cdot)$ is held constant on *random intervals* and is our approximation to $X(\cdot)$.

In Chapter 6, it will be shown that $\{\xi^h(\cdot)\}$ is tight on $D[0, \infty)$ and that if h indexes a subsequence which converges in distribution to a limit $X(\cdot)$, then $X(\cdot)$ must satisfy (2.2). As a further check on the scaling, consider the special case where $a(\cdot) \equiv 0$. Then $\dot{x} = f(x)$ replaces (2.2) and $\Delta t^h(x) = h/|f(x)| = h/(\text{velocity at } x)$. This is the correct time interval since we must hold the process at a fixed value until enough time passes so that it can jump h. This time is just h times the inverse of the velocity at x.

Define $\rho_h = t_{N_h}^h$, the escape time of $\xi^h(\cdot)$ from (0, 1). Thus (2.8) becomes

$$V^h(x) = E_x\left[\int_0^{\rho_h} k(\xi^h(s))\,ds + b(\xi^h(\rho_h))\right]. \tag{2.10}$$

Assume, for the moment:

$$\rho_h < \infty \quad \text{w.p. 1, each } h \quad \text{and} \quad \{\rho_h\} \quad \text{is uniformly integrable,} \tag{2.11}$$

$$P_x\{\tau = \tau'\} = 1, \qquad x \in (0, 1). \tag{2.12}$$

† A Lipschitz condition was assumed in Section 1.5. But the estimates are valid for any solution to the stochastic differential equation if $f(\cdot)$, $\sigma(\cdot)$ are uniformly continuous.

We shall return to these assumptions later. Under (2.11), we can "essentially" replace ρ_h in (2.10) by $\rho_h \cap T$ for large T and all h. Under (2.12), the functions on $D[0, \infty)$ given by

$$g_1(x(\cdot)) = b(x(\tau)), \qquad g_2(x(\cdot)) = \int_0^\tau k(x(s))\, ds,$$

where τ = escape time of the path $x(\cdot)$ from (0, 1), are continuous w.p. 1 relative to the $X(\cdot)$ measure. Then weak convergence (Section 2.2) implies that $E_x g_1(\xi^h(\cdot)) \to E_x g_1(X(\cdot))$ and similarly for $g_2(\cdot)$. Owing to the uniqueness of the solution to (2.2) (in the sense of distributions), the distribution of each limit is the same; it does not depend on the particular convergent subsequence that we choose. Thus $E_x g_i(X(\cdot))$ is well defined by the approximation sequence.

Condition (2.12) holds if 0 and 1 are regular (τ'). A sufficient condition for this is

$$(a(0) > 0 \quad \text{or} \quad f(0) < 0) \qquad \text{and} \qquad (a(1) > 0 \quad \text{or} \quad f(1) > 0).$$

If $a(0) > 0$ and $X(\tau) = 0$, then the "wildness" of the function of t defined by $\int_\tau^{\tau+t} [2a(X(s))]^{1/2}\, dw(s)$ forces $x(\tau + t)$ to be less than 0 infinitely often on any interval $[0, s)$, $s > 0$. If $a(0) = 0$ and $f(0) < 0$, the drift forces the process to the left.

If $k(\cdot) \equiv 0$, then (2.11) can be replaced by

$$P_x\{\tau < \infty\} = 1. \tag{2.13}$$

If (2.13) does not hold at some $x \in (0, 1)$, then the approximations—indeed any finite difference approximation—may not converge as $h \to 0$ for that value of x, although the approximations will still converge for the x values for which (2.12) and (2.13) do hold if $k(\cdot) \equiv 0$ (Chapter 6).

If (2.13) and (2.12) hold then we will have $\rho_h \to \tau$ w.p. 1 (using the Skorokhod imbedding) and so $\xi^h(\rho_h) \to X(\tau)$ w.p. 1. If (2.13) *does not hold*, then there is a nonnull set on which $X(t) \in (0, 1)$, all $t < \infty$. But for each h, the corresponding paths $\xi^h(t)$ may still hit the boundary at a finite time (although those times will go to ∞ as $h \to 0$); hence $E_x b(\xi^h(\rho_h))$ will not necessarily converge to the correct limit $E_x b(X(\tau)) I_{\{\tau < \infty\}}$.

It will be shown below that

$$\inf_{x \in (0, 1)} a(x) > 0 \tag{2.14}$$

implies (2.11). In general, for a problem in R^r, uniform positive definiteness of the matrix $a(\cdot)$ guarantees (2.11). In the degenerate case, we must check whether the problem is well defined—in particular, whether escape is possible—for otherwise the functionals may not be well defined.

Consider the special case where there is a real $\sigma^2 > 0$ such that $a(x) = \sigma^2/2$ and where $f(x) \equiv 0$. Then $X(\cdot)$ is a Wiener process with covariance $\sigma^2 t$ and $\Delta t^h(x) = h^2/\sigma^2$, a constant, and $\{\xi_n^h\}$ a random walk. Then by Section 5.1, $\xi^h(\cdot)$ converges in distribution to a Wiener process with covariance $\sigma^2 t$, which we will also denote by $X(\cdot)$. Also, (2.12) holds and so does (2.11) (see below). Hence $V^h(x) \to R(x)$.

Let us use the Skorokhod imbedding (Theorem 2.2.2). Thus, we can assume that the probability space is chosen so that (w.p. 1)

$$\xi^h(t) \to X(t)$$

uniformly on bounded t sets. Then since 0 and 1 are regular (τ'), $\rho_h \to \tau$ w.p. 1 and (w.p. 1),

$$b(\xi^h(\rho_h)) \to b(X(\tau)),$$

$$\int_0^{\rho_h} k(\xi^h(s))\, ds \to \int_0^{\tau} k(X(s))\, ds, \qquad \text{w.p. 1.}$$

The expectations of the $b(\cdot)$ terms converge also. By (2.11), the expectations of the integral terms converge.

For this simple case, more classical (and simpler) methods also work. But our intention was to illustrate the weak convergence ideas on a simple example, where the ideas are useful to prove convergence of approximations to an interesting unbounded path functional, which is the solution of a differential equation. The probabilistic approach required that we work with random variables that are elements of the abstract space $D[0, \infty)$ or that we treat convergence of processes rather than of vector-valued variables. We could not have easily treated the problem by using convergence of multivariate distributions only.

Characteristic Operator

The definition of the domain of $\tilde{A}$, the weak infinitesimal operator, involves a nonlocal calculation in the sense that $X(t)$ may not be arbitrarily close to $X(0) = x$, uniformly in ω, for small t, and we must calculate $\lim_{t\to 0} [E_x F(X(t)) - F(x)]/t$. Even if $|X(t) - x|$ were suitably and uniformly (in ω) bounded for small t, the limit may not exist. But the characteristic operator is a local operator. Define $\tau_h = \inf\{t : X(t) = x \pm h\}$ and let $X(0) = x$. It is not hard to check that if $f(\cdot) = 0$, then $E_x \tau_h = h^2/\sigma^2(x) + o(h^2)$ and if $f(x) \neq 0$ but $\sigma^2(\cdot) = 0$, then $E_x \tau_h = h/|f(x)| + o(h)$. Using (2.7), we write

$$[E_x V^h(\xi^h(\tau_h)) - V^h(x)]/\Delta t^h(x) + k(x) = 0,$$

which is just an approximation to the "local" characteristic operator. This suggests that the probabilistic approach taken here is quite natural. Indeed, the functionals of concern (in the uncontrolled case) will usually be in the domain of the characteristic operator.

PROOF THAT (2.12) AND $\inf_{x \in [0, 1]} P_x\{\tau < T\} \geq 2c > 0$ IMPLY (2.11) FOR SMALL h, WHERE T AND c ARE POSITIVE REAL NUMBERS Suppose that there are sequences $h \to 0$, $x_h \to x \in [0, 1]$, such that

$$P_{x_h}\{\rho_h < T\} \to 0 \qquad \text{as} \qquad h \to 0. \tag{2.15}$$

The sequence $\{\xi^h(\cdot), \xi^h(0) = x_h\}$ can be shown to be tight in $D[0, \infty)$ and all limits have the form (2.2) for $X(0) = x$. The measure induced by each solution (for perhaps different Wiener processes) of (2.2) is the same by our uniqueness assumption. Since, by (2.12), the escape time τ is continuous w.p. 1, and $\xi^h(t) \to X(t)$ uniformly on finite intervals (using Skorokhod imbedding), we have that $\rho_h \to \tau$ w.p. 1 as $h \to 0$. The set of paths in $D[0, \infty)$, for which† $\tau' < T$ is open. Thus (Chapter 2) $\lim_{h \to 0} P_{x_h}\{\rho_h < T\} \geq P_x\{\tau < T\} \geq 2c > 0$. This contradicts (2.15). Hence

$$\inf_{x \in [0, 1]} P_x\{\rho_h < T\} \geq c, \qquad \text{all small} \quad h. \tag{2.16}$$

Let $\tilde{\xi}^h(\cdot)$, $\tilde{\xi}_n^h$ denote the stopped processes $\xi^h(t \cap \rho_h)$ and $\xi^h_{n \cap Nh}$, resp. Then $\rho_h \geq 2T$ if and only if $\tilde{\xi}^h(T) \in (0, 1)$, and the escape time is $\geq T$ when the initial condition is $\tilde{\xi}^h(T)$. Thus (for small h)

$$P_x\{\rho_h > 2T\} = E_x I_{\{\tilde{\xi}^h(T) \in (0, 1)\}} I_{\{\tilde{\xi}^h(2T) \in (0, 1)\}}\,.$$

Using the Markov property of $\{\tilde{\xi}_n^h\}$,

$$\begin{aligned} P_x\{\rho_h \geq 2T\} &= E_x I_{\{\tilde{\xi}^h(T) \in (0, 1)\}} P_{\tilde{\xi}^h(T)}\{\rho_h \geq T\} \\ &\leq E_x I_{\{\tilde{\xi}^h(T) \in (0, 1)\}}(1 - c) \leq (1 - c)^2. \end{aligned}$$

In general (for small h) $P_x\{\rho_h \geq nT\} \leq (1 - c)^n$, which implies that there are numbers $M_n < \infty$ for which (for small h)

$$E_x \rho_h^n \leq M_n \qquad \text{uniformly in} \quad x \in (0, 1),$$

which implies the uniform integrability.

On the Finite Difference Method and the Choice (2.5)

The purpose of the choice between forward and backward differences in the approximation (2.5) is to obtain a finite difference equation whose coefficients are nonnegative. With either a forward or backward difference

† In a more formal notation, let $\tau(x(\cdot))$ denote the escape time of $x(\cdot)$ from (0, 1), where $x(\cdot)$ is an arbitrary element of $D[0, \infty)$. In this terminology, $\rho_h(\omega) = \tau(\xi^h(\omega, \cdot))$.

approximation, the coefficients would sum to unity, but they could be negative. With our choice, the dynamics help "push the particle to the right" if $f(x) > 0$ and to the left if $f(x) < 0$.

Any approximation method which provides a computationally convenient chain $\{\xi_n^h\}$ and whose interpolations converge to $X(\cdot)$ in distribution would be satisfactory. However, the finite difference procedure provides an automatic method of obtaining such processes. Neither $f(\cdot)$ nor $a(\cdot)$ need be continuous (Chapter 6) if the proper approximations are used.

CHAPTER 6

Elliptic Equations and Uncontrolled Diffusions

In this chapter, we develop in detail the basic techniques for approximation of a diffusion by a Markov chain and for showing convergence of functionals of the chain to functionals of diffusions and to weak solutions of degenerate elliptic equations.

Section 6.1 gives some assumptions and formulates part of the problem. Section 6.2 discusses a class of approximating chains. The techniques involve using certain finite difference approximations to the elliptic equations. This approach has the advantage of yielding a simple way to get the approximating chains; but it is not the only way. Section 6.3 discusses the natural interpolations for the chain, proves the necessary tightness, and shows that the limit of the interpolated chains is a diffusion. In Section 6.4, we discuss the convergence of functionals of the chain to weak solutions of the partial differential equation, and the problems that arise due to the presence of a boundary. Section 6.5 deals with various approximations for the discounted problem.

In Section 6.6, we treat a technical question concerning representing a certain term β_n^h in the form $\sigma(\xi_n^h)\, \delta W_n^h +$ "small" error, where the natural interpolations of the indefinite sum of the δW_n^h converge to a Wiener process in distribution. The representation will be very useful in the Chapters 8 and 9, which deal with approximations to nonlinear partial differential equations and to optimal control problems.

Section 6.7 discusses the use of the chains for Monte Carlo approximations. In Section 6.8, we show how to approximate invariant measures of the diffusion by a naturally weighted invariant measure for the approximating chain. Here, the space $D(-\infty, \infty)$ is used as it is very useful in dealing with stationarity properties. Various extensions are briefly discussed in Section 6.9, and in Section 6.10 some numerical results are presented. A Lipschitz condition is not used. In its place, we require only that the solution of the stochastic differential equation is unique in the sense of probability law.

The techniques developed here and in the following chapters are readily applicable for the approximation of many types of path functionals with which we do not explicitly deal. For example, we can approximate functionals such as

$$E_x \sup_{\infty > t \geq 0} F(X(t)) \exp -\lambda t, \qquad \lambda > 0.$$

In the control problems of Chapters 8 and 9, owing to the nature of weak convergence, the approximating Markov chains yield information on a great variety of functionals of the optimal process.

6.1 Problem Formulation

This chapter consists partly of a generalization of the ideas and results of Chapter 5 together with proofs. We also treat the problem of calculating invariant measures, a "discounted problem," and several related results. We will deal with the equation

$$X(t) = x + \int_0^t f(X(s))\, ds + \int_0^t \sigma(X(s))\, dw(s) \tag{1.1}$$

and again (as in Chapter 4) assume

A6.1.1† *$f(\cdot)$ and $\sigma(\cdot)$ are continuous and bounded R^r and $r \times r$ matrix-valued functions on R^r, resp., and the solution to* (1.1) *is unique in the sense of probability distributions.*

† These conditions can be weakened. See the remarks at the end of the chapter.

A6.1.2† *$k(\cdot)$ and $b(\cdot)$ are bounded continuous real-valued functions on R^r and G is a bounded open set.*

The functional (Section 4.5) $R(\cdot)$ defined by

$$R(x) = E_x \int_0^\tau k(X(s))\, ds + E_x b(X(\tau)) \tag{1.2}$$

is a weak solution of

$$\mathscr{L}V(x) + k(x) = 0, \qquad x \in G,$$

with boundary condition (1.3)

$$\lim_{y \to x} V(y) = b(x), \quad y \in G, \quad x \text{ regular } (\tau) \text{ on } \partial G.$$

As in Chapter 5, the technique is the following. For each value of an approximation parameter h (perhaps vector-valued), we seek a finite state Markov chain approximation to $X(\cdot)$. In particular, the chain should be convenient to compute the discrete analog of functionals of the form (1.2). Then we use the weak convergence theory to show that the suitably interpolated chains and the functionals converge to $X(\cdot)$ and $R(\cdot)$, resp., as $h \to 0$. There are many ways of getting such an approximating chain $\{\xi_n^h\}$. We require at least that the first- and second-order moments of $\xi_{n+1}^h - \xi_n^h$, conditioned on ξ_n^h, be consistent with those of the diffusion and that the approximation not have large jumps; i.e.,

$$\begin{aligned} E_x[\xi_{n+1}^h - \xi_n^h \mid \xi_n^h = y] &= f(y)\, \Delta t^h(y) + o(\Delta t^h(y)), \\ \operatorname{cov}_x[\xi_{n+1}^h - \xi_n^h \mid \xi_n^h = y] &= 2a(y)\, \Delta t^h(y) + o(\Delta t^h(y)), \\ P_x\{|\xi_{n+1}^h - \xi_n^h| \geq \varepsilon \mid \xi_n^h = y\} &= o(\Delta t^h(y)), \\ 2a(y) &= \sigma(y)\sigma'(y) \end{aligned} \tag{1.4}$$

for some function (depending on h) $\Delta t^h(y)$ which goes to zero as $h \to 0$ [unless, perhaps, $a(y) = f(y) = 0$].

Finding chains whose transition probabilities give these properties is not, in general, a simple task, unless one has a rather systematic approach, for otherwise it involves solving a "large" nonlinear system of equations in the unknown transition probabilities. Fortunately, a very convenient "formula" is available. Following the idea in Chapter 5, it turns out that if we approximate the operator $\mathscr{L}$ by a carefully selected finite difference approximation, then the coefficients of the resulting finite difference equation will be the

† These conditions can be weakened. See the remarks at the end of the chapter.

transition function of the desired Markov chain, where h will be the difference interval, and the $\Delta t^h(\cdot)$ will be immediately apparent from the formulas. The technique suggests other types of approximations. Some of the ideas of this chapter appeared in Kushner and Yu [K11].

6.2 The Finite Difference Method and an Approximating Markov Chain

Let h denote the finite difference interval, in each coordinate direction, and e_i the unit vector in the ith coordinate direction.

The approximations that we need to use for $V_{x_i}(x)$ and $V_{x_ix_j}(x)$, $i \neq j$, will depend on the signs of $f_i(x)$ and $a_{ij}(x)$, resp. Let R_h^r denote the finite difference grid on R^r; namely, $x \in R_h^r$ if there are integers (positive, negative, or zero) $n_1, \ldots, n_r$ such that $x = \sum_i e_i hn_i$. Define $G_h = R_h^r \cap G$ and $\partial G_h = \{$set of points on $R_h^r - G_h$ which are only one node away from G_h along a coordinate direction or diagonal$\}$. Use the approximations

$$V_{x_i}(x) \to [V(x + e_i h) - V(x)]/h \quad \text{if} \quad f_i(x) \geq 0, \tag{2.1a}$$

$$\to [V(x) - V(x - e_i h)]/h \quad \text{if} \quad f_i(x) < 0, \tag{2.1b}$$

$$V_{x_ix_i}(x) \to [V(x + e_i h) + V(x - e_i h) - 2V(x)]/h^2, \tag{2.2}$$

and for $i \neq j$ and $a_{ij}(x) \geq 0$ use

$$\begin{aligned} V_{x_ix_j}(x) \to {} & [2V(x) + V(x + e_i h + e_j h) + V(x - e_i h - e_j h)]/2h^2 \\ & - [V(x + e_i h) + V(x - e_i h) + V(x + e_j h) \\ & + V(x - e_j h)]/2h^2, \end{aligned} \tag{2.3a}$$

and for $i \neq j$, $a_{ij}(x) < 0$, use

$$\begin{aligned} V_{x_ix_j}(x) \to {} & -[2V(x) + V(x + e_i h - e_j h) + V(x - e_i h + e_j h)]/2h^2 \\ & + [V(x + e_i h) + V(x - e_i h) + V(x + e_j h) \\ & + V(x - e_j h)]/2h^2. \end{aligned} \tag{2.3b}$$

Define

$$Q_h(x) = 2 \sum_i a_{ii}(x) - \sum_{i,\, j,\, i \neq j} |a_{ij}(x)| + h \sum_i |f_i(x)|$$

and assume that

$$a_{ii}(x) - \sum_{j \neq i,\, j} |a_{ij}(x)| \geq 0, \qquad \text{all} \quad i = 1, \ldots, r. \tag{2.4}$$

Assumption (2.4) is not as restrictive as it may seem. We usually have a choice of coordinate system when doing the discretization and a transformation of coordinates often can be applied to assure (2.4) in the new system. If $a(x)$ does not depend on x, such a transformation (to the principal vectors of a) is always possible.

Define the functions $\Delta t^h(\cdot)$, $p^h(\cdot,\cdot)$, and $\bar{Q}^h(\cdot,\cdot)$ on R_h^r, $R_h^r \times R_h^r$, and $R_h^r \times R_h^r$ by $(i \neq j)$

$$
\begin{aligned}
\Delta t^h(x) &= h^2/Q_h(x),\\
p^h(x, x \pm e_i h) &= \frac{a_{ii}(x) - \sum_{j \neq i,\, j} |a_{ij}(x)| + h f_i^{\pm}(x)}{Q_h(x)} \equiv \frac{\bar{Q}_h(x, x \pm e_i h)}{Q_h(x)},\\
p^h(x, x + e_i h \pm e_j h) &= \frac{a_{ij}^{\pm}(x)}{Q_h(x)} \equiv \frac{\bar{Q}_h(x, x + e_i h \pm e_j h)}{Q_h(x)},\\
p^h(x, x - e_i h \pm e_j h) &= \frac{a_{ij}^{\mp}(x)}{Q_h(x)} \equiv \frac{\bar{Q}_h(x, x - e_i h \pm e_j h)}{Q_h(x)},\\
p^h(x, y) &= 0, \qquad \text{all other} \quad x, y \in R_h^r.
\end{aligned}
\tag{2.5}
$$

The $p^h(x, y)$ are ≥ 0, sum (over y) to unity for each x, hence are one step transition probabilities for a Markov chain, whose random variables we will denote by $\{\xi_n^h\}$.

The $p^h(x, y)$ will be coefficients in the finite difference approximation. If (2.1a) were always used to approximate $V_{x_i}(x)$, then $f_i(x)$ would replace $|f_i(x)|$ in $Q_h(x)$ and $f_i(x)$ or 0, resp., would replace $f_i^+(x)$ and $f_i^-(x)$, resp., in $\bar{Q}_h(x, x + e_i h)$ and $\bar{Q}_h(x, x - e_i h)$, resp. Then we could not always guarantee that the $p^h(x, y)$ would be ≥ 0. A similar problem would occur if (2.1b) were always used. Similarly, we allow the choice in (2.3a) and (2.3b) in order to guarantee that the diagonal transition terms $p^h(x, x \pm e_i h \pm e_j h)$ are nonnegative. If $|f_i(x)|$ is dominated from above by the other terms in the numerator of $p^h(x, x \pm e_i h)$, then either a forward or backward difference scheme could be used at x for that i.

A REMARK ON THE CHAIN Observe that, as $a_{ij}(x)$ $(i \neq j)$ increases with all other $a_{\alpha\beta}(x)$ held fixed, the probability of moving to the diagonal corners increases until some $p^h(x, x \pm e_i h)$ is zero. A further increase in $a_{ij}(x)$ would cause that "probability" to become negative, and we must transform the coordinates to preserve the probabilistic interpretation. Figure 6.1 illustrates some two-dimensional examples of the form of the transition function.

Substituting (2.1) to (2.3) into (1.3), collecting coefficients of the terms $V(x)$, $V(x \pm e_i h)$, $V(x \pm e_i h \pm e_j h)$ and multiplying each term by h^2 and

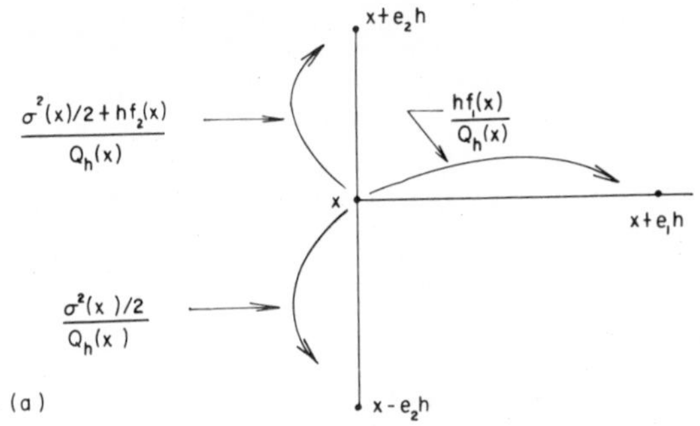

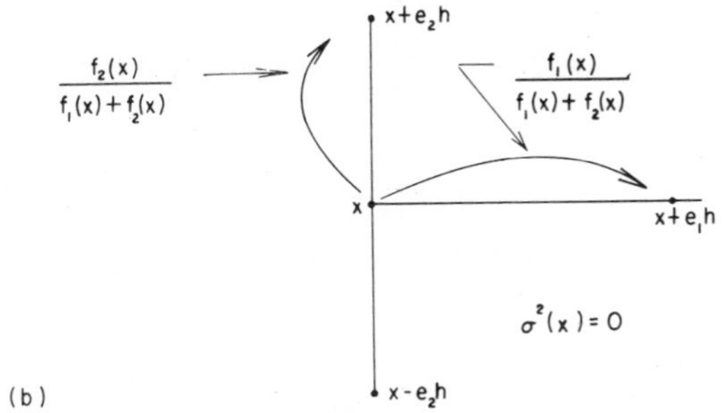

FIG. 6.1 Examples of transition probabilities for ξ_n^h. $dx_1 = f_1(x)\,dt$, $f_1(x) \geq 0$; $dx_2 = f_2(x)\,dt + \sigma(x)\,dw$, $f_2(x) \geq 0$.

using $V^h(\cdot)$ to denote the solution to the finite difference equation, yields the equation

$$O = -Q_h(x)V^h(x) + \sum_y \bar{Q}_h(x, y)V^h(y) + h^2 k(x), \qquad x \in G_h$$

or

$$V^h(x) = \sum_y p^h(x, y)V^h(y) + k(x)\,\Delta t^h(x), \qquad x \in G_h, \tag{2.6}$$

with boundary condition

$$V^h(x) = b(x), \qquad x \in \partial G_h.$$

The index y ranges over $G_h \cup \partial G_h$. Equation (2.6) can be rewritten as

$$V^h(x) = \begin{cases} E_x V^h(\xi_1^h) + k(x)\,\Delta t^h(x), & x \in G_h, \\ b(x), & x \in \partial G_h. \end{cases} \tag{2.7}$$

Define $N_h = \min\{n : \xi_n^h \notin G_h\}$, and *for the moment* assume that

$$E_x N_h < \infty. \tag{2.8}$$

Then (2.7) has a unique solution (Section 3.1.2) $R^h(\cdot)$, defined by

$$R^h(x) = E_x \sum_{n=0}^{N_h-1} k(\xi_n^h)\, \Delta t_n^h + E_x b(\xi_{N_h}^h), \tag{2.9}$$

where $\Delta t_n^h = \Delta t^h(\xi_n^h)$ is used. We will prove that $R^h(x) \to R(x)$, under some regularity conditions, as $h \to 0$.

Properties of the Chain $\{\xi_n^h\}$

Straightforward calculations yield

$$E_x[\xi_{n+1}^h - \xi_n^h \mid \xi_n^h = y] = f(y)\, \Delta t^h(y) \tag{2.10}$$

$$\begin{aligned} \text{cov}_x[\xi_{n+1}^h - \xi_n^h \mid \xi_n^h = y] &= 2a(y)\, \Delta t^h(y) + h\, \Delta t^h(y) \tilde{f}(y) - (\Delta t^h(y))^2 f(y) f'(y) \\ &\equiv \Sigma_h(y)\, \Delta t^h(y) \equiv 2a(y)\, \Delta t^h(y) + \delta\Sigma_h(y)\, \Delta t^h(y), \end{aligned} \tag{2.11}$$

where

$$\tilde{f}(y) = \begin{bmatrix} |f_1(y)| & & 0 \\ & \ddots & \\ 0 & & |f_r(y)| \end{bmatrix}.$$

Equations (2.10) and (2.11) meet the requirements of the first two lines of (1.4) [$\{\xi_n^h\}$ also has a "no jump" property corresponding to the 3rd line of (1.4)] and are consistent with the local properties of $X(\cdot)$ given in (1.5.15) and (1.5.16).

A convenient representation for $\{\xi_n^h\}$ is

$$\begin{aligned} \xi_{n+1}^h &= \xi_n^h + f(\xi_n^h)\, \Delta t_n^h + \beta_n^h \\ &= \xi_n^h + E[\xi_{n+1}^h - \xi_n^h \mid \xi_n^h] + (\xi_{n+1}^h - \xi_n^h - E[\xi_{n+1}^h - \xi_n^h \mid \xi_n^h]), \end{aligned} \tag{2.12}$$

where $\{\beta_n^h\}$ is an orthogonal sequence for each h. In particular,

$$\begin{aligned} E_x[(\beta_n^h)(\beta_n^h)' \mid \xi_i^h, i \le n] &= \Sigma_h(\xi_n^h)\, \Delta t_n^h, \\ E_x[\beta_n^h \mid \xi_i^h, i \le n] &= 0. \end{aligned} \tag{2.13}$$

A useful representation for β_n is given in Section 6.6.

An Interpolation of $\{\xi_n^h\}$ *to a Continuous Parameter Process*

So far the development has consisted of a sequence of manipulations and we have only shown that a suitably selected finite difference approximation has a probabilistic interpretation. The interpretation is quite suggestive, however. Equation (2.9) looks like a Riemann sum approximation to (1.2), and the "differential" properties of $\{\xi_n^h\}$ resemble those of $X(\cdot)$. The relationships will be made precise in the sequel. The forms of the expressions (2.9),

(2.10), and (2.11) suggest a natural continuous time interpolation. Define the time sequence $\{t_n^h\}$ and the interpolated continuous parameter process $\xi^h(\cdot)$ by

$$t_n^h = \sum_{i=0}^{n-1} \Delta t_i^h, \qquad t_0^h = 0,$$

$$\xi^h(t) = \xi_i^h \qquad \text{on} \quad [t_i^h, t_{i+1}^h).$$

Define $\rho_h = t_{N_h}^h$ = escape time of $\xi^h(\cdot)$ from G_h. Then (2.9) can be written as

$$R^h(x) = E_x \int_0^{\rho_h} k(\xi^h(s))\, ds + E_x b(\xi^h(\rho_h)). \tag{2.14}$$

The paths of $\xi^h(\cdot)$ are in $D^r[0, \infty)$. It will turn out that $\xi^h(\cdot) \to X(\cdot)$ in distribution as $h \to 0$.

A REMARK ON THE INTERPOLATION Let us examine the special case of Fig. 6.1b in more detail, when $f_1(x) = c_1, f_2(x) = c_2$, where the c_i are positive constants. Let $X(0) = 0$ and let F_ε denote an ε-neighborhood ($\varepsilon > 0$) about the path $x_1(t) = c_1 t$, $x_2(t) = c_2 t$. The random path of $\{\xi_n^h\}$ moves upward and to the right. By (2.10), the average movement (given ξ_n^h) of $\xi_{n+1}^h - \xi_n^h$ is $c\, \Delta t^h(y)$, where $\Delta t^h(y) = h/(c_1 + c_2)$, and $\text{cov}(\xi_{n+1}^h - \xi_n^h \mid \xi_n^h = y) = O(h^2) = O(\Delta t^h(y)^2)$. Also, $P_x\{\sup_{t \geq s \geq 0} |\xi^h(s) - cs| > \varepsilon\} \to 1$ as $t \to \infty$, so eventually the interpolated path wanders away from the nominal mean path ct. However, on any finite interval $[0, T]$, we can invoke a law of large numbers to show that

$$P_x\left\{\sup_{T \geq s \geq 0} |\xi^h(s) - cs| > \varepsilon\right\} \to 0$$

as $h \to 0$. So the random path becomes a very good approximation to the nominal path ct on each finite interval $[0, T]$ as $h \to 0$.

These considerations show that *for this simple case*, the values (2.9) or (2.14) are actually approximations to the integral of $k(X(t))$ along the curve $X(t)$ until exit from G, plus a boundary term; the process $\xi^h(\cdot)$ actually approximates the limit path on each fixed finite interval and the values of any functional (which is continuous w.p. 1, relative to the limit process) of the approximations actually converges to the correct limit as $h \to 0$. If the functional is discontinuous on one path (say $x_0 + ct$), then the approximations will not necessarily converge if $X(0) = x_0$, but will converge for initial values y which yield paths $y + ct$ on which the functional is continuous. There are extensions of these ideas to the more general cases, but this simple example should illustrate that the method is a very natural one, and why it should be able to handle "reasonably well" cases where $R(\cdot)$ is discontinuous.

6.3 Convergence of the Approximations to a Diffusion Process

Let $\bar{R}^+$ denote the one point compactification of $[0, \infty) = R^+$. Fix the initial condition x and define

$$B^h(t) = \sum_{t_{i+1}^h \le t} \beta_i^h, \qquad F^h(t) = \sum_{t_{i+1}^h \le t} f(\xi_i^h)\, \Delta t_i^h,$$
$$K^h(t) = \sum_{t_{i+1}^h \le t} k(\xi_i^h)\, \Delta t_i^h.$$

Then

$$\xi^h(t) = x + F^h(t) + B^h(t),$$
$$R^h(x) = E_x K^h(\rho_h) + E_x b(\xi^h(\rho_h)). \tag{3.1}$$

Let $\Phi^h(\cdot)$ denote the quadruple $(\xi^h(\cdot), F^h(\cdot), B^h(\cdot), K^h(\cdot))$.

Theorem 6.3.1 *If $a(\cdot)$, $f(\cdot)$, and $k(\cdot)$ are bounded and continuous, then $\{\Phi^h(\cdot), \rho_h, h > 0\}$ is tight on $D^{3r+1}[0, \infty) \times \bar{R}^+$. If $\{\Phi^h(\cdot)\}$ denotes any subsequence which is convergent in distribution to a process $\Phi(\cdot) = (\xi(\cdot), F(\cdot), B(\cdot), K(\cdot))$, then $\Phi(\cdot)$ has continuous paths w.p. 1.*

PROOF Since $\bar{R}^+$ is compact, $\{\rho_h\}$ is obviously tight. Let $y^h(\cdot)$ denote an arbitrary scalar component of $\Phi^h(\cdot)$. If, for each such component and real T, there is a real K such that, for all $T \ge s + \delta \ge s \ge 0$, and some $\alpha > 0$,

$$\overline{\lim_{h \to 0}}\, E_x \sup_{s+\delta \ge t \ge s} |y^h(t) - y^h(s)|^\alpha \le K\delta^2, \tag{3.2}$$

then (2.3.2) holds. Since (2.3.1a) always holds, Theorem 2.4.2 then implies the tightness and continuity assertions. Obviously, $\{\xi^h(\cdot)\}$ will be tight if $\{F^h(\cdot), B^h(\cdot)\}$ is tight. So we need only verify (3.2) for $\{F^h(\cdot)\}$ and $\{B^h(\cdot)\}$.

Fix $s > 0$ and $\delta > 0$. Define n_t by $n_0 = 0$ and (delete the h for notational simplicity)

$$n_t = \max\{i : t_i^h \le t\}, \qquad t > 0.$$

Let $I_i = 1$ if $s < t_{i+1}^h \le s + \delta$. Then we can write

$$F^h(s + \delta) - F^h(s) = \sum_{i=n_s}^{n_{s+\delta}-1} f(\xi_i^h)\, \Delta t_i^h = \sum_{i=0}^{\infty} f(\xi_i^h)\, \Delta t_i^h I_i\,.$$

Write the first sum as

$$f(\xi_{n_s}^h)\, \Delta t_{n_s}^h + \sum_{i=n_s+1}^{n_{s+\delta}-1} f(\xi_i^h)\, \Delta t_i^h.$$

The first term tends to zero uniformly (in s, ω) as $h \to 0$. Then, since

$$\sum_{i=n_s+1}^{n_{s+\delta}-1} \Delta t_i^h \le \delta,$$

(3.2) holds for $\{F^h(\cdot)\}$ with $\alpha = 2$. A similar argument works for $\{K^h(\cdot)\}$.

The sequence $\sum_{i=1}^n \beta_i$ is an R^r-valued martingale. By the discrete parameter version of the inequality (1.2.3), there is a real K_0 such that

$$E \sup_{n_{s+\delta}-1 \ge l \ge 0} \left| \sum_{i=n_s}^{l} \beta_i^h \right|^4 = E \sup_{s+\delta \ge t \ge s} |B^h(t) - B^h(s)|^4 \le K_0 E \left| \sum_{i=n_s}^{n_{s+\delta}-1} \beta_i^h \right|^4 .$$

We will show that there is a real K such that

$$E \left| \sum_{i=n_s+1}^{n_{s+\delta}-1} \beta_i^h \right|^4 \le K(\delta^2 + h^2\delta) \qquad \text{for all} \quad s, h, \tag{3.3}$$

which, together with the fact that $E|\beta_{n_s}^h|^4 \to 0$ as $h \to 0$ (uniformly in s, ω) and Chebychevs' inequality, implies (3.2) for $\{B^h(\cdot)\}$.

For notational convenience (with no loss in generality), suppose that β_i^h is scalar-valued. Define I_i' to be unity if β_i^h is in the sum on the left-hand side of (3.3), and $I_i' = 0$ otherwise. Then (3.3) equals

$$F_1 + F_2 + F_3 + F_4,$$

where

$$F_1 = E \sum_{i>j,k,l} \beta_i^h \beta_j^h \beta_k^h \beta_l^h I_i' I_j' I_k' I_l',$$

$$F_2 = 6E \sum_{i>j,k} (\beta_i^h)^2 \beta_j^h \beta_k^h I_i' I_j' I_k',$$

$$F_3 = 4E \sum_{i>j} (\beta_i^h)^3 \beta_j^h I_i' I_j', \qquad F_4 = E \sum_i (\beta_i^h)^4 I_i' .$$

$F_1 = 0$ since $E[\beta_n^h | \xi_i^h,\ i \le n] = 0$. Let K_1 be any real number $\ge \sup_{h,x} |\Sigma_h(x)|$. Write $\tilde\beta_i^h = \beta_i^h I_i'$. Then

$$F_2 = 6E \sum_{i=0}^{\infty} (\tilde\beta_i^h)^2 \left(\sum_{j=0}^{i-1} \tilde\beta_j^h \right)^2$$

$$\le 6K_1 E \sum_{i=0}^{\infty} I_i' \, \Delta t_i^h \left(\sum_{j=0}^{i-1} \tilde\beta_j^h \right)^2$$

$$\le 6K_1 \, \delta E \left(\sum_{j=0}^{\infty} \tilde\beta_j^h \right)^2 \le 6K_1^2 \delta^2.$$

Let us rewrite F_3 as

$$4E[(\tilde\beta_1^h)^3 \tilde\beta_0^h + (\tilde\beta_2^h)^3(\tilde\beta_0^h + \tilde\beta_1^h) + \cdots + (\tilde\beta_i^h)^3(\tilde\beta_0^h + \cdots + \tilde\beta_{i-1}^h) + \cdots],$$

which is dominated in absolute value by

$$4E\{(\tilde{\beta}_1^h)^2[(\tilde{\beta}_1^h)^2 + (\tilde{\beta}_0^h)^2] + \cdots + (\tilde{\beta}_i^h)^2[(\tilde{\beta}_i^h)^2 + (\tilde{\beta}_0^h + \cdots + \tilde{\beta}_{i-1}^h)^2] + \cdots\}.$$

In turn, the latter quantity equals

$$4E\sum_{i=0}^{\infty}(\tilde{\beta}_i^h)^4 + 4E\sum_{i=1}^{\infty}(\tilde{\beta}_i^h)^2(\tilde{\beta}_0^h + \cdots + \tilde{\beta}_{i-1}^h)^2.$$

The first term is of the form of F_4, and for the second term we get the bound (using $E[(\tilde{\beta}_n^h)^2 \,|\, \xi_i^h, i \le n] = \Sigma_h(\xi_n^h)\,\Delta t_n^h$)

$$4E\sum_{i=1}^{\infty}\Sigma_h(\xi_i^h)\,\Delta t_i^h I_i'(\tilde{\beta}_0^h + \cdots + \tilde{\beta}_{i-1}^h)^2 \le 4K_1\ \delta E\sup_{\infty > n \ge 0}\left(\sum_{i=0}^{n}\tilde{\beta}_i^h\right)^2 \equiv 4\delta H^h.$$

Since $Y_n^h = \sum_{i=0}^{n}\tilde{\beta}_i^h$ is a martingale, we can use the inequality (discrete parameter version of) (1.2.3) to get

$$H^h \le 4E\left(\sum_{i=0}^{\infty}\tilde{\beta}_i^h\right)^2 K_1 \le 4K_1^2\delta.$$

Since $|\beta_i^h| = |\xi_{i+1}^h - \xi_i^h - f(\xi_i^h)\,\Delta t_i^h| \le K_2 h$ for some real K_2, we can dominate F_4 by

$$F_4 \le K_2^2 h^2 E\sum_{i=0}^{\infty}(\tilde{\beta}_i^h)^2 \le K_2^2 K_1 h^2\delta.$$

Our computations imply (3.3). Q.E.D.

Let $\{\Phi^h(\cdot), \rho_h\}$ denote a subsequence which converges in distribution and let $(\Phi(\cdot), \tau)$ denote the limit process, where $\Phi(\cdot) = (\xi(\cdot), F(\cdot), B(\cdot), K(\cdot))$ has continuous paths w.p. 1.

The following lemma will be very useful when we need to characterize conditional expectations by "weak limiting" operations. The proof follows from the Radon–Nikodym definition of conditional expectation and from the fact that (see the lemma for notation) $\mathscr{B}_t(y)$ is the minimal σ-algebra over which all $g_n(y(s_1), \ldots, y(s_n))$ ($s_i \le t$, n, g_n, varying) are measurable.

Lemma 6.3.1 Let $y(\cdot)$ and z denote a vector-valued separable random process and a random vector, resp. Let $\mathscr{B}_t(y)$ denote the smallest σ-algebra over which $y(s)$, $s \le t$ is measurable. Suppose that there is a $\mathscr{B}_t(y)$ measurable function $\bar{z}$ such that, for each integer n and bounded continuous real-valued function $g_n(\cdot)$ and numbers $0 \le s_1 \le s_2 \cdots \le s_n \le t$, we have

$$Eg_n(y(s_1), \ldots, y(s_n))z = Eg_n(y(s_1), \ldots, y(s_n))\bar{z}.$$

Then, w.p. 1,

$$\bar{z} = E[z \,|\, \mathscr{B}_t(y)].$$

Theorem 6.3.2 *Assume* A6.1.1 *and* A6.1.2. *There is a Wiener process* $W(\cdot)$ *such that* $\xi(\cdot)$ *is nonanticipative with respect to* $W(\cdot)$ *and*

$$\begin{aligned}\xi(t) &= x + \int_0^t f(\xi(s))\,ds + \int_0^t \sigma(\xi(s))\,dW(s)\\ &= x + F(t) + B(t)\end{aligned}$$

and

$$K(t) = \int_0^t k(\xi(s))\,ds.$$

The multivariate distributions of $\Phi(\cdot)$ *do not depend on the particular convergent subsequence.*

PROOF *We always assume that the Skorokhod imbedding is used.* By Theorem 6.3.1, any subsequence of $\{\Phi^h(\cdot)\}$ has a further subsequence which is convergent. Fix the subsequence and index it by h. Recall that the limit $\Phi(\cdot)$ is continuous w.p. 1 and by Skorokhod imbedding we can suppose that the convergence is uniform on finite time intervals w.p. 1. Thus, (3.1) yields

$$\xi(t) = x + F(t) + B(t) \tag{3.4}$$

and it is obvious that $F(\cdot)$ and $K(\cdot)$ have the asserted forms. If we can prove that $B(\cdot)$ can be represented by the asserted stochastic integral, then the assumed uniqueness in the sense of distributions A6.1.1 implies that the subsequence is unimportant since it implies that any convergent subsequence of (hence the entire sequence) $\{\xi^h(\cdot)\}$ converges in distribution to a solution of (1.1).

To show that $B(\cdot)$ has the desired representation, we will show that it is a continuous martingale with quadratic variation $\int_0^t 2a(\xi(s))\,ds$ (Section 1.4.4).

Equation (3.4) implies that $B(t)$ is $\mathscr{B}_t(\xi)$ measurable.† Define $B^h(t, \Delta) = B^h(t + \Delta) - B^h(t)$ and $B(t, \Delta) = B(t + \Delta) - B(t)$. By the proof of Theorem 6.3.1, $E|B^h(t, \Delta)|^4$ is uniformly bounded in h for each $t, \Delta \geq 0$. Hence

$$\{B^h(t, \Delta)\}, \{B^h(t, \Delta)(B^h(t, \Delta))'\}$$

are both uniformly (in h) integrable. Let n, $g_n(\cdot)$, $s_1, \ldots, s_n \leq t$ be defined as in Lemma 6.3.1. Define C_h, D_h and C, D by

$$\begin{aligned}C_h &= Eg_n(\xi^h(s_1), \ldots, \xi^h(s_n))B^h(s, \Delta),\\ C &= Eg_n(\xi(s_1), \ldots, \xi(s_n))B(s, \Delta),\\ D_h &= Eg_n(\xi^h(s_1), \ldots, \xi^h(s_n))B^h(s, \Delta)(B^h(s, \Delta))',\\ D &= Eg_n(\xi(s_1), \ldots, \xi(s_n))B(s, \Delta)(B(s, \Delta))',\\ D^0 &= Eg_n(\xi(s_1), \ldots, \xi(s_n))\int_t^{t+\Delta} 2a(\xi(s))\,ds.\end{aligned}$$

† $\mathscr{B}_t(\xi)$ denotes the minimal σ-algebra over which $\xi(s)$, $s \leq t$, is measurable.

The uniform integrability of the integrands in C_h and D_h and the weak convergence imply that $C_h \to C$ and $D_h \to D$ as $h \to 0$. But, using the facts that $E[B^h(t, \Delta) \mid \xi^h(s), s \le t] = 0$, and the conditional covariance properties of $B^h(t, \Delta)$, yields that $C_h = 0$ (hence $C = 0$) and $D_h \to D^0$ as $h \to 0$. Hence by the Lemma (w.p. 1)

$$E[B(t, \Delta)(B(t, \Delta))' \mid \xi(s), s \le t] = E\left[\int_t^{t+\Delta} 2a(\xi(s))\, ds \mid \xi(s), s \le t\right]$$

and

$$E[B(t, \Delta) \mid \xi(s), s \le t] = 0,$$

which gives the desired martingale and quadratic covariation result. Q.E.D.

6.4 Convergence of the Cost Functionals $R^h(\cdot)$

The following theorem is a consequence of the weak convergence.

Theorem 6.4.1 *Under the conditions of Theorem 6.3.2, if $F(\cdot)$ is any real-valued bounded and measurable function on $D^r[0, \infty)$, which is a.e. continuous with respect to the measure induced on $D^r[0, \infty)$ by $\xi(\cdot)$, then*

$$E_x F(\xi^h(\cdot)) \to E_x F(\xi(\cdot)) \quad \text{as} \quad h \to 0.$$

The theorem holds if we replace boundedness by uniform integrability of $\{F(\xi^h(\cdot))\}$. The convergence is uniform on compact x sets.

We will comment on the last sentence only. Suppose that the convergence is not uniform on compact sets. Then there exists an $\varepsilon > 0$ and a bounded sequence $\{x_h\}$ such that

$$|E_{x_h} F(\xi^h(\cdot)) - E_{x_h} F(\xi(\cdot))| \ge \varepsilon \qquad \text{all} \quad h.$$

Select a subsequence such that $x_h \to x$. The sequences $\{\xi(\cdot), \xi(0) = x_h\}$ and $\{\xi^h(\cdot), \xi^h(0) = x_h\}$ are both tight on $D^r[0, \infty)$, and any convergent subsequence of either sequence converges to a solution of (1.1). This, together with the uniqueness of the probability law of the solution to (1.1), implies that ε cannot be positive.

The remaining technical difficulties in proving the desired convergence $R^h(x) \to R(x)$, concern the continuity† of the random time τ. The proof of

† By "continuity of τ," we mean that we are concerned with the continuity of a $[0, \infty]$-valued function on $D^r[0, \infty)$. The function evaluated at $x(\cdot) \in D^r[0, \infty)$, is the first exit time of $x(\cdot)$ from G. Recall that τ and τ' are the escape times of $\xi(\cdot)$ (or $X(\cdot)$) from G and $\bar{G}$, resp.

the following theorem follows from the discussion of exit times in Chapter 4, and from Theorems 6.3.1, 6.3.2, and 6.4.1.

Theorem 6.4.2 *Assume the conditions of Theorem* 6.3.2. *If*

$$P_x\{\tau = \tau' < \infty\} = 1, \tag{4.1}$$

then τ is continuous w.p. 1 with respect to $\xi(\cdot)$ measure and

$$E_x b(\xi^h(\rho_h)) \to E_x b(\xi(\tau)), \tag{4.2}$$

$$E_x \int_0^{T \cap \rho_h} k(\xi^h(s))\, ds \to E_x \int_0^{T \cap \tau} k(\xi(s))\, ds \tag{4.3}$$

for each real $T > 0$ as $h \to 0$. If, in addition,

$$\{\rho_h\} \quad \text{is uniformly integrable,} \tag{4.4}$$

then

$$E_x \int_{T \cap \rho_h}^{\rho_h} k(\xi^h(s))\, ds \to 0$$

as $T \to \infty$, uniformly in h, and we can set $T = \infty$ in (4.3).

Unfortunately, (4.1) and $E_x \tau < \infty$ do not imply (4.4), because the approximations $\xi^h(\cdot)$ may get stuck at points which are inaccessible to the process $\xi(\cdot)$. [Of course, (4.1) and $\sup_{x \in \bar{G}} E_x \tau' < \infty$ do imply (4.4).] To see this, consider the example of Fig. 6.2.

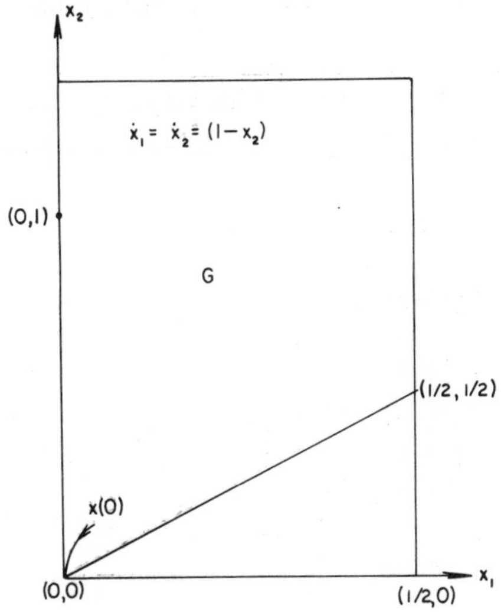

FIG. 6.2 Example where $E_x \rho_h = \infty$ and $E_x \tau < 1$.

Let $1/h$ be an integer. Starting at $X(0)$, the trajectory exits from G at time $\tau < 1$. Since $p^h(x, x + e_2 h) = \frac{1}{2} \neq 0$ for $x_2 \neq 1$, there is a nonzero probability that some ξ_n^h will hit the line $x_2 = 1$ (hence get stuck on that line since $\Delta t^h(x) = \infty$ on that line) before exiting from G. This probability goes to zero as $h \to 0$, but it is never exactly zero. Hence $E_x \rho_h = \infty$, despite the fact that $\rho_h \to \tau < 1$ w.p. 1. Thus we must somehow guarantee that the uniform integrability (4.4) holds.

Uniform Integrability

A proof similar to that used at the end of Chapter 5 shows that $\{\rho_h\}$ is uniformly integrable if there are real positive T, c such that

$$\inf_{x \in \bar{G}} P_x\{\tau' \leq T\} \geq c. \tag{4.5}$$

Property (4.5) is guaranteed if

$$\inf_{x \in \bar{G}} a_{ii}(x) > 0 \qquad \text{for some} \quad i. \tag{4.6}$$

A property implying (4.5) under the required smoothness condition [and implied by (4.6)] is (see Pinsky [Pl]): Let $\sigma(\cdot) \in C^2(R^r)$, $f(\cdot) \in C^2(R^r)$ and suppose that there is a $\delta > 0$ and an $\hat{x} \in R^r$ such that

$$(x - \hat{x})' a(x)(x - \hat{x}) \geq \delta |x - \hat{x}|^2 \qquad \text{all} \quad x \in G. \tag{4.7}$$

6.5 The Discounted Cost Problem

Consider the discounted cost functional (4.2.2) with $\lambda_0 > 0$ and associated differential equation (4.5.2). The finite difference model of Section 2 when applied to (4.5.2) yields

$$0 = -[Q_h(x) + \lambda(x)h^2]V^h(x) + \sum_{y \in R_h^r} \bar{Q}_h(x, y)\, V^h(y)$$
$$+ k(x)h^2 = 0, \qquad x \in G_h,$$

which can be rewritten as

$$V^h(x) = \left[\frac{1}{1 + \lambda(x)\,\Delta t^h(x)}\right][E_x V^h(\xi_1^h) + k(x)\,\Delta t^h(x)], \qquad x \in G_h, \tag{5.1a}$$

$$= b(x), \qquad x \in \partial G_h. \tag{5.1b}$$

Equation (5.1) has a unique solution.

Suppose that

$$\Delta t^h(x) \to 0 \qquad \text{as} \quad h \to 0, \quad \text{all } x \in \bar{G}, \tag{5.2}$$

or that there are no absorbing points in $\bar{G}$. Then it can easily be shown that the limit, as $h \to 0$, of the solutions to (5.1) is also the limit of the sequence of solutions to (5.1) with $Q_h(x)/(Q_h(x) + \lambda(x)h^2)$ replaced by the first-order approximation $[1 - \lambda(x)\,\Delta t^h(x)]$ or the limit of the unique solutions to

$$V^h(x) = (\exp - \lambda(x)\,\Delta t^h(x))[E_x V^h(\xi_1^h) + k(x)\,\Delta t^h(x)], \qquad x \in G_h, \quad (5.3)$$

or of

$$V^h(x) = \exp - \lambda(x)\,\Delta t^h(x) E_x V^h(\xi_1^h) + k(x)[1 - (\exp - \lambda(x)\,\Delta t^h(x))]/\lambda(x), \qquad x \in G_h, \quad (5.4)$$

with boundary condition (5.1b). The limit of the solutions to (5.4) as $h \to 0$ is the correct $R(x)$, whether or not (5.2) holds. Note that, in (5.4), we merely replaced $[\exp - \lambda(x)\,\Delta t^h(x)]k(x)\,\Delta t^h(x)$ by

$$k(x) \int_0^{\Delta t^h(x)} [\exp - \lambda(x)s]\, ds.$$

Let $N_h = \min\{n : \xi_n^h \notin G_h\}$. Then a slight extension of the result in Section 3.1.4 (letting the β there depend on x) yields that the unique solution to (5.3) is

$$R^h(x) = E_x \sum_{n=0}^{N_h - 1} A_n^h k(\xi_n^h)\,\Delta t_n^h + E_x A_{N_h-1}^h b(\xi_{N_h}^h), \qquad x \in G_h, \quad (5.5)$$

where

$$A_n^h = \prod_{i=0}^{n} (\exp - \lambda(\xi_i^h)\,\Delta t_i^h) = \exp - \int_0^{t_{n+1}^h} \lambda(\xi^h(s))\, ds.$$

The unique solution to (5.4) is

$$R^h(x) = E_x \sum_{n=0}^{N_h - 1} A_{n-1}^h \frac{[1 - (\exp - \lambda(\xi_n^h)\,\Delta t_n^h)]}{\lambda(\xi_n^h)} k(\xi_n^h) + E_x A_{N_h-1}^h b(\xi_{N_h}^h),$$

which can be rewritten as

$$R^h(x) = E_x \int_0^{\rho_h} [\exp - \int_0^t \lambda(\xi^h(s))\, ds] k(\xi^h(t))\, dt + E_x[\exp - \int_0^{\rho_h} \lambda(\xi^h(s))\, ds] b(\xi^h(\rho_h)). \quad (5.6)$$

The equation (5.3) discounts the cost over the interval $[t_n^h, t_{n+1}^h)$ by using the discount factor $\exp - \int_0^t \lambda(\xi^h(s))\, ds$ evaluated at $t = t_{n+1}^h$ (namely A_n^h), while the form (5.6) uses the average discount factor

$$A_{n-1}^h \int_{t_n^h}^{t_{n+1}^h} [\exp - \lambda(\xi_n^h)(s - t_n^h)]\, ds/\Delta t_n^h.$$

We can conclude that *under the conditions of Theorem* 6.3.2, *and if* $\lambda(\cdot)$ *is continuous and there is a real* $\lambda_0 > 0$ *such that* $\lambda(y) \geq \lambda_0$, *all* $y \in G$, *and if*

$$P_x\{\tau = \tau'\} = 1 \qquad (\textit{whether } \tau = \infty \textit{ or not } \text{w.p.} > 0),$$

and using (5.5) or (5.6) [only (5.6), if $\sup_{x \in G} \Delta t^h(x) \not\to 0$ as $h \to 0$], then

$$R^h(x) \to R(x)$$

as $h \to 0$. *The uniform integrability condition is not required when* $\lambda_0 > 0$.

6.6 An Alternative Representation for β_n^h and $W(\cdot)$

There is an alternative representation for β_n^h which allows us to explicitly construct the $W(\cdot)$ process, and is also very convenient to use in some of the control problems of the following chapters.

The main difficulties in the proof of the representation and limit are caused by the possibility that $a(\cdot)$ may not be positive definite. Hence, we need to introduce an auxiliary process, and to use a somewhat indirect approach.

In the chapters dealing with controlled diffusions, it will be necessary to treat a class of controls which depends on the "driving" Wiener process $w(\cdot)$. When we "discretize" these controls—in order to adapt them to the chains—the approximation $W^h(\cdot)$ which we will develop below will be very helpful. $W^h(\cdot)$ is an approximation to a Wiener process and we will be able to let the controls (for the chain) depend on the process $W^h(\cdot)$, in the same way that they depend on $w(\cdot)$ in the continuous parameter case. The comparisons obtained via this method will allow us to prove various theorems concerning convergence of minimal cost functionals for a sequence of Markov chains to the minimal cost functional for the diffusion.

Augment the probability space (whichever one we are using) by adding a R^r vector-valued Wiener process $\psi(\cdot)$, which is independent of the $\{\xi_n^h\}$. Factor $\Sigma_h(x)$ into

$$\Sigma_h(x) = P_h(x)D_h^2(x)P_h'(x),$$

and define $\sigma_h(\cdot)$ by

$$\sigma_h(x) = P_h(x)D_h(x),$$

where $P_h(\cdot)$ and $D_h(\cdot)$ are Borel measurable, and are orthonormal and diagonal matrices, resp., for each x. Define $P_{h,n} = P_h(\xi_n^h)$ and

$$D_{h,n} = \operatorname{diag}\{d_1(\xi_n^h), \ldots, d_r(\xi_n^h)\} = D_h(\xi_n^h),$$

where the $\{d_i(x)\}$ are the diagonal elements of $D_h(x)$. We can select $P_h(\cdot)$, $D_h(\cdot)$ such that $\Sigma_h(x) \to 2a(x)$, $\sigma_h(x) \to \sigma(x)$, as $h \to 0$, uniformly in x.

Define $\delta\psi_n^h = \psi(t_{n+1}^h) - \psi(t_n^h)$. Let $\alpha \in (0, 1)$ and let $D_{h,n}^+$, $D_{h,n}^{++}$, $D_{h,n}^T$ denote the diagonal matrices with ith diagonal elements

$$d_{i,n}^{-1} I_{\{d_{i,n} > 0\}}, \quad d_{i,n}^{-1} I_{\{d_{i,n} \geq h^\alpha\}}, \quad d_{i,n} I_{\{d_{i,n} \geq h^\alpha\}},$$

resp. Define

$$\begin{aligned} \delta W_n^h &= D_{h,n}^{++} P_{h,n}' \beta_n^h + (I - D_{h,n}^T D_{h,n}^{++})\, \delta\psi_n^h \\ &\equiv \gamma_n^h + \rho_n^h \end{aligned} \tag{6.1}$$

and

$$W^h(t) = \sum_{t_{n+1}^h \leq t} \gamma_n^h + \sum_{t_{n+1}^h \leq t} \rho_n^h \equiv \Gamma^h(t) + \rho^h(t).$$

In Section 1.4.4, we derived a Wiener process from a continuous square integrable martingale (and, perhaps, a Wiener process independent of the martingale) and showed that we could represent the martingale as a stochastic integral with respect to the derived Wiener process. We are trying to do something similar here and to mimic as closely as possible the derivation in Chapter 1.4.4. We are motivated by the fact that $B^h(\cdot)$ is "close" to being a square integrable martingale (Theorem 6.3.2).

It will turn out that $W^h(\cdot)$ converges in distribution to a Wiener process, with respect to which we can write

$$B(t) = \int_0^t \sigma(\xi(s))\, dW(s). \tag{6.2}$$

Theorem 6.6.1 *If* $\sup_x \Delta t^h(x) \to 0$ *as* $h \to 0$ *and if* $f(\cdot)$ *and* $\sigma(\cdot)$ *are bounded and continuous, then* $\{W^h(\cdot)\}$ *is tight on* $D^r[0, \infty)$ *and tends to an* R^r-*valued Wiener process* $W(\cdot)$, *in distribution as* $h \to 0$.

PROOF First, tightness and continuity of the limits will be shown by using the criterion of Theorem 2.4.2. Owing to the definition of $\delta\psi_n^h$ and to the fact that its coefficient in (6.1) is a nonnegative diagonal matrix dominated by I, it is clear that the interpolation $\rho^h(\cdot)$ of the $\{\rho_n^h\}$ satisfies the criterion of Theorem 2.4.2. Criterion (2.3.3) will be verified for $\Gamma^h(\cdot)$, as it was for the $B^h(\cdot)$ in Theorem 6.3.1. We can suppose, in this paragraph, that the γ_i^h are scalar-valued. Let G_i be defined like the F_i were defined in Theorem 6.3.1, but with γ_i^h replacing β_i^h. Since

$$E[\gamma_n^h \mid \xi_i^h,\ i \leq n] = 0,$$

$G_1 = 0$. Let K denote a real number, whose value may vary from usage to usage but which does not depend on either s, δ or h. The calculation which showed that $F_2 \le K\delta^2$ can also be used to show that $G_2 \le K\delta^2$ since

$$E[|\gamma_n^h|^2 | \xi_i^h, i \le n] \le \Delta t_n^h.$$

Also, we can show that $G_3 \le K(\delta^2 + G_4)$. Using the fact that $|\gamma_n^h| \le Kh^{1-\alpha}$, we get that $G_4 \le Kh^{2-2\alpha} E \sum_{i=0}^{\infty} |\gamma_i^h|^2 I_i' \le Kh^{2-2\alpha}\delta \to 0$ as $h \to 0$. Thus (3.3) holds for $\Gamma^h(\cdot)$ with h^2 replaced by $h^{2-2\alpha}$, and $\{\Gamma^h(\cdot), \rho^h(\cdot)\}$ are tight, and all limits are continuous w.p. 1.

Observe that $\{\delta W_n^h\}$ is an orthogonal sequence and, in particular, satisfies

$$\begin{aligned} E[\delta W_n^h | \xi_i^h, \delta\psi_{i-1}^h, i \le n] &= 0 \\ \operatorname{cov}[\delta W_n^h | \xi_i^h, \delta\psi_{i-1}^h, i \le n] &= \Delta t_n^h I. \end{aligned} \tag{6.3}$$

Let $\{W^h(\cdot)\}$ denote any subsequence which converges in distribution. A proof like that used in Theorem 6.3.2 [to show that $B(\cdot)$ was a continuous martingale with a certain quadratic variation can be used to show that $W(\cdot)$, the limit of $\{W^h(\cdot)\}$, is a continuous martingale with quadratic variation It. Hence it is a Wiener process. Q.E.D.

The only place where the assumption $\sup_x \Delta t^h(x) \to 0$ as $h \to 0$ is required, is in showing that the quadratic variation is It; i.e., that there are no gaps—as there would be if $\Delta t^h(x) \equiv \infty$, say. If the assumption does not hold, then we need to modify $W^h(\cdot)$ and $\{\delta W_n^h\}$ slightly. Let K be a large real number. Define δW_n^h as in (6.1), and define $\delta \tilde{W}_n^h$ and $\tilde{W}_n^h(\cdot)$ by $\tilde{W}_n^h = \sum_{i=0}^{n-1} \delta \tilde{W}_i^h$

$$\delta \tilde{W}_n^h = \delta W_n^h I_{\{\Delta t_n^h \le Kh^\alpha\}}, \qquad \tilde{W}^h(t) = \tilde{W}_n^h \qquad \text{on} \quad [t_n^h, t_{n+1}^h).$$

Let $I^h(\cdot)$ denote the function which has the value 1 on $[t_i^h, t_{i+1}^h)$ if $\Delta t_i^h > Kh^\alpha$ and is zero otherwise. Define $\bar{W}^h(\cdot)$ by

$$\bar{W}^h(t) = \tilde{W}^h(t) + \int_0^t I^h(s)\, d\psi(s). \tag{6.4}$$

$\{\bar{W}^h(\cdot)\}$ is also tight and all limits are Wiener processes, but we will not go into the details. The process $\bar{W}^h(\cdot)$ can be used instead of $W^h(\cdot)$ below.

Henceforth, we assume that $\sup_x \Delta t^h(x) \to 0$ as $h \to 0$. All the results can be extended to the general case at the cost of extra detail. A direct calculation (in which we use the fact that $P_{h,n} D_{h,n} D_{h,n}^+ P_{h,n}' \beta_n^h = \beta_n^h$ w.p. 1) show that we can write β_n^h as

$$\beta_n^h = P_{h,n} D_{h,n}\, \delta W_n^h + \varepsilon_n^h \equiv \sigma_h(\xi_n^h)\, \delta W_n^h + \varepsilon_n^h, \qquad \text{w.p. 1}, \tag{6.5}$$

where

$$\begin{aligned} -\varepsilon_n^h = {} & [P_{h,n} D_{h,n}(D_{h,n}^{++} - D_{h,n}^+) P_{h,n}'] \beta_n^h \\ & + [P_{h,n} D_{h,n}(D_{h,n} D_{h,n}^+ - D_{h,n}^T D_{h,n}^{++})]\, \delta\psi_n^h. \end{aligned}$$

The continuous parameter process $\varepsilon^h(\cdot)$, obtained by interpolating $\{\varepsilon_n^h\}$, can be verified to converge to the identically zero process in distribution. We can write

$$\xi_{n+1}^h = \xi_n^h + f(\xi_n^h)\,\Delta t_n^h + \sigma_h(\xi_n^h)\,\delta W_n^h + \varepsilon_n^h. \tag{6.6}$$

Let $\{\xi^h(\cdot), W^h(\cdot), \varepsilon^h(\cdot)\}$ denote a convergent subsequence whose limit is denoted by $\{\xi(\cdot), W(\cdot), 0\}$. By using the technique of the proof of Theorem 8.2.1 and Theorem 8.2.2, it can be verified that $\xi(\cdot)$ is nonanticipative with respect to the Wiener process $W(\cdot)$ and that

$$\xi(t) = x + \int_0^t f(\xi(s))\,ds + \int_0^t \sigma(\xi(s))\,dW(s). \tag{6.7}$$

Only an outline of the proof of the latter assertion will be given.

For each $\Delta > 0$, we can rewrite $\xi^h(\cdot)$ in the form

$$\begin{aligned}\xi^h(m\Delta + \Delta) = \xi^h(m\Delta) &+ \int_{m\Delta}^{m\Delta+\Delta} f(\xi^h(s))\,ds + O(h)\\ &+ \sigma_h(\xi^h(m\Delta))[W^h(m\Delta + \Delta) - W^h(m\Delta)]\\ &+ [\varepsilon^h(m\Delta + \Delta) - \varepsilon^h(m\Delta)]\\ &+ [\tilde{\varepsilon}^{h,\Delta}(m\Delta + \Delta) - \tilde{\varepsilon}^{h,\Delta}(m\Delta)],\end{aligned} \tag{6.8}$$

where $\tilde{\varepsilon}^{h,\Delta}(m\Delta + \Delta) - \tilde{\varepsilon}^{h,\Delta}(m\Delta)$ is an error due to the approximation of

$$\sum_{i=n_{m\Delta}}^{n_{m\Delta+\Delta}-1} \sigma_h(\xi_i^h)\,\delta W_i^h,$$

by the 3rd term on the right, which term equals

$$\sigma_h(\xi^h(m\Delta)) \sum_{i=n_{m\Delta}}^{n_{m\Delta+\Delta}-1} \delta W_i^h = \sigma_h(\xi^h(m\Delta))[W^h(m\Delta + \Delta) - W^h(m\Delta)].$$

Now, $\xi^h(\cdot) \to \xi(\cdot)$, a continuous process, uniformly on finite intervals, the second term tends to $\int_{m\Delta}^{m\Delta+\Delta} f(\xi(s))\,ds$ as $h \to 0$, the third term tends to $\sigma(\xi(m\Delta))[W(m\Delta + \Delta) - W(m\Delta)]$ and the $\varepsilon^h(\cdot)$ term tends to zero as $h \to 0$. It can be shown that, as $h \to 0$, $\Delta \to 0$, the $\tilde{\varepsilon}^{h,\Delta}(\cdot)$ terms tends to the zero process in distribution. It is obvious that the indefinite sum of the third term tends to the stochastic integral in (6.7) as $h \to 0$, $\Delta \to 0$. These facts, taken all together, imply that $\xi(\cdot)$ satisfies (6.7).

6.7 Monte Carlo

The chain $\{\xi_n^h\}$ can be used for Monte Carlo type estimations of any functional for which the weak convergence Theorems 6.3.1 and 6.3.2 hold. To generate a sample path of the chain, we simply need to generate a sequence of random variables, each drawn from a simple discrete distribution; e.g., if $\xi_n^h = x$, to get a sample value of ξ_{n+1}^h, we need only to draw a sample according to the distribution $\{p^h(x, y)\}$.

Suppose we wish to calculate $E_x \int_0^\tau k(X(s))\, ds$. Using $\{p^h(x, y)\}$, generate, say, N sequences of sample paths $\{\xi_n^{i,h}, n = 1, 2, \ldots\}$, $i = 1, \ldots, N$, with $\xi_0^{i,h} = x$ and each sequence being generated independently. The ith such sequence need be generated only until N_h^i, the time that $\{\xi_n^{i,h}, n = 1, 2, \ldots\}$ first leaves G_h. Then use

$$\frac{1}{N} \sum_{i=1}^{N} \sum_{j=0}^{N_h^i - 1} k(\xi_j^{i,h})\, \Delta t^h(\xi_j^{i,h})$$

for the estimate.

We do not know how the efficiency compares with other approaches, e.g., with that which uses a discrete time approximation to $X(\cdot)$. For the latter method, we choose a $\Delta > 0$, generate a sample sequence $\{\delta w_n, n = 1, 2, \ldots\}$ of independent "almost" Gaussian random vectors with covariance ΔI and use the iteration

$$X_{n+1}^\Delta = X_n^\Delta + f(X_n^\Delta)\, \Delta + \sigma(X_n^\Delta)\, \delta w_n^\Delta.$$

Perhaps the natural time scaling will work in favor of our method. Also, it is somewhat easier to generate a sample from our discrete distribution, than to generate a good approximation to the Gaussian random vector.

6.8 Approximation of Invariant Measures

The development will require the following limit theorem for Markov chains (Chung [Cl, Section I.15, Theorems 1, 2, 3]). Let $\{Y_n\}$ denote a recurrent finite state Markov chain with stationary transition probabilities $\{p_{ij}\}$, and whose states form a connected class. Define

$$\pi_j = \lim_{n\to\infty} \frac{1}{n} \sum_{m=1}^{n} p_{ij}^{(m)}.$$

Theorem 6.8.1 *For any real-valued function $F(\cdot)$,*

$$\lim_{n\to\infty} \frac{1}{n} \sum_{i=0}^{n-1} F(Y_i) = \sum_j \pi_j F(j) \tag{8.1}$$

w.p. 1 *and in expectation.*

In order to do computations with the chain $\{\xi_n^h\}$, the state space must be finite for each $h > 0$. The state space will be finite if the process varies over a finite region with boundary reflections. Although we will not emphasize such problems in this book (except for the material in Chapter 10), much the same development would be used—with the proper model for a reflected diffusion (Strook and Varadhan [S4], Kushner [K7]). The range will also be bounded if the diffusion is on a manifold such as a hypersphere. On the other hand, the dynamics for large values of the state are often somewhat arbitrary, and do not—or should not with a proper model—seriously affect the behavior for moderate values of x.

Consider the special two-dimensional case.

$$dX_1 = X_2\,dt,$$
$$dX_2 = f_2(X)\,dt + \sigma(X)\,dw, \qquad X = (X_1, X_2).$$

If the model is to be used to represent some type of phenomenon arising in a physical situation, it will usually be approximate in some sense. In particular, the values of $f_2(\cdot)$ and $\sigma(\cdot)$ for large values of x are somewhat arbitrary. If the process is of interest over a very long time, one would usually try modelling the functions so that values such as $E_x k(X(t))$ do not differ much from their true (or experimental) values for large t. With this idea in mind, there are several ways by which the range space of $X(\cdot)$ can be made finite. We only need to assume that there is a bounded open set G so that the range of the process $X(\cdot)$ is contained in $\bar{G}$, when $X(0) \in \bar{G}$. In order to avoid details concerned with the specification of the finite difference procedure in an arbitrary region $\bar{G}$, we assume that $\bar{G}$ is a hyperrectangle, whose sides are all integral multiples of h, for each h which we actually use as a finite difference interval. Except for the difficulty in specifying the approximation, the treatment of more general regions will be the same as for the hyperrectangle. Let $\bar{G}_h \equiv \bar{G} \cap R_h^r$ denote the state space. Under our assumptions on $\bar{G}$, $\{\xi_n^h\} \in \bar{G}$ if $\xi_0^h \in \bar{G}$.

Formally, if μ is an invariant measure for (1.1), then $\gamma \equiv E_\mu k(X)$ satisfies Eq. (4.5.5), for some bounded function $V(\cdot)$. Let us proceed formally and discretize (4.5.5) using the approximations of Section 6.2. Denoting the finite difference solution by $(V^h(\cdot), \gamma^h)$, we have

$$V^h(x) = E_x V^h(\xi_1^h) + k(x)\,\Delta t^h(x) - \gamma^h\,\Delta t^h(x). \tag{8.2}$$

If (8.2) has a solution γ^h which does not depend on x, then (8.2) implies that

$$\gamma^h = \lim_{n\to\infty} E_x \sum_{i=0}^{n-1} k(\xi_i^h)\,\Delta t_i^h / E_x \sum_{i=0}^{n-1} \Delta t_i^h. \tag{8.3}$$

If $\bar{G}_h$ is an ergodic class, then there is a unique constant γ^h which satisfies (8.2), for some function $V^h(\cdot)$. The function will not be unique. In general, there will be transient states, perhaps several ergodic classes [it is not clear whether even uniqueness of invariant measure of $X(\cdot)$ implies a similar uniqueness for the chain, for small h] and perhaps periodicity. But there must be at least one recurrent state for each $h > 0$.

Suppose that x is a recurrent state on $\bar{G}_h$: Define $\bar{G}_h(x)$ to be the recurrence class which communicates with x, and define

$$\pi^h(y) = \lim_{n\to\infty} \frac{1}{n} \sum_{i=0}^{n-1} P\{\xi_i^h = y \mid \xi_0^h = x\}.$$

Then the term defined by the right-hand side of (8.3) takes the value (by Theorem 6.8.1)

$$\gamma^h(x) = \sum_{y \in \bar{G}_h(x)} k(y)\mu^h(y), \tag{8.4}$$

where $\mu^h(y)$ is defined by

$$\mu^h(y) = \begin{cases} \Delta t^h(y)\pi^h(y) \Big/ \sum_{z \in \bar{G}_h(x)} \Delta t^h(z)\pi^h(z), & y \in \bar{G}_h(x), \\ 0, & y \notin \bar{G}_h(x). \end{cases} \tag{8.5}$$

Note that $\pi^h = \{\pi^h(y),\ y \in \bar{G}_h\}$ is an invariant measure for $\{\xi_i^h\}$, and that it has support on $\bar{G}_h(x)$ only.

Since the relative time that $\xi^h(\cdot)$ spends in a state y is proportional to $\Delta t^h(y)$, we have

$$P\{\xi^h(t) = y \mid \text{initial measure for the chain, at } n = -\infty, \text{ is either } \pi^h \text{ or } \xi_{-\infty}^h = x\} = \mu^h(y).$$

Thus for $t \in (-\infty, \infty)$

$$E[k(\xi^h(t)) \mid \xi_{-\infty}^h = x \text{ or has measure } \pi^h] = E_{\mu^h} k(X) \equiv \int k(y)\, d\mu^h(y). \tag{8.6}$$

By Theorem 6.8.1 and the definition of μ^h we have the following relationships (initial condition, at $n = t = -\infty$, is x)

$$\begin{aligned} \gamma^h(x) &= \sum_{y \in \bar{G}_h(x)} k(y)\mu^h(y) = \lim_{n\to\infty,\ \text{w.p. } 1} \left[\frac{\sum_{i=0}^n k(\xi_i^h)\,\Delta t_i^h}{\sum_{i=0}^n \Delta t_i^h}\right] \\ &= \lim_{n\to\infty,\ \text{w.p. } 1} \int_0^{t_n^h} k(\xi^h(s))\, ds/t_n^h = \lim_{t\to\infty,\ \text{w.p. } 1} \int_0^t k(\xi^h(s))\, ds/t \\ &= \lim_{t\to\infty} \int_0^t E_x k(\xi^h(s))\, ds/t. \end{aligned} \tag{8.7}$$

These calculations suggest the following theorem.

Theorem 6.8.2 *Assume* A6.1.1 *and that* $X(\cdot)$ *has a unique invariant measure* μ. *Let* x *denote a recurrent state in* $\bar{G}_h$ *(which can vary with* h*). Then the measure* $\mu^h(\cdot)$ *given by* (8.5) *is a weak sense approximation to* μ *in the sense that*

$$\sum_{y \in \bar{G}_h(x)} k(y)\mu^h(y) \to E_\mu k(X) \qquad \text{as} \quad h \to 0, \tag{8.8}$$

for any bounded and continuous function $k(\cdot)$.

PROOF Let $\{\xi_n^h\}$ and $\xi^h(\cdot)$ denote the chain and interpolated process, both defined on $(-\infty, \infty)$, with initial measure (at $t = n = -\infty$)π^h. Via arguments like those used in Section 3 of this chapter, it can be shown that $\{\xi^h(\cdot)\}$ is tight on $D^r(-\infty, \infty)$ and that any limit $\xi(\cdot)$ satisfies: there is a Wiener process $W(\cdot)$ with respect to which $\xi(\cdot)$ is nonanticipative and such that

$$\xi(t) = \xi(0) + \int_0^t f(\xi(s))\, ds + \int_0^t \sigma(\xi(s))\, dW(s), \qquad t \in (-\infty, \infty). \tag{8.9}$$

Suppose that we choose a subsequence for which both $\{\xi^h(\cdot)\}$ and $\{\mu^h(\cdot)\}$ converge (in distribution and weakly, resp.). Let μ_0 denote the weak limit of $\{\mu^h(\cdot)\}$.

By (8.6) and the weak convergence, $\xi(\cdot)$ is a stationary process with measure μ_0. Hence μ_0 is an invariant measure for (1.1) and, by uniqueness, $\mu_0 = \mu$. The theorem follows since

$$E_{\mu^h} k(X) \to E_{\mu_0} k(X),$$

as $h \to 0$. Q.E.D.

REMARKS By uniqueness of μ, all convergent subsequences of $\{\mu^h\}$ have the same weak limit, namely μ. Also, the invariant measure need not be unique. Let B be a subset of $\bar{G}$ such that there is a unique invariant measure μ with support in B. If $P_x\{\xi_n^h \in B\} = 1$ for $x \in B$, then we can use B in place of G in the theorem.

REMARK ON THE PARABOLIC CASE OF CHAPTER 7 There is a natural analog of Theorem 6.8.2 for the chain $\{\xi_n^{h,\Delta}\}$ of Chapter 7. We only need to let Δt_i^h be replaced by Δ, and to replace $\pi^h(y)$ by (for $y \in \bar{G}_{h,\Delta}(x)$, the recurrence set associated with a recurrent state $x \in \bar{G}_h$)

$$\lim_{n \to \infty} \frac{1}{n} \sum_{i=0}^{n-1} P\{\xi_i^{h,\Delta} = y \mid \xi_0^{h,\Delta} = x\} = \pi^{h,\Delta}(y).$$

6.9 Remarks and Extensions

We have concentrated on diffusion processes on R^r. The same ideas work when the diffusion is confined to any smooth manifold, provided that an appropriate discretization procedure is used. Some special cases of current importance deal with diffusions on circles or hyperspheres (see Khazminskii [K2] or Mitchell and Kozin [M4] for an example in stability theory, and Lo and Willsky [L2] for an example in estimation theory). This type of example is particularly easy to discretize since the state variable is the "angle" and the state space is inherently finite.

The conditions on $f(\cdot)$, $\sigma(\cdot)$, $k(\cdot)$, and $b(\cdot)$ can be weakened. We assumed that they were continuous only for convenience since otherwise we would have had to take somewhat greater care with the details in the specification of the discretizations and in the proofs. For example, let $\bar{D}(g)$ denote the closure of the discontinuity set of a bounded R^r-valued measurable function $g(\cdot)$. Then the continuity of $f(\cdot)$ in A6.1.1 can be replaced by

$$P_x\{X(t) \in \bar{D}(f)\} = 0, \qquad \text{almost all} \quad t > 0.$$

In lieu of using $f(\cdot)$ or $\sigma(\cdot)$, we could use some local average of these functions in the construction of $p^h(\cdot, \cdot)$. For example, let $N_\varepsilon(y)$ denote an ε-neighborhood of y and let $\varepsilon(h) \to 0$ as $h \to 0$. Then use $f^h(\cdot)$ defined by

$$f^h(y) = \int_{N_\varepsilon(y)} f(u)\, du \Big/ \int_{N_\varepsilon(y)} du, \qquad \varepsilon = \varepsilon(h),$$

in place of $f(\cdot)$ in $p^h(\cdot, \cdot)$. It is likely that some such alteration would yield faster convergence but the question has not been investigated. Indeed, there are probably better ways than the ones used here of approximating the transition probabilities—or even of constructing the state space for the approximating chain (it need not be G_h or even a linear transformation of G_h)—and more work needs to be done on the question. We have used the discretization of the partial differential equations by finite differences largely because it yields a suitably convergent approximating process with relatively little work. Yet, once the basic ideas of the approximation method and convergence proofs are clear, one can exercise considerable freedom in altering the transition functions and either experimentally or theoretically checking rates of convergence and other properties.

Suppose that we solve for the $R^h(x)$, $x \in G_h \cup \partial G_h$, given by (2.14). An area needing much work concerns the problem of smoothing $R^h(\cdot)$ in some way to obtain a "better" approximation to $R(\cdot)$. Almost nothing is known about this. Yet in several simple numerical experiments for functionals of the form (2.14) and for other functionals as well, simple types of smoothings seem to yield improved approximations of the limit.

6.10 Numerical Data

Some numerical data of interest in stochastic stability will be presented. Let $\xi_1(\cdot)$ and $\xi_2(\cdot)$ denote two independent wide band zero mean ergodic processes, with spectral densities of heights σ_1^2 and σ_2^2, resp., at 0. The equation (10.1) arises frequently in applications and, as the band width goes to infinity, the solution processes converge weakly [K2, M2, W3] to the solution of (10.2) in $C^2[0, \infty)$.

$$\left.\begin{aligned} \dot{Y}_1 &= Y_2\ dt \\ \dot{Y}_2 &= -a_1 Y_1 - a_2 Y_2 + Y_1 \xi_1(t) + Y_2 \xi_2(t) \end{aligned}\right\} y(0) = x. \tag{10.1}$$

$$\begin{aligned} dX_1 &= X_2\ dt \\ dX_2 &= -a_1 X_1\ dt - (a_2 - \sigma_2^2/2) X_2\ dt \\ &\qquad + \sigma_1 X_1\, dw_1 + \sigma_2 X_2\ dw_2, \qquad X(0) = x, \end{aligned} \tag{10.2}$$

$w_1(\cdot)$, $w_2(\cdot)$ independent Wiener processes.

One way to get information on the stability properties of (10.1) is to calculate various quantities for (10.2), say the average time required for the path to leave some bounded set or similar quantities. Consider the set

$$G = \{x : |x_i| < 3,\ i = 1, 2\} - \{0\} \text{ and quantity}$$
$$R(x) = E_x \exp -\lambda\tau, \qquad \tau = \inf\{t : X(t) \notin G\}.$$

The values of $R(\cdot)$ yield some "stability" information. Indeed, if it can be calculated for several (small) values of λ, then the characteristic function of the distribution of τ can be estimated. Denote the solution to the finite difference approximation (5.3) by $V^h(\cdot)$, where we set $k(\cdot) = 0$ and $b(\cdot) = 1$ ($b(0) = 0$). Set $a_1 = a_2 = 1$, $\sigma_1^2 = \sigma_2^2 = \sigma^2$. A necessary and sufficient condition for the mean value (resp., second moment) of (10.2) to go to zero as $t \to \infty$ is $\sigma^2 < 4$ (resp., $\sigma^2 < \frac{4}{3}$) [K2, M4].

Refer to Fig. 6.3, where $\lambda = 0.1$, $\sigma^2 = 3$, and $h = 0.3$ and 0.15. The level curves are close, suggesting that the approximations are "reasonably" good.

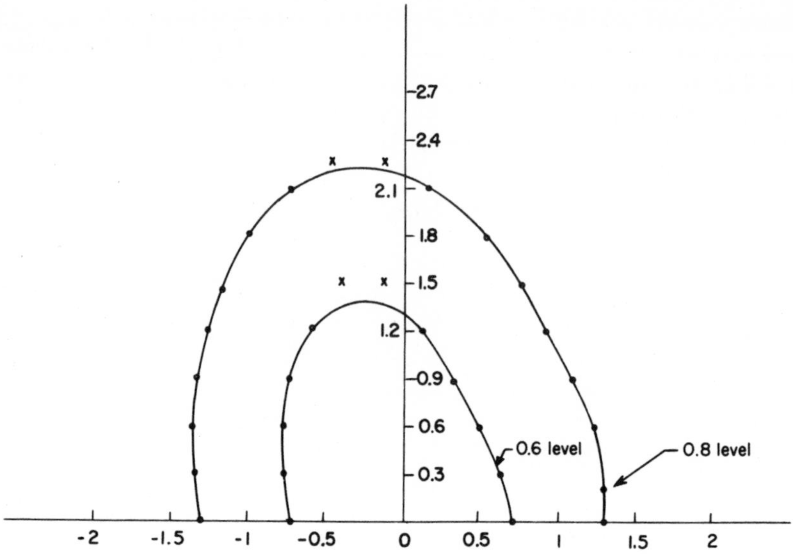

FIG. 6.3 Level surfaces for discrete approximation to $E_x \exp - \lambda\tau$. $\sigma^2 = 3$, $\lambda = 0.1$; – or -∘-, $h = 0.3$; -•- or -×-, $h = 0.15$.

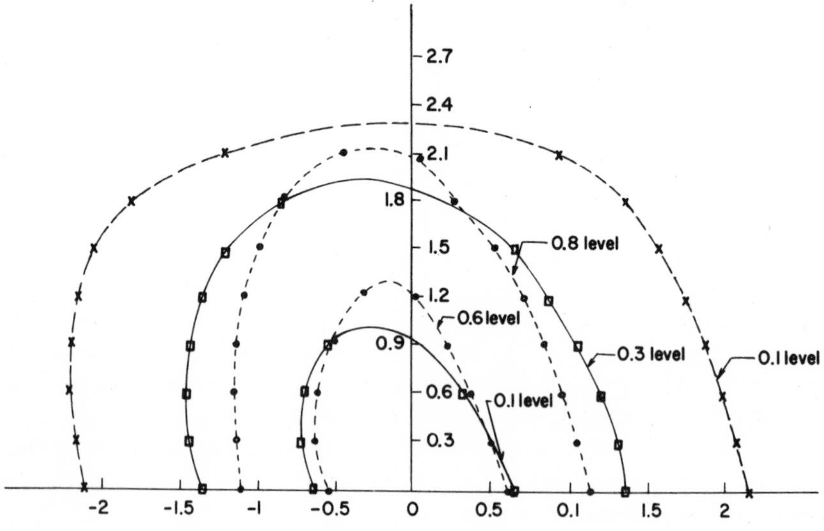

FIG. 6.4 Level surfaces for discrete approximation to $E \exp - \lambda\tau$. $\lambda = 0.05$, $h = 0.3$, (-∘-) $\sigma^2 = 3$, (—□—) $\sigma^2 = 1$, (-×-) $\sigma^2 = \frac{1}{2}$.

Similar results hold for $\lambda = 0.05$. Refer to Figure 6.4, where $\lambda = 0.05$, $h = 0.3$, and several level curves for several values of σ^2 are plotted. The results for $h = 0.15$ are fairly close to the plotted results. The curves give information on the path excursions of the unstopped process (10.2). In particular, they illustrate how rapidly the path excursions grow as σ increases, even when it is well below the upper limit for mean square stability.

CHAPTER 7

Approximations for Parabolic Equations and Nonlinear Filtering Problems

The chapter treats the nonhomogeneous finite time version of the problem of Chapter 6. A degenerate parabolic equation replaces the degenerate elliptic equation. As in Chapter 6, a suitable finite difference approximation to the partial differential equation yields a Markov chain approximation to the diffusion. The interpolations of the Markov chains converge to the diffusion in distribution, as the finite difference intervals go to zero. With the use of the chain, many functionals of the diffusion can be approximated. Equivalently, we can approximate the solution to the degenerate parabolic equation by simple calculations with the chain. The explicit, implicit, and explicit–implicit methods are discussed, together with their probabilistic interpretations.

An iterative method is discussed for solving for the weak-sense solution to the Fokker–Planck equation; i.e., to approximate the evolution of the weak-sense density of the diffusion as time increases. A standard problem in nonlinear filtering is also treated; we give a method analogous to that for (approximately) solving the Fokker–Planck equation for approximating the conditional density of the diffusion, given "noise-corrupted" data.

In the last section, there is some numerical data concerning the approximation of the invariant measure for a simple problem.

7.1 Problem Statement

We will treat computational problems associated with the nonhomogeneous Eq. (1.1) and the cost functional (4.2.4).

$$X(s) = x + \int_t^s f(X(v), v)\, dv + \int_t^s \sigma(X(v), v)\, dw(v), \qquad X(t) = x. \quad (1.1)$$

Assume A7.1.1 and A7.1.2.

A7.1.1 $f(\cdot, \cdot)$, $\sigma(\cdot, \cdot)$ *are bounded and continuous, and* (1.1) *has a unique solution* (*in the sense of probability law*) *on the finite interval* $[t, T]$, *for each* $0 \le t \le T$, *and each* $x \in R^r$. *The solution is a strong Markov and a Feller process.*

(Again, as in Chapter 4, the Markov and Feller properties actually are consequences of the uniqueness.)

A7.1.2 $k(\cdot, \cdot)$, $b_1(\cdot, \cdot)$, *and* $b_T(\cdot)$ *are bounded and continuous functions on* $R^r \times [0, T]$, $R^r \times [0, T]$ *and* R^r *resp.*

In the next section we will apply a finite difference technique to the parabolic partial differential equation (4.5.4) which, formally, is satisfied by the functional (4.2.4). The finite difference method will yield a Markov chain, whose interpolations will converge to $X(\cdot)$ in distribution as the finite difference intervals Δ, $h \to 0$. Again, the interpolation will be piecewise constant, but over nonrandom time intervals Δ.

7.2 The Finite Difference Approximation and Weak Convergence

In the last chapter it was necessary to choose the finite difference approximations with some care if we wanted the coefficients in the finite difference equations to be one-step transition probabilities for some Markov chain. We must make analogous choices for the current problem. Let Δ and h denote the finite difference intervals to be used to approximate the derivatives with respect to time and with respect to each of the spatial directions, resp. The quantity h could vary with the coordinate direction but the formulas then become much longer. The approximations (2.1)–(2.4) are for the explicit method of approximation for parabolic equations. The implicit method is treated in the next section.

We will use the finite difference approximations (2.1)–(2.4). In these formulas, we do let h depend on the direction, although in the subsequent development we will set $h = h_i$.

$$V_t(x, t) \to [V(x, t + \Delta) - V(x, t)]/\Delta, \tag{2.1}$$

$$\begin{aligned} V_{x_i}(x, t) &\to [V(x + e_i h_i, t + \Delta) - V(x, t + \Delta)]/h_i, \qquad \text{if } f_i(x, t) \geq 0, \\ &\to [V(x, t + \Delta) - V(x - e_i h_i, t + \Delta)]/h_i, \qquad \text{if } f_i(x, t) < 0, \end{aligned} \tag{2.2}$$

$$V_{x_i x_i}(x, t) \to [V(x + e_i h_i, t + \Delta) + V(x - e_i h_i, t + \Delta) - 2V(x, t + \Delta)]/h_i^2, \tag{2.3}$$

$$\begin{aligned} V_{x_i x_j}(x, t) \to [&2V(x, t + \Delta) + V(x + e_i h_i + e_j h_j, t + \Delta) \\ &+ V(x - e_i h_i - e_j h_j, t + \Delta)]/2h_i h_j \\ &- [V(x + e_i h_i, t + \Delta) + V(x - e_i h_i, t + \Delta) + V(x + e_j h_j, t + \Delta) \\ &+ V(x - e_j h_j, t + \Delta)]/2h_i h_j, \qquad a_{ij}(x, t) \geq 0, \quad i \neq j, \end{aligned} \tag{2.4a}$$

$$\begin{aligned} V_{x_i x_j}(x, t) \to -&[2V(x, t + \Delta) + V(x + e_i h_i - e_j h_j, t + \Delta) \\ &+ V(x - e_i h_i + e_j h_j, t + \Delta)]/2h_i h_j \\ &+ [V(x + e_i h_i, t + \Delta) + V(x - e_i h_i, t + \Delta) + V(x + e_j h_j, t + \Delta) \\ &+ V(x - e_j h_j, t + \Delta)]/2h_i h_j, \qquad a_{ij}(x, t) < 0, \quad i \neq j. \end{aligned} \tag{2.4b}$$

Analogously to (6.2.4), we assume that (for $x \in R^r$, $t < T$)

$$a_{ii}(x, t) - \sum_{i \neq j,\, j} |a_{ij}(x, t)| \geq 0, \qquad i = 1, \ldots, r, \tag{2.5}$$

and that the quantity defined below by $p^{h,\Delta}(x, x)$ is nonnegative:

$$p^{h,\Delta}(x, x) = 1 - \frac{\Delta}{h^2}\left[h \sum_i |f_i(x, t)| + 2 \sum_i a_{ii}(x, t) - \sum_{i \neq j,\, i,\, j} |a_{ij}(x, t)|\right] \geq 0. \tag{2.6a}$$

Both (2.5) and (2.6a) are needed to assure that the finite difference coefficients will be nonnegative. The conditions (2.5) and (2.6a) can be considerably weakened if we allow h to depend on the coordinate direction or by use of a coordinate rotation before the substitution of (2.1)–(2.4) into the differential equations. Before proceeding, we briefly examine a simple two-dimensional case where h_i is allowed to depend on i.

Example Consider

$$V_t(x, t) + 9V_{x_1x_1}(x, t) + 6V_{x_1x_2}(x, t) + V_{x_2x_2}(x, t) + k(x, t) = 0, \qquad x \in G, \quad t \le T.$$

Then $a_{11} = 9$, $a_{12} = a_{21} = 3$, and $a_{22} = 1$. A rotation of coordinates can clearly be used to transform the problem into one in which the $a(\cdot)$ matrix is diagonal. For numerical purposes this may be useful. But if we restrict ourselves to using (2.1)–(2.4) without a prior rotation and allow the possibility that $h_1 \neq h_2$, then (2.5) must be replaced by

$$a_{11}/h_1^2 - a_{12}/h_1 h_2 \ge 0, \qquad a_{22}/h_2^2 - a_{12}/h_1 h_2 \ge 0.$$

Any pair (h_1, h_2) satisfying $h_1/h_2 = 3$ will satisfy these two inequalities.

The Approximating Markov Chain

Define M_Δ by $M_\Delta \Delta = T$, where we assume (an unimportant assumption, for notational convenience only) that M_Δ is an integer. For $x, y \in R_h^r$, $t = 0, \Delta, \ldots, \Delta(M_\Delta - 1)$, define the function $p_t^{h,\Delta}(x, y)$ by (2.6a) for $y = x$ and otherwise by

$$p_t^{h,\Delta}(x, x \pm e_i h) = \frac{\Delta}{h^2}\left[a_{ii}(x, t) - \sum_{i \neq j,\, j} |a_{ij}(x, t)| + hf_i^{\pm}(x)\right], \tag{2.6b}$$

$$p_t^{h,\Delta}(x, x + e_i h - e_j h) = p_t^{h,\Delta}(x, x - e_i h + e_j h) = (\Delta/h^2)a_{ij}^-(x, t), \tag{2.6c}$$

$$p_t^{h,\Delta}(x, x + e_i h + e_j h) = p_t^{h,\Delta}(x, x - e_i h - e_j h) = (\Delta/h^2)a_{ij}^+(x, t), \tag{2.6d}$$

$$p_t^{h,\Delta}(x, y) = 0, \qquad \text{for all } x, y \text{ pairs not covered in (2.6a–c).}$$

Note that the $p_t^{h,\Delta}(x, y)$ are ≥ 0 and sum (over y) to unity for each x, t. Hence, $\{p_t^{h,\Delta}(\cdot, \cdot)\}$ is a one-step transition function for a Markov chain, whose random variables we denote by $\{\xi_n^{h,\Delta}\}$. By definition, $P\{\xi_{n+1}^{h,\Delta} = y \mid \xi_n^{h,\Delta} = x\} = p_{n\Delta}^{h,\Delta}(x, y)$. If either $a(\cdot, \cdot)$ or $f(\cdot, \cdot)$ actually do depend on t, then the chain is nonhomogeneous.

Next, to obtain a finite difference equation, let us substitute (2.1)–(2.4) into (4.5.4), where we replace R by V. Denote the finite difference solution by $V^{h,\Delta}(\cdot, \cdot)$, and set $V^{h,\Delta}(x, T) = b_T(x)$, $x \in G_h$, and $V^{h,\Delta}(x, t) = b_1(x, t)$ for $t \le T$, $x \notin G_h$. By multiplying all terms in the finite difference equation by Δ we get the equation

$$V^{h,\Delta}(x, n\Delta) = \sum_y p_{n\Delta}^{h,\Delta}(x, y)V^{h,\Delta}(y, n\Delta + \Delta) + k(x, n\Delta)\Delta, \quad n < M_\Delta, \quad x \in G_h, \tag{2.7a}$$

with the above boundary conditions. In terms of the random process $\{\xi_n^{h,\Delta}\}$, we can write (2.7a) as

$$V^{h,\Delta}(x, n\Delta) = E_{x,n} V^{h,\Delta}(\xi_{n+1}^{h,\Delta}, n\Delta + \Delta) + k(x, n\Delta)\Delta, \qquad n < M_\Delta, \quad x \in G_h, \tag{2.7b}$$

where we use $t = n\Delta$ and $\xi_n^{h,\Delta} = x$. The notation $E_{x,n}$ denotes the expectation given that $\xi_n^{h,\Delta} = x$.

Let $N(h, \Delta) = \min\{n : \xi_n^{h,\Delta} \notin G_h\}$. Since we have defined the chain only up to time M_Δ, set $N(h, \Delta) = \infty$ if $\xi_n^{h,\Delta} \in G_h$ for all $n \le M_\Delta$. According to Section 3.1.3, the unique solution to (2.7b) is

$$V^{h,\Delta}(x, n\Delta) = E_{x,n}\Bigg[\sum_{i=0}^{(M_\Delta \cap N(h,\Delta))-1} k(\xi_i^{h,\Delta}, i\Delta)\Delta + b_T(\xi_{M_\Delta}^{h,\Delta}) I_{\{M_\Delta < N(h,\Delta)\}}$$

$$+ b_1(\xi_{N(h,\Delta)}^{h,\Delta}, N(h, \Delta)\Delta) I_{\{M_\Delta \ge N(h,\Delta)\}}\Bigg]. \tag{2.8}$$

$V^{h,\Delta}(\cdot, \cdot)$ satisfies the required boundary conditions.

Fix $x \in R_h^r$ *and* $t = n\Delta$. Define the interpolated process $\xi^{h,\Delta}(\cdot)$ on $[t, T]$ by $\xi^{h,\Delta}(t) = x$ and $\xi^{h,\Delta}(s) = \xi_m^{h,\Delta}$ on $[m\Delta, m\Delta + \Delta)$, $m \ge n$, and let $\rho(h, \Delta) = N(h, \Delta)\Delta$ denote the escape time of $\xi^{h,\Delta}(\cdot)$ from G_h. Then (2.8) can be rewritten as

$$V^{h,\Delta}(x, t) = E_{x,t} \int_t^{T \cap \rho(h,\Delta)} k(\xi^{h,\Delta}(s), s)\, ds + E_{x,t} b_T(\xi^{h,\Delta}(T)) I_{\{T < \rho(h,\Delta)\}}$$

$$+ E_{x,t} b_1(\xi^{h,\Delta}(\rho(h, \Delta)), \rho(h, \Delta)) I_{\{T \ge \rho(h,\Delta)\}}. \tag{2.9}$$

The notation $E_{x,t}$ denotes the expectation under the condition that $\xi^{h,\Delta}(t) = x$. The chain has the local properties $(M_\Delta > m \ge n)$

$$E_{x,n}[\xi_{m+1}^{h,\Delta} - \xi_m^{h,\Delta} \mid \xi_m^{h,\Delta} = y] = f(y)\Delta, \tag{2.10}$$

$$\begin{aligned} \operatorname{cov}_{x,n}[\xi_{m+1}^{h,\Delta} - \xi_m^{h,\Delta} \mid \xi_m^{h,\Delta} = y] &= 2a(y)\Delta + h\Delta\tilde{f}(y) - \Delta^2 f(y) f'(y) \\ &= \Sigma_h(y)\Delta, \end{aligned} \tag{2.11}$$

which are precisely the same expressions as (6.2.10) and (6.2.11) with Δ replacing $\Delta t^h(y)$. The term $\tilde{f}(\cdot)$ is defined below (6.2.11). We can write

$$\xi_{m+1}^{h,\Delta} = \xi_m^{h,\Delta} + f(\xi_m^{h,\Delta})\Delta + \beta_m^{h,\Delta}, \quad m \ge n, \quad \xi_n^{h,\Delta} = x, \tag{2.12}$$

where $\{\beta_m^{h,\Delta}\}$ is an orthogonal sequence whose conditional covariance is given by the right-hand side of (2.11).

From here on, the development is almost exactly like that in Chapter 6. It is a little simpler since we need only consider the processes on a finite interval $[0, T]$ and since the interpolation interval is a constant, not a random variable. Define (where $i \geqslant n$)

$$\left.\begin{aligned} F^{h,\Delta}(s) &= \sum_{t \le i\Delta + \Delta \le s} f(\xi_i^{h,\Delta})\Delta \\ B^{h,\Delta}(s) &= \sum_{t \le i\Delta + \Delta \le s} \beta_i^{h,\Delta} \end{aligned}\right\} \quad s \in [t, T].$$

Then, analogously to (6.3.1), we have

$$\xi^{h,\Delta}(s) = x + F^{h,\Delta}(s) + B^{h,\Delta}(s), \qquad s \in [t, T].$$

The proofs of Theorems 6.3.1 and 6.3.2 immediately yield

Theorem 7.2.1 *Fix x, t and let $f(\cdot, \cdot)$, $\sigma(\cdot, \cdot)$ be bounded and continuous. Then $\{\xi^{h,\Delta}(\cdot), B^{h,\Delta}(\cdot), t \le T, \rho(h, \Delta) \cap T\}$ is tight on $D^{2r}[0, T] \times [0, T]$. If h, Δ indexes any convergent subsequence with limit $\xi(\cdot)$, $B(\cdot)$, $\rho \cap T$, then $\xi(\cdot)$, $B(\cdot)$ have continuous paths* w.p. 1. *There is a Wiener process $W(\cdot)$, with respect to which $\xi(\cdot)$ is nonanticipative and*

$$\begin{aligned} \xi(s) &= x + \int_t^s f(\xi(v), v)\, dv + \int_t^s \sigma(\xi(v), v)\, dW(v) \\ &= x + \int_t^s f(\xi(v), v)\, dv + B(s), \qquad s \in [t, T]. \end{aligned} \tag{2.13}$$

Assume A7.1.1. *Then the multivariate distributions of $\xi(\cdot)$, $B(\cdot)$ do not depend on the particular subsequence.*

Define τ and τ' as in Section 4.2. Then Theorems 7.2.1, 6.4.1, and 6.4.2 immediately yield

Theorem 7.2.2 *Assume* A7.1.1, A7.1.2, *and fix x, t and suppose that*

$$P_{x,t}\{\tau = T\} = 0 \tag{2.14a}$$

$$P_{x,t}\{\tau \cap T = \tau' \cap T\} = 1. \tag{2.14b}$$

Then

$$V^{h,\Delta}(x, t) \to R(x) \tag{2.15}$$

as $h, \Delta \to 0$ [always assuming that (2.5) *and* (2.6a) *hold as $h, \Delta \to 0$]. If $g(\cdot)$ is any bounded measurable function: $D^r[0, T] \to R^r$, which is continuous* w.p. 1 *with respect to the measure induced by* (1.1) *on $D^r[0, T]$, then $E_{x,t}\, g(\xi^{h,\Delta}(\cdot)) \to E_{x,t}\, g(X(\cdot))$ as $h, \Delta \to 0$.*

REMARKS For each x, t (2.14a) must hold except for a countable set of T. In fact (2.14) holds except for a set of (x, T) of zero measure. The problem of showing (2.14b) [which is more difficult than showing (2.14a)] is precisely the problem of showing the analogous result $P_x\{\tau = \tau'\} = 1$ for the problem of Chapter 6. The condition (2.14a) is not required if the boundary function has no discontinuity at $t = T$, $x \in \partial G$.

Consider the classical problem of the r-dimensional heat equation, where $f(\cdot) \equiv 0$, $a_{ij} = 0$ for $i \neq j$ and $a_{ii}(x) = \sigma^2/2$, where $\sigma > 0$ is a real number. Then the nonnegativity condition (2.6a) reduces to $1 - r\Delta\sigma^2/h^2 \geq 0$ or $r\Delta \leq h^2/\sigma^2$, which is precisely the classical criterion of von Neumann for stability of the finite difference approximation [using (2.1) and (2.3)] to the heat equation (Garabedian [G1, p. 476], for $r = 1$).

The chain $\{\xi_m^{h,\Delta}\}$ can be used for Monte Carlo simulation of the $X(\cdot)$ process (1.1). Interpolations over random time intervals are not required and owing to the weak convergence Theorem 7.2.2, the distributions of the values of a large class [the $g(\cdot)$ functions of the theorem statement] of functions of $X(\cdot)$ can be approximated using simulations of the chain $\{\xi_m^{h,\Delta}\}$. In this context we should note that piecewise constant interpolations are not required. The interpolations can be taken piecewise linear, for example, without losing the convergence property; perhaps this would yield "better" estimates of the functions. The best form for the interpolation has not been investigated. Whether it is possible to (or how one may) interpolate between several points is not at all clear. In certain cases, one must exercise great care in doing this since we may get convergence of the resulting $\{\xi^{h,\Delta}(\cdot)\}$ to some process other than (1.1). See Wong and Zakai [W3] or McShane [M2] for a discussion of limits corresponding to various naive or not "approximations" to diffusions.

Let $\psi(\cdot)$ be a R^r-valued Wiener process which is defined on the same probability space on which $\{\xi_m^{h,\Delta}\}$ is defined, and let the two processes be independent of each other. Then, using $\{\xi_m^{h,\Delta}\}$ instead of $\{\xi_n^h\}$ in the constructions of Section 6.6, we define the random variables $\{\delta W_m^{h,\Delta}\}$ and process $W^{h,\Delta}(\cdot)$, exactly as $\{\delta W_n^h\}$ and $W^h(\cdot)$ were defined. Then, as in Section 6.6, we have that $W^{h,\Delta}(\cdot) \to W(\cdot)$, as $h, \Delta \to 0$, where $W(\cdot)$ is the Wiener process used in (2.13), and we can also write (see (6.6.6))

$$\xi_{m+1}^{h,\Delta} = \xi_m^{h,\Delta} + f(\xi_m^{h,\Delta})\Delta + \sigma_h(\xi_m^{h,\Delta})\,\delta W_m^{h,\Delta} + \varepsilon_m^{h,\Delta},$$

where the interpolation $\varepsilon^{h,\Delta}(\cdot)$ of $\{\varepsilon_m^{h,\Delta}\}$ tends to the zero process as $h, \Delta \to 0$.

The Fokker–Planck and Kolmogorov Equations

Under various assumptions (which we will not be concerned with here—see Section 4.3), the density $p(x, t; y, s)$ of the transition function $P(x, t; \Gamma, s)$ $(s > t)$ satisfies the backward p.d. Kolmogorov equation in the initial variables (x, t), and the forward Kolmogorov or Fokker–Planck equation (4.3.2) in the forward variables (y, s). There are many problems that are associated with the usual approximations to the solution of the backward equation, even if $f(\cdot, \cdot)$, $\sigma(\cdot, \cdot)$ are smooth and $a(\cdot, \cdot)$ is positive definite. It is not uncommon that "numerical solutions" actually take negative values or have other pathologies. Furthermore, even if a "good" solution can be obtained, if it is only an approximation to the fundamental solution of (4.3.2), then it may be of use for approximating expectations of functionals of $X(\cdot)$ at some fixed time s, but may not be useful for approximating expectations of more general functionals of $X(\cdot)$. It should be clear that we can get an approximation to the solution of both the forward and backward equations directly from the $p_t^{h,\Delta}(\cdot, \cdot)$. We will develop the details for the nonhomogeneous case. Suppose that $T = \infty$, and order the points on R_h^r and† $G_h \cup \partial G_h$ in some way.

Define the matrices

$$
\begin{aligned}
P_n^{h,\Delta} &\equiv \{p_{n\Delta}^{h,\Delta}(x, y),\ x, y \in R_h^r\}, \\
D_{n,N}^{h,\Delta} &\equiv P_n^{h,\Delta} \cdots P_{N-1}^{h,\Delta} \equiv \{d_{n,N}^{h,\Delta}(x, y),\ x, y \in R_h^r\},
\end{aligned}
\tag{2.16}
$$

where we let $n\Delta = t$ and $N\Delta = s$. The empty product is defined to be the identity matrix. The matrix $D_{n,N}^{h,\Delta}$ can be considered to be a function of four variables, the row and column entries x, y and the initial and final times n, N. Considered as such, it is a weak-sense approximation to the weak-sense density $p(x, t; y, s)$ (as a function of x, t, y, s) in the following sense. Let $g(\cdot)$ denote a bounded and continuous real-valued function on R^r. As $h, \Delta \to 0$, but with $n\Delta$ and $N\Delta$ fixed at t and s, resp., Theorem 7.2.2 yields that

$$
\sum_{y \in R_h^r} g(y) d_{n,N}^{h,\Delta}(x, y) \to \int g(y) P(x, t; dy, s). \tag{2.17}
$$

The approximation $D_{n,N}^{h,\Delta}$ is on the infinite state space R_h^r. We can stop the process on leaving a bounded open set G. Let $\tilde{P}(x, t; \Gamma, s)$ denote the transition function for the stopped process, and $\tilde{p}_{n\Delta}^{h,\Delta}(\cdot, \cdot)$ the one-step transition function for the chain, stopped on first hitting ∂G_h. Define $\tilde{P}_{n\Delta}^{h,\Delta} = \{\tilde{p}_n^{h,\Delta}(x, y),\ x, y \in G_h \cup \partial G_h\}$, and define

$$
\tilde{D}_{n,N}^{h,\Delta} \equiv \tilde{P}_n^{h,\Delta} \cdots \tilde{P}_{N-1}^{h,\Delta} \equiv \{\tilde{d}_{n,N}^{h,\Delta}(x, y),\ x, y \in G_h \cup \partial G_h\}. \tag{2.18}
$$

Assume that $P_{x,t}\{\tau = \tau'\} = 1$ and $P_{x,t}\{\tau = s\} = 0$. Theorem 7.2.2 yields

$$
\sum_{y \in G_h \cup \partial G_h} g(y) \tilde{d}_{n,N}^{h,\Delta}(x, y) \to \int_G g(y) \tilde{P}(x, t; dy, s) \qquad \text{as} \quad h, \Delta \to 0. \tag{2.19}
$$

The limit in Eq. (2.19) is valid if we sum and integrate only over ∂G_h and ∂G, resp., or over G_h and G, resp. Using (2.18), the expectation can be iterated

† Recall that ∂G_h is the set of points in $R_r - G_h$ which are only one node away from G_h.

forward or backward in time, yielding weak-sense approximations to the weak-sense density $\tilde{p}(x, t; y, s)$ as s or t vary, resp.

If we are only interested in approximating (weakly) the weak-sense solution to the forward equation (for the stopped diffusion) for some particular fixed initial value $x \in G_h$, then the iteration in (2.18) can be simplified by proceeding as follows. Let $\tilde{D}^{h,\Delta}_{n,N}(x)$ denote the xth row of $\tilde{D}^{h,\Delta}_{n,N}$. Define $\tilde{D}^{h,\Delta}_{n,n}(x) = (0, 0, \ldots, 1, 0, \ldots)$ where the only nonzero element is in the xth place. Define, recursively

$$\tilde{D}^{h,\Delta}_{n,N+1}(x) = \tilde{D}^{h,\Delta}_{n,N}(x)\tilde{P}^{h,\Delta}_{N}, \qquad N = n, \ldots. \tag{2.20}$$

The $\tilde{D}^{h,\Delta}_{n,N}(x)$ constructed in (2.20) is precisely the xth row of $\tilde{D}^{h,\Delta}_{n,N}$. The amount of computation required to calculate (2.20) is of the same order as that required with the use of any finite difference approach to the forward equation. Our approximation is a transition density for a process $\{\xi^{h,\Delta}_m\}$ which is very closely related to the diffusion $X(\cdot)$.

7.3 Implicit Approximations

The approximations (2.1)–(2.4) constitute an explicit type of approximation, in the sense that they yield explicit recursive equations for the evaluation of the solution, in the classical sense of numerical analysis (Forsythe and Wasow [F2]). This is made evident by the iteration (2.7a or b), which starts at $t = T = N\Delta$ (where the boundary condition is given), and moves backward one Δ step at a time, and $V^{h,\Delta}(x, n\Delta)$ is computed as an *explicit* linear combination of the $V^{h,\Delta}(y, n\Delta + \Delta)$, $y \in G_h \cup \partial G_h$.

To get the standard type of implicit approximation (Forsythe and Wasow [F2]), we need to approximate the parabolic equation by considering it as a degenerate elliptic equation and use the method of Chapter 6. To do this, we make time into a state variable by defining a new variable $X_0(\cdot)$, with $\dot{X}_0(t) = 1$. Then we have (4.5.4) with $\partial/\partial x_0$ and x_0, resp., replacing $\partial/\partial t$ and t, resp. The set G is replaced by $G \times [0, T)$, from which the escape time is always bounded above by T. Let τ and τ' be defined as in Chapter 4.2 and assume (2.14). Then we can apply the various theorems in Chapter 6. Since there are no second derivatives in t appearing in (4.5.4), even if we use a difference interval Δ for the approximation of the derivative $\partial/\partial t$ and h for the approximation of the spatial derivativatives, where h is not necessarily equal to Δ, we still get tractable formulas. The method is straightforward. Here we will give an illustration of the idea, by means of a simple one-dimensional problem.

We consider the example

$$V_t(x, t) + a(x, t)V_{xx}(x, t) + f(x, t)V_x(x, t) + k(x, t) = 0, \quad (x, t) \in G \times [0, T),$$
$$V(x, T) = b_T(x) \quad x \in G, \tag{3.1}$$
$$V(x, t) = b_1(x, t), \quad x \in \partial G, \quad t \leq T.$$

In Chapter 6, we required that the boundary function $b(\cdot)$ be continuous. But, if $P_{x,t}\{\tau = T\} = 0$, then the discontinuity at the corner $\partial G \times \{T\}$ does not affect the convergence since the function with values (at arbitrary $x(\cdot) \in D^r[0, T]$, and where $\tau(x(\cdot)) = \tau$ is the escape time of $x(\cdot)$ from $G \times [0, T)$)

$$b_T(x(\tau))I_{\{\tau > T\}} + b_1(x(\tau), \tau)I_{\{\tau \leq T\}}$$

is still continuous w.p. 1 with respect to the measure induced by (1.1).

Define

$$Q_{h,\Delta}(x, t) = 2a(x, t)\Delta/h^2 + 1 + |f(x, t)|\Delta/h,$$
$$\Delta t^h(x, t) = \Delta/Q_{h,\Delta}(x, t).$$

since $\dot{X}_0 = 1 > 0$, we use the forward difference approximation for $V_t(x, t)$, namely

$$V_t(x, t) \to [V(x, t + \Delta) - V(x, t)]/\Delta. \tag{3.2}$$

Substituting (6.2.1), (6.2.2), and (3.2) into (3.1) and rearranging terms yields the finite difference approximation

$$\begin{aligned} V^{h,\Delta}(x, t) = Q_{h,\Delta}^{-1}(x, t)\{&V^{h,\Delta}(x, t + \Delta) + V^h(x + h, t)[a(x, t)(\Delta/h^2) \\ &+ f^+(x, t)(\Delta/h)] + V^h(x - h, t)[a(x, t)(\Delta/h^2) + f^-(x, t)(\Delta/h)]\} \\ &+ k(x, t)\Delta t^{h,\Delta}(x, t), \quad x \in G_h, \quad t < T. \end{aligned} \tag{3.3}$$

Define the function $p^{h,\Delta}(x, t; y, s)$ by letting $p^{h,\Delta}(x, t; x, t + \Delta)$ and $p^{h,\Delta}(x, t; x \pm h, t)$ be the coefficients of $V^{h,\Delta}(x, t + \Delta)$ and $V^{h,\Delta}(x \pm h, t)$, resp., in (3.3). Define $p^{h,\Delta}(x, t; y, s) = 0$ for all values of (y, s) besides $(x, t + \Delta)$, $(x \pm h, t)$. The $p^{h,\Delta}(z; z')$ ($z = (x, t)$ pair) are ≥ 0 and sum to unity over z' for each z. They are one-step transition probabilities for a two-dimensional Markov chain whose random variables we denote by $\{\xi_n^{h,\Delta}\}$. We can rewrite (3.3) in the form

$$\begin{aligned} V^{h,\Delta}(x, t) = &V^h(x, t + \Delta)p^{h,\Delta}(x, t; x, t + \Delta) \\ &+ V^h(x + h, t)p^{h,\Delta}(x, t; x + h, t) \\ &+ V^h(x - h, t)p^{h,\Delta}(x, t; x - h, t) \\ &+ k(x, t)\Delta t^{h,\Delta}(x, t), \quad x \in G_h, \quad t < T. \end{aligned} \tag{3.4a}$$

If we write $z = (x, t)$, then (3.4a) can be rewritten using the notation for the homogeneous case; namely, as

$$V^{h,\Delta}(z) = E_z V^{h,\Delta}(\xi_1^{h,\Delta}) + k(z)\Delta t^{h,\Delta}(z), \tag{3.4b}$$

where $\xi_1^{h,\Delta}$ is the successor state to z. Define the process $\xi^{h,\Delta}(\cdot)$ to be the interpolation of $\{\xi_n^{h,\Delta}\}$, in the usual fashion, with interpolation intervals $\Delta t_i^{h,\Delta} = \Delta t^{h,\Delta}(\xi_i^{h,\Delta})$. Then the results of Chapter 6 imply that $V^{h,\Delta}(z) \to R(z) = R(x, t)$ as $h, \Delta \to 0$.

The flow of time is implicit in (3.4). Strictly speaking, (3.4a) cannot be solved in the same simple backwards iteration fashion as can (2.7). In a

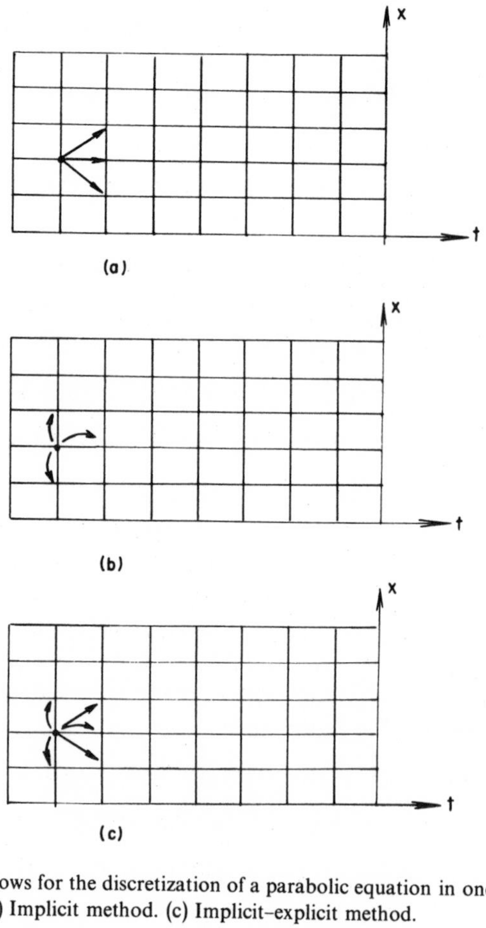

FIG. 7.1 State flows for the discretization of a parabolic equation in one state variable. (a) Explicit method. (b) Implicit method. (c) Implicit–explicit method.

sense, "real" time (the zeroth component of $\tilde{\xi}_n^{h,\Delta}$) advances in any step only with probability $p^{h,\Delta}(x, t; x, t + \Delta)$. When the zeroth component advances, the "real" state, namely x, remains the same. Figures 7.1a and 7.1b illustrate the state flows for $\{\xi_m^{h,\Delta}\}$ and for $\{\tilde{\xi}_n^{h,\Delta}\}$.

To get the implicit–explicit method, we simply randomize between the implicit and the explicit methods. For any $\alpha \in (0, 1)$, choose an explicit step with probability α and an implicit step probability $1 - \alpha$ (transitions illustrated in Fig. 7.1c).

The implicit, or implicit–explicit, method can also be used for Monte Carlo (see Section 6.7).

Real time increases at a rate of Δ per step of $\{\xi_m^{h,\Delta}\}$. For the implicit scheme, Δ can take greater values than it can take for the explicit scheme—in fact, it is usually of the order of h rather than of h^2. But the average increase in real time, namely $\Delta t^{h,\Delta}(z) = \Delta/Q_{h,\Delta}(z)$, is of the order of h^2 also. The time increment $\Delta t^{h,\Delta}(z)$ is "adaptive" to local dynamical conditions; perhaps this helps to explain the frequently observed numerical superiority of the implicit method.

7.4 Discounted Cost: Explicit Method

Consider the cost functional (4.2.4b) with associated differential equation (4.5.4), with $k(x)$ replaced by $k(x) - \lambda(x)R(x, t)$. Applying the approximations (2.1)–(2.4) to this equation yields the finite difference equation (compare with the elliptic case of Section 6.5)

$$V^{h,\Delta}(x, t) = \frac{1}{(1 + \lambda(x, t)\Delta)} [E_{x,t} V^{h,\Delta}(\xi_{n+1}^{h,\Delta}, t + \Delta) + k(x, t)\Delta],$$
$$(x, t) \in G_h \times [0, T) \quad (4.1)$$

instead of (2.7). Of course the boundary conditions remain the same. The limit of $V^{h,\Delta}(x, t)$ as $h, \Delta \to 0$ can be shown to be the same as the limit (as $h, \Delta \to 0$) of both the solutions to (4.2) or (4.3) [compare with (6.5.3) or (6.5.4) in the elliptic case]

$$V^{h,\Delta}(x, t) = (\exp -\lambda(x, t)\Delta)[E_{x,t} V^{h,\Delta}(\xi_{n+1}^{h,\Delta}, t + \Delta) + k(x, t)\Delta], \quad (4.2)$$

$$V^{h,\Delta}(x, t) = [\exp -\lambda(x, t)\Delta] E_{x,t} V^{h,\Delta}(\xi_{n+1}^{h,\Delta}, t + \Delta)$$
$$+ \frac{[1 - \exp - \lambda(x, t)\Delta]}{\lambda(x, t)} k(x, t), \quad (x, t) \in G_h \times [0, T). \quad (4.3)$$

The unique solution to (4.3), with the correct boundary conditions, is

$$V^{h,\Delta}(x,t) = E_{x,t}\int_t^{T\cap\rho(h,\Delta)}\left[\exp - \int_t^s \lambda(\xi^h(v))\,dv\right]k(\xi^{h,\Delta}(s),s)\,ds$$

$$+ E_{x,t}\left[\exp - \int_t^T \lambda(\xi^h(s))\,ds\right]b_T(\xi^{h,\Delta}(T))I_{\{T<\rho(h,\Delta)\}}$$

$$+ E_{x,t}\left[\exp - \int_t^{\rho(h,\Delta)} \lambda(\xi^h(s))\,ds\right]$$

$$\times\, b_1(\xi^{h,\Delta}(\rho(h,\Delta)),\rho(h,\Delta))I_{\{T\geq\rho(h,\Delta)\}}\,. \tag{4.4}$$

It is not hard to see that Theorem 7.2.2 holds for the above $V^{h,\Delta}(\cdot,\cdot)$ and the functional $R(\cdot,\cdot)$ given by (4.2.4b). Of course, the same convergence result holds if we discretize by either the implicit or the implicit–explicit method.

7.5 Nonlinear Filtering

One of the basic and difficult problems of current stochastic control and communication theory is the nonlinear filtering problem for diffusion models. *Let* $X(\cdot)$ *be defined by* (1.1), *assume* A7.1.1, *and suppose that for some integer* u, $z(\cdot)$ *is a* R^u*-valued Wiener process which is independent of* $w(\cdot)$. *Let* $g(\cdot,\cdot)$ *and* $b(\cdot)$ *denote bounded continuous* R^u*-valued and* R*-valued, resp., functions on* $R^r\times[0,\infty)$ *and* R^r*, resp. At each time* $T\geq t$ (t = *initial time*), *let the observations* $Y(s)$, $s\in[t,T]$, *be available, where we define* $Y(\cdot)$ *by*

$$dY(s) = g(X(s),s)\,ds + dz(s), \qquad Y(t) = 0.$$

In many applications, we wish to calculate or approximate either the conditional probability

$$P_{x,t}\{X(T)\in\Gamma\mid Y(s),\ s\in[t,T]\},$$

or the conditional expectation

$$E_{x,t}\{b(X(T))\mid Y(s),\ s\in[t,T]\},$$

where $x = X(t)$ is either a constant or a random variable.

Using the approximations $\{\xi_n^{h,\Delta}\}$ or $\xi^{h,\Delta}(\cdot)$, a recursive solution to the approximation problem can be obtained (which will be similar to the recursive solution to the approximation problem for the Fokker–Planck equation). For computational feasibility, it is necessary that the state space $G_h\cup\partial G_h$ be finite. So we will assume that $X(\cdot)$ stops on reaching the boundary ∂G. Next, a precise formulation of the problem, which is convenient for our purposes (and also very convenient for other theoretical purposes—such as the derivation of Itô-type equations for the evolution of conditional moments) will be given.

Let $\bar{X}(\cdot)$, $\tilde{X}(\cdot)$ be solutions of the Itô equations

$$\bar{X}(s) = x + \int_t^{s \cap \bar{\tau}} f(\bar{X}(s), s)\, ds + \int_t^{s \cap \bar{\tau}} \sigma(\bar{X}(s))\, d\bar{w}(s), \tag{5.1a}$$

$$\tilde{X}(s) = x + \int_t^{s \cap \tilde{\tau}} f(\tilde{X}(s), s)\, ds + \int_t^{s \cap \tilde{\tau}} \sigma(\tilde{X}(s))\, d\tilde{w}(s), \tag{5.1b}$$

where $(\bar{w}(\cdot), \bar{X}(\cdot))$ and $(w(\cdot), X(\cdot))$ are independent processes and $\bar{\tau}$ and $\tilde{\tau}$ are the escape times of $\bar{X}(\cdot)$ and $\tilde{X}(\cdot)$, resp., from G. By A7.1.1, both $\bar{X}(\cdot)$ and $\tilde{X}(\cdot)$ have the same multivariate distributions. Let $z(\cdot)$ be a Wiener process which is independent of $\bar{w}(\cdot)$ and $\tilde{w}(\cdot)$ and define the process $\tilde{Y}(\cdot)$ by

$$d\tilde{Y}(s) = g(\tilde{X}(s), s)\, ds + dz(s), \qquad s > t, \quad \tilde{Y}(t) = 0.$$

Define $R(\cdot, \cdot)$ by

$$R(t, T) = \exp\left[\int_t^T g'(\bar{X}(s), s)\, d\tilde{Y}(s) - \tfrac{1}{2}\int_t^T |g(\bar{X}(s), s)|^2\, ds\right], \tag{5.2}$$

and define

$$\tilde{\mathscr{B}}_t^T = \text{minimal } \sigma\text{-algebra over which } \tilde{Y}(s),\ s \in [t, T], \text{ is measurable,}$$

and the term

$$V_b(x, t, T, \omega) = E_{x,t}[R(t, T) b(\bar{X}(T)) \,|\, \tilde{\mathscr{B}}_t^T].$$

Note that the conditioning serves to fix the $\tilde{Y}(\cdot)$ process; the expectation above is only over the $\bar{X}(\cdot)$ process.

One of the fundamental results of nonlinear filtering theory is the representation

$$E_{x,t}[b(\tilde{X}(T)) \,|\, \tilde{\mathscr{B}}_t^T] = \frac{V_b(x, t, T, \omega)}{V_1(x, t, T, \omega)} \quad \text{w.p. } 1, \tag{5.3}$$

where V_1 uses the function which is identically unity for $b(\cdot)$ (Kushner [K4], Zakai [Z1]).

In (5.3) the initial condition is $X(t) = x$, a constant. However, if $X(t)$ is a random variable with distribution $P_t(\cdot)$ supported on $\bar{G}$, we have (w.p. 1)

$$E[b(\tilde{X}(T) \,|\, \tilde{\mathscr{B}}_t^T] = \int V_b(x, t, T, \omega) P_t(dx) \Big/ \int V_1(x, t, T, \omega) P_t(dx).$$

Thus, to compute the conditional expectation, we must compute V_b for arbitrary b. The continuity of $b(\cdot)$ is not required for the validity of (5.3), where $b(\cdot)$ may be an arbitrary Borel function with finite expectation but it is used in the proof of convergence of the approximation procedure.

The exponential in the definition of $R(\cdot, \cdot)$ suggests that an approximation to V_b can be computed in a manner similar to the way we computed an approximation to the discounted cost problem in Section 4, where the exponent in (5.2) would be related to a discount factor such as $\lambda(\cdot)$. In fact, if we note that ($d_t =$ Itô differential)

$$d_t\left[\exp - \int_t^T \lambda(s)\, ds\right] = \left[\exp - \int_t^T \lambda(s)\, ds\right]\lambda(t)\, dt$$

and that (use Itô's lemma)

$$-d_t R(t, T) = R(t, T) g'(\bar{X}(t), t)\, d\tilde{Y}(t),$$

we might guess that the proper replacement for $\lambda(x)\, dt$ in the differential equation (4.5.4) (with $k(x) - \lambda(x)R(x, t)$ replacing $k(x)$ there) would be $-g'(x, t)\, d\tilde{Y}(t)$. The subscript t denotes the variable with respect to which we are calculating the differential. In fact, it can be shown that the unnormalized conditional expectation $V_b(x, t, T, \omega)$ formally satisfies the backward equation (in the initial variables x, t) (5.4), although we will not go through the formal calculations; our only aim is the development of a computational procedure for approximation of (5.3) via the use of the chain $\{\xi_m^{h,\Delta}\}$, and observed data $\tilde{Y}(\cdot)$.

$$d_t V_b(x, t, T, \omega) + \mathscr{L} V_b(x, t, T, \omega)\, dt + g'(x, t)\, d\tilde{Y}(t) V_b(x, t, T, \omega) = 0,$$

$$(x, t) \in G \times [0, T), \quad (5.4)$$

$$V_b(x, T, T, \omega) = b(x), \qquad x \in G,$$

$$V_b(x, t, T, \omega) = \exp\left[\int_t^T g'(x, s)\, d\tilde{Y}(s) - \tfrac{1}{2}\int_t^T |g(x, s)|^2\, ds\right] b(x),$$

$$t \le T, \quad x \in \partial G.$$

Equation (5.4) is a backward equation. Equations of the forward Kolmogorov or Fokker–Plank type for the conditional density have been discussed by Kushner [K3] and Mortenson [M5]. The backward equation for the unnormalized conditional expectation is better behaved and will serve, as it did for the ordinary Fokker–Plank equation, to yield a *recursive procedure* for approximating the conditional expectation or weak-sense conditional density, as T increases, and new observational data become available. However, we start the development by actually doing an iteration in the direction of decreasing t.

In what follows, $\tilde{X}(\cdot)$ is the signal process, and $\tilde{Y}(\cdot)$ is the actual process of observations taken on the diffusion $\tilde{X}(\cdot)$. We can assume, with no loss of generality, that $\tilde{X}(\cdot)$, $\tilde{Y}(\cdot)$ are the original diffusion and observation process, respectively. Let $\{\bar{\xi}_m^{h,\Delta}\}$ be a Markov chain on the state space $G_h \cup \partial G_h$, which is stopped on first exit from G_h and which has the same transition function that $\{\xi_m^{h,\Delta}\}$ has in the set G_h. The process $\{\bar{\xi}_m^{h,\Delta}\}$ is taken to be independent of the given process $\tilde{Y}(\cdot)$. Also, assume that†

$$P_{x,t}\{\tau = \tau'\} = 1, \qquad \text{all } x \in G, \quad 0 \le t < \infty. \tag{5.5}$$

Let $T = M_\Delta \Delta$, where M_Δ is an integer, and define $\delta_\Delta \tilde{Y}(t) = \tilde{Y}(t+\Delta) - \tilde{Y}(t)$, and define $V_b^{h,\Delta}(x, t, T, \omega)$ by the iteration

$$\begin{aligned} V_b^{h,\Delta}(x, n\Delta, T, \omega) = \exp&\left[g'(x, n\Delta)\delta_\Delta \tilde{Y}(n\Delta) - \tfrac{1}{2}|g(x, n\Delta)|^2\Delta\right] \\ &\times E_{x,n} V_b^{h,\Delta}(\bar{\xi}_{n+1}^{h,\Delta}, n\Delta + \Delta, T, \omega), \qquad x \in G_h, \quad n < M_\Delta, \end{aligned} \tag{5.6}$$

with the same boundary conditions as in (5.4). Note that in (5.6) we never take expectations over the $\tilde{Y}(\cdot)$ process, only over the $\{\bar{\xi}_m^{h,\Delta}\}$ process. Equation (5.6) has a unique solution, which can be written in the form

$$V_b^{h,\Delta}(x, n\Delta, T, \omega) = E_{x,n}[R^{h,\Delta}(n\Delta, T)b(\bar{\xi}^{h,\Delta}(T)) \mid \tilde{\mathscr{B}}_{n\Delta}^T], \tag{5.7}$$

where

$$R^{h,\Delta}(n\Delta, T) = \prod_{i=n}^{M_\Delta - 1} \exp[g'(\bar{\xi}_i^{h,\Delta}, i\Delta)\delta_\Delta \tilde{Y}(i\Delta) - \tfrac{1}{2}|g(\bar{\xi}_i^{h,\Delta}, i\Delta)|^2\Delta].$$

The conditioning serves to remind us that we do not take expectations over the $\tilde{Y}(\cdot)$ process. Skorokhod imbedding will be used where useful.

Now let us fix x, t and use the weak convergence results of Section 7.2. Under the conditions (5.5) and A7.1.1, Theorems 7.2.1 and 7.2.2 imply that

$$\bar{\xi}^{h,\Delta}(s) \to \bar{\xi}(s), \qquad s \ge t,$$

uniformly in each finite time interval, as $h, \Delta \to 0$, where $\bar{\xi}(\cdot)$ is a solution to (5.1a) [for some Wiener process $W(\cdot)$] with $\bar{\xi}(t) = x$. This implies that, w.p. 1, as $h, \Delta \to 0$,

$$R^{h,\Delta}(t, T) \to \exp\left[\int_t^T g'(\bar{\xi}(s), s)\, d\tilde{Y}(s) - \tfrac{1}{2}\int_t^T |g(\bar{\xi}(s), s)|^2\, ds\right].$$

Since $g(\cdot, \cdot)$ is bounded and (x, t) are arbitrary, there is also convergence in the mean for each (x, t). Hence, w.p. 1,

$$V_b^{h,\Delta}(x, t, T, \omega) \to V_b(x, t, T, \omega) \tag{5.8}$$

† The condition can be weakened. We only need that it holds for the value of t which is the initial value in the original filtering problem. Similarly, if the initial distribution is concentrated on $x \in$ all G_h, it need only hold for x.

as $h, \Delta \to 0$. Thus, via the finite iteration (5.6), we can calculate $V_b^{h,\Delta}(x, t, T, \omega)/V_1^{h,\Delta}(x, t, T, \omega)$, an approximation to the conditional expectation (5.3). In the paper (Kushner [K6]), a similar convergence is shown if the exponential in (5.6) is replaced simply by the first-order approximation $[1 + g'(x, n\Delta)\delta_\Delta \tilde{Y}(n\Delta)]$.

Recursive Equation for the Approximate Conditional Density

We can derive a recursive formula for an approximation to the weak-sense conditional density from (5.6), (5.7). (Compare with the development for the Fokker–Planck equation in Section 7.2.)

Fix h, Δ and let H denote the number of points on the grid $G_h \cup \partial G_h$. Order the points in some way, and recall the definition of the one-step transition matrix $\tilde{P}_n^{h,\Delta}$ from Section 7.2.

For arbitrary H vectors V, X and an $H \times H$ matrix F, define the $\circ$ operator by

$$V \circ F = \begin{bmatrix} V_1 F_{11} & \cdots & V_1 F_{1H} \\ \vdots & & \vdots \\ V_H F_{H1} & & V_H F_{HH} \end{bmatrix}, \qquad V \circ X = \begin{bmatrix} V_1 X_1 \\ \vdots \\ V_H X_H \end{bmatrix}.$$

Henceforth, the variables x, y range over $G_h \cup \partial G_h$ and are ordered as the points of $G_h \cup \partial G_h$ are ordered.

Let B, E_m, and $V_b(m, \omega)$ denote the vectors $\{b(y)\}$, $\{\exp[g'(y, m\Delta)\delta_\Delta \tilde{Y}(m\Delta) - \frac{1}{2}|g(y, m\Delta)|^2\Delta]\}$, and $\{V_b^{h,\Delta}(y, m\Delta, T, \omega)\}$, resp. *We will suppress some of the h, Δ notation.* Equation (5.6) can now be rewritten in *vector* form as

$$V_b(n, \omega) = E_n \circ \tilde{P}_n^{h,\Delta} V_b(n + 1, \omega), \qquad n < M, \tag{5.9}$$

where

$$V_b(M_\Delta, \omega) = B.$$

Now, for each pair of integers n, m, with $m \geq n$, define the $H \times H$ matrix $Q(n, m, \omega)$ with components $q_{x,y}(n, m, \omega)$ by

$$Q(n, m, \omega) = I, \quad \text{the identity}, \qquad \text{for} \quad m = n,$$

and, for $m > n$, by the recursive formula

$$\begin{aligned} Q(n, m, \omega) &= E_n \circ \tilde{P}_n^{h,\Delta} E_{n+1} \circ \tilde{P}_{n+1}^{h,\Delta} \cdots E_{m-1} \circ \tilde{P}_{m-1}^{h,\Delta} \\ &= E_n \circ \tilde{P}_n^{h,\Delta} Q(n + 1, m, \omega) \\ &= Q(n, m - 1, \omega) E_{m-1} \circ \tilde{P}_{m-1}^{h,\Delta}. \end{aligned}$$

Now let $n\Delta = t$, $M_\Delta \Delta = T$. Then by iterating (5.9),

$$V_b(n, \omega) = Q(n, M_\Delta, \omega) B. \tag{5.10}$$

In terms of components, (5.10) can be written as

$$V_b^{h,\Delta}(x, t, T, \omega) = \sum_y q_{xy}(n, M_\Delta, \omega)b(y). \tag{5.11}$$

By (5.3), (5.11), and the weak convergence,

$$\begin{aligned} V_b^{h,\Delta}(x, t, T, \omega)/V_1^{h,\Delta}(x, t, T, \omega) &= \sum_y q_{xy}(n, M_\Delta, \omega)b(y)\Big/\sum_y q_{xy}(n, M_\Delta, \omega) \\ &\to E_{x,t}[b(\tilde{X}(T)) \mid \tilde{Y}(s), s \in [t, T]], \\ &\qquad \text{w.p. } 1, \end{aligned} \tag{5.12}$$

as $h, \Delta \to 0$. As we will see next, only one row of (5.9) needs to be iterated.

Suppose that t is the initial time and that $\tilde{X}(t)$ is not concentrated at x but has a weak-sense density $p_t(\cdot)$. Let the row vector $\{p_t^h(x)\}$ denote an approximation to $p_t(\cdot)$, with support on $G_h \cup \partial G_h$, and which converges weakly to $p_t(\cdot)$, as $h \to 0$. Let $Q(n, m, \omega, x)$ denote the xth row of $Q(n, m, \omega)$. Then

$$Q(n, m, \omega, x) = Q(n, m-1, \omega, x)E_{m-1} \circ P_{m-1}^{h,\Delta}. \tag{5.13}$$

For each pair of integers $n, m, m \geq n$, define the row vector $\tilde{Q}(n, m, \omega)$ with components $\tilde{q}_y(n, m, \omega)$ by $\tilde{Q}(n, m, \omega) = \{p_t^h(y)\}$ for $m = n$, and for $m > n$, by the recursive formula

$$\tilde{Q}(n, m, \omega) = \tilde{Q}(n, m-1, \omega)E_{m-1} \circ \tilde{P}_{m-1}^{h,\Delta}. \tag{5.14}$$

Then we have that

$$\sum_y \tilde{q}_y(n, M_\Delta, \omega)b(y)\Big/\sum_y \tilde{q}_y(n, M_\Delta, \omega) \to E[b(\tilde{X}(T)) \mid \tilde{Y}(s), s \in [t, T]], \qquad \text{w.p. } 1,$$

as $h, \Delta \to 0$. Thus the vector with components

$$\tilde{q}_y(n, m, \omega)\Big/\sum_u \tilde{q}_u(n, m, \omega) \equiv \tilde{p}_y(n, m, \omega) \tag{5.15}$$

is simply a weak-sense approximation to the weak-sense density of $\tilde{X}(m\Delta)$, conditioned on $\tilde{Y}(s)$, $n\Delta \leq s \leq m\Delta$, in the sense that it converges weakly w.p. 1, as $h, \Delta \to 0$. Furthermore, it can be computed recursively via (5.14).

The quantity $\tilde{p}(n, m, \omega)$ in Eq. (5.15) is simply the *Bayes-rule conditional density for the chain* $\{\xi_m^{h,\Delta}\}$, namely (with an obvious abuse of notation) $\tilde{p}_y(n, m, \omega) = P\{\bar{\xi}_m^{h,\Delta} = y \mid \delta_\Delta \tilde{Y}(i) = g(\bar{\xi}_i^{h,\Delta}, i\Delta)\Delta + \delta_\Delta z(i\Delta), n \leq i < m\}$, where $\{\bar{\xi}_m^{h,\Delta}\}$ is independent of $z(\cdot)$ and has the same law that $\{\xi_m^{h,\Delta}\}$ has. We can see this as follows. It is true for $m = n$. Suppose that it is true for $m = l - 1 \geq n$. By formal manipulations with Bayes's rule, we show it to be true for $m = l$.

Let $E_l(y) =$ yth component of the vector E_l. We have, by the induction hypothesis,

$$\begin{aligned} P\{\xi_{l-1}^{h,\Delta} = y \mid &\text{current observation} = \delta_\Delta \tilde{Y}(l\Delta - \Delta) = g(\xi_{l-1}^{h,\Delta}, l\Delta - \Delta)\Delta \\ &+ \delta_\Delta z(l\Delta - \Delta);\ \delta_\Delta \tilde{Y}(i\Delta - \Delta),\ n+1 \le i < l\} \\ &= P\{\text{current observation has value } \delta_\Delta \tilde{Y}(l\Delta - \Delta) \mid \xi_{l-1}^{h,\Delta} = y\} \\ &\quad \cdot P\{\xi_{l-1}^{h,\Delta} = y \mid \delta_\Delta \tilde{Y}(i\Delta - \Delta),\ n+1 \le i < l\} K_l \\ &= (\exp -\tfrac{1}{2} |\delta_\Delta \tilde{Y}(l\Delta - \Delta) - g(y, l\Delta - \Delta)\Delta|^2) \cdot \tilde{q}_y(n, l-1, \omega) K_l' \\ &= E_{l-1}(y)\tilde{q}_y(n, l-1, \omega),\ K_l'', \end{aligned}$$

where K_l, K_l', K_l'' are normalizing factors which do not depend on y. The last equation implies that, modulo normalizing factors,

$$\begin{aligned} P\{\xi_l^{h,\Delta} &= x \mid \delta_\Delta \tilde{Y}(i\Delta),\ n \le i < l\} \\ &= \sum_y E_{l-1}(y)\tilde{q}_y(n, l-1, \omega) P\{\xi_l^{h,\Delta} = x \mid \xi_{l-1}^{h,\Delta} = y\}, \\ &\qquad x \in G_h \cup \partial G_h, \qquad (5.16) \end{aligned}$$

which equals $\tilde{q}_x(n, l, \omega)$ by the definition (5.14). Thus, the normalized value (5.15) is the conditional probability, as asserted.

If we used the first-order approximation $\{1 + g'(y, m\Delta)\delta_\Delta \tilde{Y}(m\Delta)\}$ in lieu of E_m, we would not have the Bayes-rule interpretation and, indeed, it is possible that some $\tilde{q}_x(n, m, \omega)$ would be negative even though we would still get the correct conditional expectation as $h, \Delta \to 0$.

7.6 Numerical Data: Estimation of an Invariant Measure

Tables 7.1–7.3 give data on a numerical approximation to the invariant measure for the simple scalar case

$$dx = -(x/2)\,dt + \sqrt{2}\,dw,$$

using both the elliptic and parabolic approximations. The unique invariant measure is normal with mean zero and variance 2. Since its support is the entire real line, the state space was truncated to either the interval $[-3, 3]$ (Tables 7.1 and 7.2) or to the interval $[-1\tfrac{1}{2}, 1\tfrac{1}{2}]$ (Table 7.3). Let $x =$ left endpoint of the truncation interval. In order to account for the truncation, we selected a number $q \in (0, 1)$ and used $p^h(x, x+h) = 1 - q$, $p^h(x, x) = q$. Similarly, $p^h(x, x-h) = 1 - p^h(x, x) = 1 - q$ when $x =$ right endpoint. The tables give the *cumulatives*, i.e., the calculated distribution functions. Since these are antisymmetric about $x =$ zero, only half the points are plotted.

TABLE 7.1

ESTIMATE OF AN INVARIANT MEASURE. SPACE TRUNCATED TO $[-3, 3]$, $h = \frac{1}{5}$

x	Gaussian	Approximate Gaussian	E $q = \frac{1}{2}$	P $B = \frac{1}{2}, q = \frac{1}{2}$	E $q = .75$
.2	.556	.558	.559	.559	.558
.4	.610	.614	.616	.617	.614
.6	.665	.669	.670	.672	.668
.8	.713	.720	.721	.723	.718
1.0	.761	.767	.768	.770	.764
1.2	.805	.810	.810	.812	.805
1.4	.841	.846	.847	.850	.842
1.6	.873	.881	.880	.880	.874
1.8	.900	.910	.907	.910	.901
2.0	.921	.933	.931	.934	.924
2.2	.940	.953	.950	.950	.943
2.4	.962	.968	.966	.970	.959
2.6	.967	.980	.978	.982	.971
2.8	.976	.989	.988	.992	.981
3.0	.983	.997	.996	.998	.993

Refer to Table 7.1. Column 1 gives the value of the state, and column 2 the cumulative Gaussian (variance 2). We calculated the Gaussian *density* at each point $0, \pm h, \ldots,$ on $[-3, 3]$ or $[-1\frac{1}{2}, 1\frac{1}{2}]$, interpolated this by use of a piecewise linear interpolation, and then used the interpolation to calculate an approximation to the Gaussian distribution. This distribution is tabulated in column 3. The value at $x = 3$ is not unity; the error is $\frac{1}{2}$ the density at $x = 3$, which mass we assigned to $(3, \infty)$. The other columns give the numerical data using either the elliptic (E) or parabolic (P) schemes. In all cases (and all tables), we took a piecewise linear interpolation between the value obtained at $0, \pm h, \ldots,$ and plotted the cumulative from that. The

TABLE 7.2

ESTIMATE OF AN INVARIANT MEASURE. SPACE TRUNCATED TO $[-3, 3]$, $h = \frac{3}{5}$

P	Gaussian	Approximate Gaussian	Table 7.1 column 4	E $q = .3$	P $B = \frac{1}{2}, q = \frac{3}{4}$	P $B = \frac{1}{2}, q = .3$
.6	.665	.665	.670	.669	.667	.673
1.2	.805	.804	.810	.804	.799	.811
1.8	.900	.903	.907	.889	.982	.907
2.4	.962	.961	.966	.957	.949	.967
3.0	.983	.991	.996	.789	.985	.994

TABLE 7.3

ESTIMATE OF AN INVARIANT MEASURE. SPACE TRUNCATED TO $[-1\frac{1}{2}, 1\frac{1}{2}]$, $h = \frac{1}{4}$.

	Approximate Gaussian	P $B = 1, q = \frac{1}{2}$	P $B = \frac{1}{2}, q = \frac{1}{2}$	P $B = \frac{1}{2}, q = .3$
.25	.593	.595	.601	.602
.5	.684	.686	.697	.700
.75	.769	.770	.786	.791
1	.846	.846	.866	.873
1.25	.915	.913	.937	.945
1.5	.973	.972	.985	.989

errors were smaller with the use of the interpolation. The value of h is given in the table; to calculate Δ (for the parabolic case), we chose a number $B \in [0, 1]$ and [see (2.6a)] selected Δ by the formula

$$B = (\Delta/h^2) \max_x [h|x/2| + \sigma^2],$$

where $|x| \le 3$ or $|x| \le 1\frac{1}{2}$ according to the case and $\sigma^2 = 2$.

It is not at all clear whether column 2 or 3 should be taken as the basis of comparison. The values with the parabolic method and $B = 1$ were about identical to the values with the elliptic method. The approximations calculated by both the elliptic and parabolic methods seem to be rather close to the true values (columns 1 or 2). The main problem is the choice of q. The value should reflect the average amount of real time that the original process spends outside the truncation region.

CHAPTER 8

Optimal Stopping and Impulsive Control Problems

In Section 8.1, we discuss the discretization of the optimal stopping problem by use of the finite difference methods of Chapter 6 and show that the discretization is an optimal stopping problem for a Markov chain.

In Section 8.2, we discuss the optimality of the limits of the costs and of the "interpolated" optimal stopping times for the approximating chains. With a fixed initial condition (x), the interpolated Markov chain converges to a solution $\xi(\cdot)$ of (1.5.1) with respect to some Wiener process $W(\cdot)$ and initial condition x. Also ρ, the limit of the discretized stopping times, is nonanticipative with respect to $W(\cdot)$. We then define a Markov process [solution to (1.5.1)] $(\Omega, \mathscr{B}, \mathscr{B}_t, X(\cdot), P_y, y \in R^r)$ such that a stopping time $\bar{\rho}$ (relative to $\{\mathscr{B}_t\}$) is defined on Ω, and under P_x, $(X(\cdot), \bar{\rho})$ has the same law and cost as has $(\xi(\cdot), \rho)$. Thus we can suppose that $\rho \in \overline{\mathscr{T}}_{\mathrm{R}}^{+}(x)$. This theorem (8.2.4) is proved under a Lipschitz condition. If there is uniqueness (A6.1.1) but not necessarily a Lipschitz condition then $\bar{V}(x) =$ limit of discretized costs $\leq V(x)$. The method of proof involves a "discretization" of stopping times $\tau \in \overline{\mathscr{T}}_{\mathrm{W}}$. Subsequently, the Lipschitz condition is dropped and the results strengthened.

In Theorem 8.2.5 we show that if $V(\cdot)$ is continuous, then $\bar{V}(x) = V(x)$ and in Corollary 8.2.2, the continuity is removed. Here, we approximate and discretize the pure Markov stopping rules. The methods of Theorems 8.2.4 and 8.2.5 are of independent interest, for they illustrate different approximation techniques which are useful in other problems also. Then various extensions to the discounted problem or to the problem where we are forced to stop on first hitting a set are discussed.

The Lipschitz condition is helpful for two reasons. It implies that the $\bar{\mathscr{T}}_{\mathrm{W}}$ class is as "good as" the $\bar{\mathscr{T}}_{\mathrm{PM}}$ or $\bar{\mathscr{T}}_{\mathrm{R}}$ class, because the path can be represented as an explicit $w(\cdot)$ function in a convenient way. Second, it facilitates the proof that certain approximations to diffusions converge as the approximation parameter goes to zero. In fact, these proofs use standard techniques (requiring the Lipschitz condition) and most of the details are left out.

In Section 8.3, we discuss the optimal stopping problem with side constraints, a topic on which (as with all stochastic control problems with side constraints) there is relatively little known. In Sections 8.4 and 8.5, the impulse control problem is discretized and it is shown that the discretization is an impulse control problem for a Markov chain. Again, under a Lipschitz condition, it is shown that the limiting sequence of action times and values is minimal. The removal of the Lipschitz condition is also discussed in Section 8.5, and in Section 8.6, some numerical results for the optimal stopping problem are given.

In Chapters 5, 6, and 7, linear partial differential equations were dealt with. The optimal stopping and impulsive control problems of this chapter relate to a class of nonlinear partial differential equations.

8.1 Discretization of the Optimal Stopping Problem

The treatment of the optimal stopping problem of Section 4.7.1 is a relatively simple extension of the discretization and convergence ideas in Chapter 6. We use the notation of Section 4.7.1 and of Chapter 6. The main problem concerns verification of the *minimality* of the limiting cost function. *Assumptions* A6.1.1 *and* A6.1.2 *are assumed to hold throughout Sections* 8.1–8.3. *We also suppose (unless otherwise mentioned), that there is used a real* $k_0 > 0$ *such that* $\inf_x k(x) \geq k_0$.

According to the formal optimality Eq. (4.7.6), if we find a smooth function $V(\cdot)$ which satisfies

$$\begin{aligned} V(x) &\leq b(x), \\ \mathscr{L}V(x) + k(x) &= 0 \qquad \text{on} \quad R^r - B, \end{aligned} \tag{1.1}$$

where $B = \{x : V(x) = b(x)\}$, then B is an optimal stopping set and $V(x)$ is the minimal cost. We will use the finite difference approximations of Section 6.2, and let $V^h(\cdot)$ denote the finite difference solution. Substituting those finite difference approximations into (1.1) and keeping in mind that $V(x) \le b(x)$, we get that (6.2.7) holds if $V^h(x) < b(x)$. Thus since we must use the condition $V^h(x) \le b(x)$, the correct discretized equation is

$$V^h(x) = \min[b(x), E_x V^h(\xi_1^h) + k(x)\, \Delta t^h(x)]. \tag{1.2}$$

Equation (1.2) is the dynamic programming equation for an optimal stopping problem on $\{\xi_n^h\}$. Let $B(h)$ denote the optimal stopping set $\{x : V^h(x) = b(x)\}$ for $\{\xi_n^h\}$. We will not be explicitly concerned with boundedness of the state space of $\{\xi_n^h\}$, but we implicitly assume that it is bounded. Boundaries can be added (see below) with little added difficulty or $b(\cdot)$ can grow fast enough as $|x| \to \infty$ to assure that for each initial condition there is a bounded region in which the processes will stay. Also, the state space may be bounded by a suitable selection of the dynamics $f(\cdot)$ and $\sigma(\cdot)$ for large x, by the addition of a reflecting boundary, or by a combination of the four methods.

Let $\mathscr{T}_{\mathrm{PM}}(h)$ and $\mathscr{T}_{\mathrm{R}}(h)$ (with or without superscripts + or 0 and additional argument x) denote the set of *pure Markov and randomized strategies, resp., for* $\{\xi_n^h\}$. For a stopping time m, define the cost $R^h(x, m)$ by

$$R^h(x, m) = E_x\left[\sum_{i=0}^{m-1} k(\xi_i^h)\, \Delta t_i^h + b(\xi_m^h)\right]. \tag{1.3}$$

Note that $\inf_x k(x)\, \Delta t^h(x) > 0$. Hence by Section 3.2, (1.2) always has a unique solution, which is the minimal cost, namely,

$$V^h(x) = \inf_{m \in \mathscr{T}_{\mathrm{PM}}(h)} R^h(x, m) = \inf_{m \in \mathscr{T}_{\mathrm{R}}(h)} R^h(x, m) \tag{1.4}$$

and a minimizing pure Markov stopping rule exists.

Let m_h denote the *optimal stopping time* (in $\mathscr{T}_{\mathrm{PM}}(h)$ or $\mathscr{T}_{\mathrm{R}}(h)$) *for the chain, and define* $\rho_h = \sum_{i=0}^{m_h-1} \Delta t_i^h$, the *optimal stopping time* for the *interpolated process*, $\xi^h(\cdot)$. When dealing with the $\xi^h(\cdot)$ process, we will sometimes use the notation $R^h(x, \rho_h)$ for $R^h(x, m_h)$. Also, a *bar* over $\mathscr{T}_{\mathrm{PM}}(h)$, etc., implies that the *interpolation* is referred to.

If ρ is a stopping time for $\xi^h(\cdot)$, define $R^h(x, \rho)$ by

$$R^h(x, \rho) = E_x\left[\int_0^\rho k(\xi^h(s))\, ds + b(\xi^h(\rho))\right].$$

$R^h(x, \rho)$ obviously makes sense for ρ of the form

$$\rho = \sum_{i=0}^{m-1} \Delta t_i^h = t_m^h,$$

where $m \in \mathcal{T}_R(h)$; but since $\xi^h(\cdot)$ is a continuous parameter process, the values of ρ need not be restricted to t_i^h, $i = 0, 1, \ldots$. Observe that†

$$E_x \rho_h \leq 2 \max_x |b(x)|/k_0 . \tag{1.5}$$

The upper bound in (1.5) also holds for the optimal stopping time for the diffusion.

Let us use the representation of Section 6.6 for β_n^h and W_n^h. Define

$$\hat{\Phi}^h(\cdot) = (\xi^h(\cdot), B^h(\cdot), F^h(\cdot), W^h(\cdot), K^h(\cdot); \xi^h(0) = x).$$

It follows from Theorems 6.3.1 and 6.6.1 that the sequence $\{\hat{\Phi}^h(\cdot), \rho_h\}$ is tight on $D^{4r+1}[0, \infty) \times \bar{R}^+$. The limit $(\hat{\Phi}(\cdot), \rho) = (\xi(\cdot), B(\cdot), F(\cdot), W(\cdot), K(\cdot), \rho; \xi(0) = x)$ of any convergent subsequence satisfies

$$\begin{aligned} \xi(t) &= x + \int_0^t f(\xi(s))\, ds + \int_0^t \sigma(\xi(s))\, dW(s) \\ &\equiv x + F(t) + B(t) \end{aligned} \tag{1.6}$$

$$K(t) = \int_0^t k(\xi(s))\, ds.$$

Henceforth, fix a convergent subsequence. By (1.5), the $\{\rho_h\}$ are also tight on $R^+ = [0, \infty)$, and $E_x \rho < \infty$. If we also assume that the Skorokhod imbedding is used, then $\rho_h \to \rho$ w.p. 1. The multivariate distributions of $\xi(\cdot)$ do not depend on the subsequence (uniqueness) but the distribution of ρ may possibly depend on the subsequence. Also,

$$R^h(x, m_h) = E_x\left[\int_0^{\rho_h} k(\xi^h(s))\, ds + b(\xi^h(\rho_h))\right],$$

$$\begin{aligned} R^h(x, m_h) &= V^h(x) \to R(x, \rho) \\ &= E_x\left[\int_0^{\rho} k(\xi(s))\, ds + b(\xi(\rho))\right]. \end{aligned} \tag{1.7}$$

8.2 Optimality of the Limiting Stopping Time ρ

We must show that ρ is a legitimate stopping time in $\bar{\mathcal{T}}_R^+(x)$ in some sense and that $R(x, \rho) = V(x)$. So far, we know very little about ρ. In order for ρ to be a stopping time in $\bar{\mathcal{T}}_R^+$, it is necessary to show at least that the function with values $I_{\{\rho \leq t\}}$ is nonanticipative with respect to $W(\cdot)$. [Then we say that

† Equation (1.5) follows from $|b(x)| \geq V^h(x) \geq -\max_x |b(x)| + k_0 E_x \rho_h$.

the stopping time ρ is *nonanticipative* with respect to $W(\cdot)$.] Our formulation of the optimal stopping problem actually requires more: namely that ρ be a stopping time with respect to a sequence of nondecreasing σ-algebras on a space which supports a Markov process. This is not necessarily the case with ρ. However, we will find a Markov diffusion process $X(\cdot)$ and an appropriate stopping time $\bar{\rho}$ such that $(X(\cdot), \bar{\rho}, X(0) = x)$ and $(\xi(\cdot), \rho)$ have the same probability law. Hence $R(x, \bar{\rho}) = R(x, \rho)$. Since we are mainly concerned with the cost functionals, this will eventually allow us to show that $R(x, \rho)$ is minimal.

The next theorem will be helpful in showing the nonanticipative property of ρ.

Theorem 8.2.1 *Let $w(\cdot)$ denote an R^r-valued Wiener process, and Y a real-valued random variable, and s a nonnegative real number. Suppose that for each integer n and each pair of bounded and continuous functions $F(\cdot): R \to R$, $F_n(\cdot): R^{nr} \to R$, and each set of real numbers $(t, s_n, \ldots, s_1)$ satisfying $t \geq s \geq s_n \geq s_{n-1} \geq \cdots \geq s_1$, we have*

$$EF(Y)F_n(w(s_1), \ldots, w(s_n))(w(t) - w(s)) = 0. \tag{2.1}$$

Then Y is independent of $\{w(t) - w(s), t \geq s\}$.

PROOF Let $q(\cdot)$ denote an arbitrary bounded continuous R^r-valued function on $[0, \infty)$, and let λ denote an arbitrary real number. By Itô's lemma,

$$\exp i\left(\lambda Y + \int_s^t q'(\sigma)\, dw(\sigma)\right) = \exp i\lambda Y + \int_s^t (\exp i\lambda Y)\left[\exp i \int_s^\sigma q'(\mu)\, dw(\mu)\right] \times \left[iq'(\sigma)\, dw(\sigma) - \frac{|q(\sigma)|^2}{2}\, d\sigma\right].$$

Note that

$$E \exp i \int_s^t q'(\sigma)\, dw(\sigma) = \exp -\tfrac{1}{2} \int_s^t |q(\sigma)|^2\, d\sigma.$$

Taking expectations in the first equation above and using (2.1) and the second equation and the definition

$$C_t = E \exp i\left(\lambda Y + \int_s^t q'(\sigma)\, dw(\sigma)\right)$$

yields

$$C_t = C_s - \int_s^t C_\sigma\, |q(\sigma)|^2\, d\sigma/2,$$

which has the unique solution

$$C_t = E(\exp i\lambda Y)E\left(\exp i \int_s^t q'(\sigma)\, dw(\sigma)\right).$$

The last equation yields the cited independence since it implies that the characteristic functional of $\{Y, w(t) - w(s), t \geq s\}$ is the product of the characteristic functionals of Y and of $\{w(t) - w(s), t \geq s\}$, resp. Q.E.D.

Theorem 8.2.2 *Let ρ denote the limit of $\{\rho_h\}$ and let $F(\cdot)$, $F_n(\cdot)$ and $s, t, s_1, \ldots, s_n$ be as in Theorem 8.2.1. If t_0 is such that $t_0 \leq s$ and $P_x\{\rho = t_0\} = 0$ and $Y = I_{\{\rho \leq t_0\}}$, then*

$$EF(Y)F_n(W(s_1), \ldots, W(s_n))(W(t) - W(s)) = 0. \tag{2.2}$$

Proof We use the Skorokhod imbedding. By a direct evaluation, with $Y^h = I_{\{\rho_h \leq t_0\}}$,

$$EF(I_{\{\rho_h \leq t_0\}})F_n(W^h(s_1), \ldots, W^h(s_n))(W^h(t) - W^h(s)) = 0 \tag{2.3}$$

since $E[W^h(t) - W^h(s) \mid \xi^h(\sigma), \sigma \leq s] = 0$ and Y^h is a functional of $\xi^h(\sigma), \sigma \leq s$. Recall that (proof of Theorem 6.3.1.) $\{W^h(t) - W^h(s)\}$ is uniformly (in h) integrable and note that $I_{\{\rho_h \leq t_0\}} \to Y$ w.p. 1 since $P_x\{\rho = t_0\} = 0$. Thus by the weak convergence, the left-hand side of (2.3) tends to the left-hand side of (2.2) as $h \to 0$. Hence (2.2) holds. Q.E.D.

Remark If $P_x\{\rho = t_0\} > 0$, then we would not necessarily have the convergence $I_{\{\rho_h \leq t_0\}} \to Y$ as $h \to 0$, w.p. 1.

Corollary 8.2.1 *The random variable ρ is nonanticipative with respect to $W(\cdot)$, in the sense that $I_{\{\rho \leq s\}}$ is independent of $\{W(t) - W(s), t \geq s\}$ for all $s \geq 0$.*

Proof Let s be fixed. By Theorems 8.2.1 and 8.2.2, $I_{\{\rho \leq t_0\}}$ is independent of $\{W(t) - W(s), t \geq s\}$ if $P_x\{\rho = t_0\} = 0$ and $t_0 \leq s$. There are at most a countable number of values of t_0 for which $P_x\{\rho = t_0\} > 0$. Let $t_n \downarrow s$ with $P_x\{\rho = t_n\} = 0$. Let $\varepsilon > 0$ denote a positive real number. Note that $I_{\{\rho \leq t_n\}}$ is independent of $S_\varepsilon \equiv \{W(t) - W(s + \varepsilon), t \geq s + \varepsilon\}$ for large n by Theorems 8.2.1 and 8.2.2. Thus $I_{\{\rho \leq s\}}$ is also independent of S_ε since $I_{\{\rho \leq s\}} = \lim_n I_{\{\rho \leq t_n\}}$ w.p. 1. Since the independence is for each $\varepsilon > 0$, and $W(s^+) = W(s)$, the proof is concluded. Q.E.D.

Next, we discuss the sense in which ρ is an optimal stopping time, and $R(x, \rho)$ is minimal $[= V(x)]$. Indeed, $R(x, \rho)$ always equals $V(x)$. There are several different types of approximation techniques which can be applied to prove the result. These techniques seem to be of some independent interest, and can be applied to the study of other problems in stochastic control theory. In order to illustrate them, several different proofs, under different conditions, will be given.

Theorem 8.2.3 *Fix* $x = X(0)$, *and let* $f(\cdot)$ *and* $\sigma(\cdot)$ *satisfy a uniform Lipschitz condition. Then* $\rho \in \bar{\mathcal{T}}_R^+(x)$.

REMARK In order to assure that the main idea is apparent, the theorem is worded loosely. A more formal statement (which is the actual assertion to be proved) is the following.

Let $f(\cdot)$ *and* $\sigma(\cdot)$ *satisfy a uniform Lipschitz condition. There is a strong Markov process* $(\Omega, \mathcal{B}, \mathcal{B}_t, P_y, y \in R^r, X(\cdot))$, *and a Wiener process* $w(\cdot)$ *defined on each* $(\Omega, \mathcal{B}, P_y)$, *where* $w(\cdot)$ *is adapted to* $\{\mathcal{B}_t\}$ *and for each* $y \in R^r$, $(w(\cdot), X(\cdot))$ *solve* (1.5.1) (with $X(0) = y$) w.p. 1 *under* P_y. There is a random variable $\bar{\rho}$ *defined on each* $(\Omega, \mathcal{B}, P_y)$, *which is a stopping time relative to* $\{\mathcal{B}_t\}$. *Also,* $(\xi(\cdot), \rho)$ *and* $(X(\cdot), \bar{\rho})$ *(under* P_x*) have the same distributions. In addition* $\bar{\rho} \in \bar{\mathcal{T}}_R^+(x)$. *The costs* $R(x, \bar{\rho})$ *and* $R(x, \rho) = \lim_{h\to 0} R^h(x, \rho_h) = \lim_{h\to 0} V^h(x)$ *are the same.*

PROOF We have shown that $\xi(\cdot)$ satisfies (1.6) and that ρ is nonanticipative with respect to the $W(\cdot)$ in (1.6). Owing to the Lipschitz condition, we can define a solution to (1.6) for any initial condition $y \in R^r$ on the same sample space on which $W(\cdot)$ is defined since $W(\cdot)$ can be used to constructively generate the solution for any initial condition (Section 1.5). In fact, the solution [denoted by $\tilde{X}(\cdot)$] to (1.6) which we generate from $W(\cdot)$ by the iterative method of Section 1.5 equals $\xi(\cdot)$ w.p. 1, when $\tilde{X}(0) = x$. However, we will now change the sample space.

Define $\Omega' = C^r[0, \infty)$, $\mathcal{B}[0, t] =$ Borel algebra over $[0, t]$, $\mathcal{B}' =$ Borel algebra over Ω', $\mathcal{B}'_t =$ smallest sub-σ-algebra of $\mathcal{B}'$ which contains $\{x(\cdot): x(s) < \alpha\}$, all $s \le t$ and all $\alpha \in R^r$. Define $\Omega = \Omega' \times \Omega' \times R^+$, and let $\mathcal{B}$ and $\mathcal{B}_t$ denote the product σ-algebras $\mathcal{B}' \times \mathcal{B}' \times \mathcal{B}[0, \infty)$ and $\mathcal{B}'_t \times \mathcal{B}'_t \times (\mathcal{B}[0, t] \cup [t, \infty))$, resp. Let P_y denote the measure which the triple $(\tilde{X}(\cdot), W(\cdot), \rho; \tilde{X}(0) = y)$ induces on $(\Omega, \mathcal{B})$, and let the triple $(X(\cdot), w(\cdot), \bar{\rho})$ denote the corresponding† random processes on $(\Omega, \mathcal{B})$. Clearly, $X(\cdot)$ and the solution to (1.6) or to (1.5.1) all have the same probability law, if their initial conditions are the same.

The random variable $\bar{\rho}$ is obviously a stopping time with respect to $\{\mathcal{B}_t\}$ since $\{\bar{\rho} \le t\} = \Omega' \times \Omega' \times [0, t] \in \mathcal{B}_t$ for each t. Since ρ was nonanticipative with respect to $W(\cdot)$, and since $\tilde{X}(s)$, $s \ge t$, was a functional of $\{\tilde{X}(t), W(u) - W(t), u \ge t\}$, we have that $\{I_{\{\rho \le t\}}, w(u), u \le t\}$, is conditionally independent of $X(s)$, $s \ge t$, given $X(t)$, under each P_y. Thus we still have the Markov

† That is, we identify the processes $X(\cdot)$, $w(\cdot)$ and random variable $\bar{\rho}$ with the three components of ω.

property [i.e., the "presence" of $w(\cdot)$ and $\bar{\rho}$ and their measure theoretic supporting structure does not ruin the Markov property of $X(\cdot)$: namely, for each $\Gamma \in \mathscr{B}(R^r)$ and $y \in R^r$ and $t,\ s \geq 0$, $P_y\{X(t+s) \in \Gamma \mid \mathscr{B}_t\} = P_{X(t)}\{X(s) \in \Gamma\}$ w.p. 1]. Since $X(\cdot)$ is a Feller process under the P_y, $(\Omega, \mathscr{B}, \mathscr{B}_t, P_y, y \in R^r, X(\cdot))$ is a strong Markov process. Hence $\bar{\rho} \in \bar{\mathscr{T}}_{\mathrm{R}}^+(x)$. The costs $R(x, \rho)$ and $R(x, \bar{\rho})$ are the same since $(X(\cdot), \bar{\rho})$ under P_x and $(\xi(\cdot), \rho)$ have the same probability law. Q.E.D.

The proof that $R(x, \rho)$ is the minimal cost requires that we compare $R(x, \rho)$ to $R(x, \tau)$ for a class of stopping times τ. To do this, we must somehow use the properties of ρ. The most important property is its definition as a limit of optimal times for the $\xi^h(\cdot)$. Hence we must somehow "discretize" τ to obtain a policy which can be compared to ρ_h, and then get the comparison between $R(x, \rho)$ and $R(x, \tau)$, by taking limits of the costs for the discretized policies in a suitable way. There are several ways in which the discretization can be done. Theorem 8.2.4 illustrates a technique for discretizing stopping times which are functionals of $w(\cdot)$ and Theorem 8.2.5 and Corollary 8.2.2 illustrate a method for a discretization based on approximating the optimal stopping sets B or B_ε. In Chapter 9, some different methods are developed. Those methods are actually more general than the discretization techniques of this chapter.

Theorem 8.2.4 *Under the conditions of Theorem 8.2.3, $R(x, \rho) = V(x)$. Hence, since the limit is independent of the subsequence, $V^h(y)$ converges to the minimal cost as $h \to 0$, for each $y \in R^r$. If we assume only uniqueness of the solution to (1.5.1) in the sense of distributions (and not a Lipschitz condition), then $R(y, \rho) \leq R(y, \tau)$ for every $\tau \in \bar{\mathscr{T}}_{\mathrm{W}}$ and $y \in R^r$.*

PROOF We will only prove the first sentence. The second and third sentences are actually implied by the proof. Under the Lipschitz condition, $X(\cdot)$ can be represented as a functional of $w(\cdot)$ for each $x(0) = x$. Thus, by Section 4.7.1,

$$\inf_{\tau \in \bar{\mathscr{T}}_{\mathrm{PM}}} R(x, \tau) = \inf_{\tau \in \bar{\mathscr{T}}_{\mathrm{W}}} R(x, \tau) = \inf_{\tau \in \bar{\mathscr{T}}_{\mathrm{R}}} R(x, \tau) = V(x).$$

So by Theorem 8.2.3, we only need to show that

$$R(x, \rho) \leq R(x, \tau) \qquad \text{for all} \quad \tau \in \bar{\mathscr{T}}_{\mathrm{W}}. \tag{2.4}$$

The proof is divided into several steps. Owing to (1.5), (2.4) will hold if, for each $T < \infty$, it holds for all $\tau \in \bar{\mathscr{T}}_{\mathrm{W}}$ for which $\tau \leq T$. We approximate each such τ by a stopping time which takes only finitely many values and which depends on the values of $w(\cdot)$ at only a finite number of times. We then use a similar stopping rule for the chain, and show that the cost is no less than $V^h(x) = R^h(x, m_h)$. A suitable limiting procedure then yields the desired result.

PART 1 Let $0 < \delta \ll \Delta < T$ denote real numbers such that T/Δ and Δ/δ are integers, and let $\mathscr{B}^{\delta}_{n\Delta}(w)$ denote the minimal σ-algebra over which $\{w(i\delta),$ all $i\delta \le n\Delta\}$ is measurable. Let $\tau \in \bar{\mathscr{T}}_W$, $\tau \le T$. All $i\delta$ or $n\Delta$ terms are assumed to be no larger than T. Define the sequence

$$\rho_i(\delta, \Delta) = E_x[\tau \mid \mathscr{B}^{\delta}_{i\Delta}(w)], \; i = 1, 2, \ldots,$$

and the stopping time

$$\tau(\delta, \Delta) = \sum_{i=1}^{T/\Delta} (i\Delta) I_{\{\rho_i(\delta, \Delta) \le i\Delta, \, \rho_j(\delta, \Delta) > j\Delta, \, j < i\}} .$$

If $\tau(\delta, \Delta)$ or any other stopping time is not defined at some ω, set it equal to T there.

As $\delta \to 0$, $\tau(\delta, \Delta)$ converges w.p. 1 to a stopping time [which we denote by $\tau(\Delta)$] which equals (w.p. 1) either $i\Delta$ or $i\Delta + \Delta$ on $\{i\Delta - \Delta < \tau \le i\Delta\}$. As $\delta \to 0$ and $\Delta \to 0$, $\tau(\delta, \Delta) \to \tau$ w.p. 1.

Define $q = \Delta/\delta$, $Q = T/\Delta$. For each Δ, δ, i with $i \le Q$, there is a Lebesgue set $A_i(\delta, \Delta) \in R^{iqr}$ such that $\tau(\delta, \Delta) = i\Delta$ if $\{w(j\delta), j\delta \le i\Delta\} \in A_i(\delta, \Delta)$. Let $I_i(\delta, \Delta; \cdot)$ denote the indicator function of $A_i(\delta, \Delta)$. Note, for use below, that for any integer α and any Lebesgue set $A \in R^{\alpha}$, there is a sequence of continuous functions $f_n(\cdot) : R^{\alpha} \to [0, 1]$, which converges to $I_A(\cdot)$ almost everywhere as $n \to \infty$. Also, if ϕ is a vector of α independent nondegenerate scalar-valued Gaussian random variables, then, since the Gaussian measure on R^{α} is absolutely continuous with respect to Lebesgue measure on R^{α}, we have that $f_n(\phi) \to I_A(\phi)$ w.p. 1 (ϕ measure) as $n \to \infty$.

Let us now fix δ, Δ and approximate the stopping time $\tau(\delta, \Delta)$ by a stopping time $\tau_n(\delta, \Delta)$ as follows. Let $\{f_i^n(\cdot)\}$ denote a continuous sequence of functions with values in $[0, 1]$ which converges to $I_i(\delta, \Delta; \cdot)$ almost everywhere (Lebesgue, hence Gaussian measure) as $n \to \infty$. Now, let us drop the δ, Δ indices where no ambiguity will result. Define the set

$$B_1^n = \{v \in R^{qr} : f_1^n(v) > \tfrac{1}{2}\}$$

and, recursively, for $i = 2, \ldots, Q$,

$$B_i^n = \{v \in R^{iqr} : f_i^n(v) > \tfrac{1}{2}\} - \overline{\bigcup_{l=1}^{i-1} B_l^n \times R^{qr(i-l)}}.$$

We have $I_{B_i^n}(\cdot) \to I_{A_i}(\cdot)$ almost everywhere as $n \to \infty$. The $\{B_i^n \times R^{q(Q-i)r}$, $i = 1, 2, \ldots\}$ are disjoint and open. Hence, there is at most one value of i for which $\{w(j\delta), j\delta \le i\Delta\} \in B_i^n$. Finally, let $\tau_n(\delta, \Delta)$ denote the stopping rule:

$$\tau_n(\delta, \Delta) = \begin{cases} i\Delta & \text{if } \{w(j\delta), j\delta \le i\Delta\} \in B_i^n, \\ T & \text{if not otherwise defined.} \end{cases} \tag{2.5}$$

Let $\tau_n(\delta, \Delta; w(\cdot))$ denote the value of the stopping time.

By construction, $\tau_n(\delta, \Delta) \to \tau(\delta, \Delta)$ as $n \to \infty$ (w.p. 1). Hence since $T < \infty$, for each $\varepsilon > 0$, there is an $n_\varepsilon < \infty$ such that $n \geq n_\varepsilon$ implies that

$$R(x, \tau_n(\delta, \Delta)) \leq R(x, \tau(\delta, \Delta)) + \varepsilon. \tag{2.6}$$

Furthermore, as $\delta, \Delta \to 0$,

$$R(x, \tau(\delta, \Delta)) \to R(x, \tau). \tag{2.7}$$

PART 2 Now, we must find a stopping rule, for the $\{\xi^h(\cdot)\}$ process, which "corresponds" to $\tau_n(\delta, \Delta)$. Let us denote this stopping time by $\tau_n^h(\delta, \Delta)$ and define it by [see (2.5)]

$$\tau_n^h(\delta, \Delta) = \begin{cases} i\Delta & \text{if } \{W^h(j\delta),\, j\delta \leq i\Delta\} \in B_i^n \\ T & \text{if not otherwise defined.} \end{cases} \tag{2.8}$$

Let $\tau_n^h(\delta, \Delta; W^h(\cdot))$ denote the value of the stopping time. This rule may yield stopping times which take values outside of the set $\{t_i^h, i = 0, 1, \ldots\}$, but that is unimportant. By the optimality of ρ_h,

$$R^h(x, \tau_n^h(\delta, \Delta)) \geq R^h(x, \rho_h) \to R(x, \rho). \tag{2.9}$$

Fix $x = \xi_0^h = X(0)$ and let h index a weakly convergent subsequence of $\{\xi^h(\cdot), B^h(\cdot), W^h(\cdot), \rho_h\}$, and suppose that the Skorokhod imbedding is used. Since the boundaries of the sets B_i^n have zero probability [$W(\cdot)$ measure], the weak convergence yields that $\tau_n^h(\delta, \Delta; W^h(\cdot)) \to \tau_n(\delta, \Delta; W(\cdot))$ w.p. 1, as $h \to 0$, where $W(\cdot)$ denotes the weak limit of $\{W^h(\cdot)\}$. Also, $\xi^h(\cdot) \to \xi(\cdot)$ uniformly on $[0, T]$ w.p. 1, and by virtue of the Lipschitz condition [which allows us to write $\xi(\cdot)$ and $X(\cdot)$ as explicit $W(\cdot)$ and $w(\cdot)$ functions, resp.], the laws of $(\xi(\cdot), \tau_n(\delta, \Delta; W(\cdot)))$ and of $(X(\cdot), \tau_n(\delta, \Delta; w(\cdot)))$ are the same, if $X(0) = x$. Hence,

$$R^h(x, \tau_n^h(\delta, \Delta)) \to R(x, \tau_n(\delta, \Delta)) \tag{2.10}$$

as $h \to 0$. Equations (2.6), (2.7), (2.9), and (2.10) imply that $R(x, \rho) \leq R(x, \tau)$ for all $\tau \in \bar{\mathscr{T}}_{\mathrm{W}}$. Q.E.D.

Theorem 8.2.5 *Let $V(\cdot)$ be continuous, and suppose that* A6.1.1 *holds. Fix $x = X(0)$. Let ρ denote the limit of $\{\rho_h\}$ corresponding to a convergent subsequence of $\{\xi^h(\cdot), B^h(\cdot), \rho_h\}$. Then*† *$\rho \in \bar{\mathscr{T}}_{\mathrm{R}}^{+}$. The value of $R(x, \rho)$ does not depend on the subsequence and $R(x, \rho) = V(x)$.*

REMARK In the proof, we do not assume that $X(\cdot)$ can be constructed from $w(\cdot)$. The approach is different from that in the previous theorem.

PROOF Since A6.1.1 holds, we can suppose that there is an Ω', σ-algebras $\mathscr{B}_t'$, $\mathscr{B}'$ and measures P_y' such that $\{\Omega', \mathscr{B}', \mathscr{B}_t', X(\cdot), P_y', y \in R^r\}$ is a strong

† In the sense of the remark under Theorem 8.2.3.

Markov and Feller process, where under P'_y, $X(\cdot)$ has the probability law of the diffusion (1.5.1) with initial condition y. The proof will be divided into several parts. First, we enlarge the above probability space to add ρ, and show that the process is still a strong Markov and Feller process. Then we show that ρ is as "good" as any ε-optimal stopping time for $X(\cdot)$. We suppose that the Ω', $\mathscr{B}'$, $\mathscr{B}'_t$, and P'_y are the same as those used in Theorem 8.2.3.

PART 1 Let h index a convergent subsequence of $\{\xi^h(\cdot), B^h(\cdot), \rho_h\}$ with limit $(\xi(\cdot), B(\cdot), \rho)$. We use the construction $B(t) = \int_0^t \sigma(\xi(s))\, dW(s)$ of Theorem 6.3.2, where $W(t)$ is given directly as a functional of $\xi(s)$, $s \le t$ and $\psi(s)$, $s \le t$, where $\psi(\cdot)$ is a Wiener process which is independent of $\xi(\cdot)$. Define $\rho_t \equiv I_{\{\rho \le t\}}$. We will first show that (for any Borel set Γ, and $t \ge 0$, $s \ge 0$, and $x \in R^r$)

$$P_x\{\xi(t+s) \in \Gamma \mid \xi(u), \rho_u, u \le t\} = P_x\{\xi(t+s) \in \Gamma \mid \xi(u), u \le t\} \quad \text{w.p. 1.} \tag{2.11}$$

We will actually prove (2.12a) [the proof of (2.12b) is similar to that of (2.12a)]

$$\begin{aligned} E_x[\xi(t+\Delta) - \xi(t) \mid \xi(u), \rho_u, u \le t] &= f(\xi(t))\Delta + o(\Delta) \\ &= E_x[\xi(t+\Delta) - \xi(t) \mid \xi(t)] + o(\Delta), \end{aligned} \tag{2.12a}$$

$$\begin{aligned} \operatorname{cov}_x[\xi(t+\Delta) - \xi(t) \mid \xi(u), \rho_u, u \le t] &= 2a(\xi(t))\Delta + o(\Delta) \\ &= \operatorname{cov}_x[\xi(t+\Delta) - \xi(t) \mid \xi(t)] + o(\Delta), \end{aligned} \tag{2.12b}$$

where all $o(\cdot)$ terms are uniform in t. First, we will show that (2.12) implies (2.11). Let t_i, $i = 1, \ldots, n$, be real numbers $\le t$. Let $F(\cdot)$ be a bounded infinitely differentiable function from R^r to R and $G_{1n}(\cdot)$, $G_{2n}(\cdot)$ be real-valued bounded continuous functions on R^{nr} and R^n, resp. Then (2.12a) and (2.12b) imply that

$$\begin{aligned} &E_x[F(\xi(t+s)) - F(\xi(t))]G_{1n}(\xi(t_1), \ldots, \xi(t_n))G_{2n}(\rho_{t_1}, \ldots, \rho_{t_n}) \\ &= E_x G_{1n}(\xi(t_1), \ldots, \xi(t_n))G_{2n}(\rho_{t_1}, \ldots, \rho_{t_n}) \\ &\quad \times E_x\left[\int_t^{t+s} \mathscr{L}F(\xi(u))\, du \mid \xi(u), u \le t\right]. \end{aligned} \tag{2.13}$$

Equation (2.13) and the arbitrariness of n, $t_1, \ldots, t_n$, $F(\cdot)$ and $G_{1n}(\cdot)$, $G_{2n}(\cdot)$ imply that

$$E_x[F(\xi(t+s)) \mid \xi(u), \rho_u, u \le t] = E_x[F(\xi(t+s)) \mid \xi(u), u \le t]$$

w.p. 1 for all bounded infinitely differentiable $F(\cdot)$, hence for all Borel $F(\cdot)$, from which (2.11) follows for each Borel set Γ.

Although we will not explicitly use it, we note that Eq. (2.11) implies that $\{\xi(t+s), s \geq 0\}$ and $\{\rho_u, u \leq t\}$ are conditionally independent, given $\xi(u), u \leq t$. Thus (Loeve [L3, p. 351]) for all Borel sets $B_1 \in R^1$ and $B_2 \in R^r$, $\sigma \leq t$, and $s > 0$, we have w.p. 1 $(T \geq t)$

$$P_x\{\rho_\sigma \in B_1 \,|\, \xi(s), s \leq T\} = P_x\{\rho_\sigma \in B_1 \,|\, \xi(u), u \leq t\},$$

$$P_x\{\rho_\sigma \in B_1, \xi(t+s) \in B_2 \,|\, \xi(u), u \leq t\} = P_x\{\rho_\sigma \in B_1 \,|\, \xi(u), u \leq t\} \cdot P_x\{\xi(t+s) \in B_2 \,|\, \xi(u), u \leq t\}.$$

Next, we will prove (2.12a). In (2.13), let $s = \Delta$ and suppose that $P_x\{\rho = t_i\} = 0$, $i = 1, \ldots, n$, and that $F(\xi) = \xi$, and note that (where $\rho^h_{ti} = I_{\{\rho_h \leq ti\}}$)

$$E_x G_{1n}(\xi^h(t_1), \ldots, \xi^h(t_n)) G_{2n}(\rho^h_{t_1}, \ldots, \rho^h_{t_n})[\xi^h(t+\Delta) - \xi^h(t)] \tag{2.14}$$

converges to the left-hand side of (2.13) by weak convergence and uniform (in h) integrability of $\{\xi^h(t+\Delta) - \xi^h(t)\}$. Also, by using the weak convergence of $\xi^h(\cdot)$ to $\xi(\cdot)$ and the continuity of the limit $\xi(\cdot)$, we can show that the value of (2.14) changes by at most $o(\Delta)$ (uniformly in h, t) if $[\xi^h(t+\Delta) - \xi^h(t)]$ is replaced by $f(\xi^h(t))\Delta$. These arguments yield

$$0 = E_x G_{1n}(\xi(t_1), \ldots, \xi(t_n)) G_{2n}(\rho_{t_1}, \ldots, \rho_{t_n})[\xi(t+\Delta) - \xi(t) - f(\xi(t))\Delta + o(\Delta)], \quad \text{small } h, \text{ and } t \text{ in a bounded interval.}$$

This last equation, the arbitrariness of n, $G_{1n}(\cdot)$, $G_{2n}(\cdot)$, and $t_1, \ldots, t_n$ (excluding a countable set) implies (2.12a).

PART 2 Let $\mathscr{B}'$, $\mathscr{B}'_t$ and $\mathscr{B}[0, t]$ be defined as in Theorem 8.2.3. Define $\Omega = \Omega' \times R^+$, $\mathscr{B} = \mathscr{B}' \times \mathscr{B}[0, \infty)$ and $\mathscr{B}_t = \mathscr{B}'_t \times (\mathscr{B}[0, t] \cup [t, \infty))$.

For each $y \in R^r$, select a convergent subsequence of $\{\xi^h(\cdot), \rho_h, \xi^h(0) = y\}$ with limit $(\xi(\cdot), \rho, \xi(0) = y)$, and let P_y denote the measure that the limit induces on $(\Omega, \mathscr{B})$. The corresponding† processes on $(\Omega, \mathscr{B})$ will be denoted by $\bar{X}(\cdot)$, $\bar{\rho}$. Under P_x, the pair $\bar{X}(\cdot)$, $\bar{\rho}$ has the same law as does $(\xi(\cdot), \rho)$, when $\xi(0) = x$. By uniqueness A6.1.1, the measure P_y reduces to P'_y on $\mathscr{B}' \times [0, \infty)$; that is, if $B \in \mathscr{B}'$, then $P_y\{B \times [0, \infty)\} = P'_y\{B\}$. The last two sentences and the conditional independence (2.11) imply that $(\Omega, \mathscr{B}, \mathscr{B}_t, P_y, y \in R^r, \bar{X}(\cdot))$ is a strong Markov diffusion and a Feller process. [Obviously, the laws of $\bar{X}(\cdot)$ and $X(\cdot)$ are the same (under each P_y and P'_y, resp.). It is only necessary to check the Markov property $P_y\{\bar{X}(t+s) \in \Gamma \,|\, \mathscr{B}_t\} = P(\bar{X}(t), s, \Gamma)$ w.p. 1.] Also, $\bar{\rho} \in \mathscr{T}^+_{\mathbf{R}}(x)$.

† That is, we identify the process $\bar{X}(\cdot)$ and random variable ρ with the two coordinates of ω.

To prove the theorem, we only need to show that for each $\varepsilon > 0$,

$$R(x, \rho) \le R(x, \tau(\varepsilon)), \tag{2.15}$$

where $\tau(\varepsilon)$ is the ε-optimal stopping time defined by

$$\tau(\varepsilon) = \min\{t : \bar{X}(t) \in B_\varepsilon\},$$

where

$$B_\varepsilon = \{x : V(x) \ge b(x) - \varepsilon\}.$$

This will be shown next. Define $\tau'(\varepsilon) = \min\{t : \bar{X}(t) \in B_\varepsilon - \partial B_\varepsilon\}$.

NOTE It is always possible to define $\{\Omega, \mathscr{B}_t, \mathscr{B}, P_y, y \in R^r\}$ as we have done, whether or not the resulting $\bar{X}(\cdot)$ process is Markov on that space, and by construction, $\bar{\rho}$ will always be a stopping time relative to $\{\mathscr{B}_t\}$ and $R(x, \rho) = R(x, \bar{\rho})$. However, unless $\bar{X}(\cdot)$ is Markov on that space [or at least unless the conditional independence (2.11) holds], the basic stopping problem is altered and, in fact, $R(x, \rho) < V(x)$ is possible. See the example in the remark after Corollary 8.2.2.

PART 3 Fix x and let $\rho_h(\varepsilon)$ denote the stopping time

$$\rho_h(\varepsilon) = \min\{t : \xi^h(t) \in B_\varepsilon\}.$$

Assume that ∂B_ε satisfies the regularity condition $P_x\{\tau'(\varepsilon) = \tau(\varepsilon)\} = 1$. Otherwise, we can perturb B_ε slightly to get a set with that property without invalidating the following argument. The fact that

$$\inf_y P_y\{\tau(\varepsilon) \le T\} \ge c > 0$$

for a real T, c, implies (see Chapter 5) that the $\{\rho_h(\varepsilon)\}$ are uniformly (in h) integrable. Let h index a convergent subsequence of $\{\xi^h(\cdot), B^h(\cdot), \rho_h(\varepsilon), \rho_h\}$. By the weak convergence and Skorokhod imbedding, $\xi^h(\cdot) \to \xi(\cdot)$ w.p. 1 uniformly on finite intervals and (using the regularity assumption) $\rho_h(\varepsilon) \to \rho(\varepsilon)$ w.p. 1, where $\rho(\varepsilon) = \inf\{t : \xi(t) \in B_\varepsilon\} = \inf\{t : \xi(t) \in B_\varepsilon - \partial B_\varepsilon\}$ w.p. 1. The pairs $(\bar{X}(\cdot), \tau(\varepsilon))$ and $(\xi(\cdot), \rho(\varepsilon))$ have the same probability law [under P_x and with $\xi(0) = x$, resp.]. By the optimality of ρ_h and by weak convergence and uniform integrability,

$$R^h(x, \rho_h) \le R^h(x, \rho_h(\varepsilon)) \to R(x, \rho(\varepsilon)) = R(x, \tau(\varepsilon)),$$

$$R^h(x, \rho_h) \to R(x, \rho),$$

which proves (2.15). Q.E.D.

It is not necessary that $V(\cdot)$ be continuous, and Theorem 8.2.5 holds without that assumption. We made the assumption to simplify the details. Continuity of $V(\cdot)$ was used only to assure that $B_\varepsilon^0 \equiv B_\varepsilon - \partial B_\varepsilon \supset \bar{B}$ for $\varepsilon > 0$. We needed this fact in order to assure that $\xi^h(\cdot)$ entered B_ε^0 for small h. This latter fact allowed us to compare the costs for the limit $\rho(\varepsilon) = \lim_{h\to 0} \rho_h(\varepsilon)$ and $\tau(\varepsilon)$, and ultimately, to get the optimality of ρ. We can approximate $V(\cdot)$ by a continuous function, even in the absence of the Lipschitz condition, and accomplish the same comparison.

This can be done as follows. Let $X(\cdot)$ denote the process given by (1.5.1). Fix Δ and n and define $V_\Delta^0(x) = b(x)$ and for $0 < i \leq n$,

$$V_\Delta^i(x) = \min\left\{b(x), E_x V_\Delta^{i-1}(X(\Delta)) + E_x \int_0^\Delta k(X(s))\, ds\right\}. \tag{2.16}$$

Since $X(\cdot)$ is a Feller process, each $V_\Delta^i(\cdot)$ is a continuous function. Let $\bar{\mathcal{T}}_{\mathrm{PM}}(n, \Delta)$ denote the set of stopping times which take values $\{0, \Delta, \ldots, n\Delta\}$ and where whether or not a stopping time is $\leq i\Delta$ can depend only on $\{X(0), \ldots, X(i\Delta)\}$. $V_\Delta^i(x)$ is the minimal cost for the stopping problem when the stopping time is allowed to vary only over $\bar{\mathcal{T}}_{\mathrm{PM}}(i, \Delta)$. Also, $V_\Delta^i(x) \geq V(x)$ and $V_\Delta^i(x) \to V(x)$ as $i \to \infty$, $\Delta \to 0$ (Shiryaev [S1]). Define the sets

$$B(i, \Delta) = \{x : V_\Delta^i(x) = b(x)\},$$

$$B_\varepsilon(i, \Delta) = \{x : V_\Delta^i(x) \geq b(x) - \varepsilon\}.$$

Within the class $\bar{\mathcal{T}}_{\mathrm{PM}}(n, \Delta)$, the optimal rule is to stop at the time $i\Delta$, $n \geq i$, if

$$i\Delta = \min\{j\Delta : X(j\Delta) \in B(n - j, \Delta)\}.$$

An ε-optimal policy [stopping time denoted by $\tau(\varepsilon, n, \Delta)$], within the class $\bar{\mathcal{T}}_{\mathrm{PM}}(n, \Delta)$ of stop rules is given by: stop at $i\Delta$ if

$$i\Delta = \min\{j\Delta : X(j\Delta) \in B_\varepsilon(n - j, \Delta)\}.$$

Define $\rho_h(\varepsilon, n, \Delta)$ as follows: $\rho_h(\varepsilon, n, \Delta) = i\Delta$ if

$$i\Delta = \min\{j\Delta : \xi^h(j\Delta) \in B_\varepsilon(n - j, \Delta)\}.$$

Let $x = \xi_0^h = X(0)$.

By the optimality of ρ_h,

$$V^h(x) = R^h(x, \rho_h) \leq R^h(x, \rho_h(\varepsilon, n, \Delta)).$$

Let h index a convergent subsequence of $\{\xi^h(\cdot), B^h(\cdot), \rho_h, \rho_h(\varepsilon, n, \Delta)\}$. Then, under a regularity assumption,† $\rho_h(\varepsilon, n, \Delta) \to \rho(\varepsilon, n, \Delta)$ as $h \to 0$, where $\rho(\varepsilon, n, \Delta) = \min\{j\Delta : \xi(j\Delta) \in B_\varepsilon(n - j, \Delta)\}$. The pairs $(\xi(\cdot), \rho(\varepsilon, n, \Delta))$ and

† As in the proof of the previous theorem, if the regularity assumption does not hold, we need only perturb the $\partial B_\varepsilon(i, \Delta)$ slightly.

$(X(\cdot), \tau(\varepsilon, n, \Delta))$ have the same probability law. We also have that

$$R^h(x, \rho_h(\varepsilon, n, \Delta)) \to R(x, \rho(\varepsilon, n, \Delta)) = R(x, \tau(\varepsilon, n, \Delta)) \qquad \text{as} \quad h \to 0,$$

and (τ = optimal for $X(\cdot)$)

$$R(x, \tau(\varepsilon, n, \Delta)) \to R(x, \tau)$$

as $\varepsilon \to 0$, $\Delta \to 0$, $n \to \infty$.

A formalization of these arguments yields

Corollary 8.2.2 *Theorem 8.2.5 is true if we drop the continuity condition on $V(\cdot)$.*

Remark on the importance of (2.11) Consider the two-stage Markov process (Y_0, Y_1), where Y_i takes values a_1 or a_2, and we are to select a stopping rule $\rho = 0$ or 1. Let $k(x) = 1$, $b(a_1) = 1$, $b(a_2) = -2$, $p_{a_i a_j} = \frac{1}{2}$, all i, j. Let $Y_0 = a_1$ and let the stopping rule be (the cheating rule) $\rho = 0$ if $Y_1 = a_1$ and $\rho = 1$ if $Y_1 = a_2$. The corresponding cost (for $Y_0 = a_1$) is 0, while the minimal cost over all Markov stopping rules is $\frac{1}{2}$.

Analogously to the construction of $(\Omega, \mathscr{B}_t, P_x)$ in the proof of the last theorem, define $\Omega = \{(a_i, a_j, 0), (a_i, a_j, 1); i, j = 1, 2\}$. Let $\mathscr{B}_0$ be the σ-algebra over the sets $\{(a_i, (a_1, a_2), 0), (a_i, (a_1, a_2), 1); i = 1, 2\}$, and let $\mathscr{B}_1$ be the σ-algebra containing "all" subsets of Ω. We can easily define measures P_x $(x = a_1, a_2)$ and random variables $\{\bar{Y}_0, \bar{Y}_1, \bar{\rho}\}$ on this new space which are consistent with the random triple $Y_0, Y_1, \rho\}$ and its measures. The random variable $\bar{\rho}$ will be a stopping time relative to $\mathscr{B}_i$, $i = 0, 1$. Also $R(a_1, \bar{\rho}) = R(a_1, \rho) = 0$ by construction. But even though $\bar{\rho}$ is a stopping time, the process $\{\bar{Y}_0, \bar{Y}_1\}$ is not Markov on the new space. In this example, $P_{a_i}\{\bar{\rho} = 1 \mid \bar{Y}_0 = a_1\} = \frac{1}{2}$. But we cannot select $\bar{\rho} = 0$ or 1 "randomly," each with probability $\frac{1}{2}$ (which would give a randomized Markov, hence a legitimate strategy), and then expect the process to evolve from $\bar{Y}_0$ to $\bar{Y}_1$ according to the law of the process. If $\bar{\rho} = 1$, then we *must* have $\bar{Y}_1 = a_2$. The property (2.11) eliminates such possibilities.

Counterexample to Convergence if $k(\cdot)$ Is Not Strictly Positive

Consider the two-dimensional deterministic example where $x = (x_1, x_2)$, $\dot{x}_1 = \dot{x}_2 = 1$, $k(x) = 0$ and

$$b(x) = \frac{\min(x_1 + \frac{1}{2}, 1)}{1 + |x_1 - x_2|}.$$

Let us restrict our attention to $A = \{x : x_1 \geq 0, x_2 \geq 0\}$. We have

$$p^h(x + e_1 h) = p^h(x + e_2 h) = \tfrac{1}{2}, \qquad x \in R_h^r \cap A.$$

The point $x = 0$ is in B, the stopping set. But if $\xi_0^h = 0$, the chain $\{\xi_n^h\}$ still wanders infinitely often to points x where $b(x)$ is arbitrarily small ($|x_1 - x_2|$ arbitrarily large), as $n \to \infty$. In fact no finite point will be in the stopping set for the chain.

The difficulty arises from the fact that the time horizon is infinite. The $\xi^h(\cdot)$ will approximate $X(\cdot)$ arbitrarily closely over any finite time interval as $h \to 0$ [if $\xi^h(0) = X(0)$]. But, for any fixed h, the approximations eventually "wander away" from the $X(\cdot)$ path as $t \to \infty$. Thus, we would expect good results only when the "essential" behavior of the system—as far as the costs are concerned—takes place on a perhaps large, but finite, time interval. When $k(x) \geq k_0 > 0$, the stopping times that we need to consider satisfy $E_x \tau \leq 2 \sup_x |b(x)|/k_0$, and we have the "finiteness" condition.

Discounted Cost

An alternative way to guarantee the "finiteness" is to introduce a discount factor. Assume that $k(\cdot)$ is bounded and continuous, but not necessarily positive. Let $\lambda(\cdot)$, λ_0, $A(t)$ and A_n^h be defined as in Section 6.5, and define

$$R(x, \tau) = E_x\left[\int_0^\tau A(t)k(X(s))\, ds + A(\tau)b(X(\tau))\right], \qquad \tau \in \mathcal{T}_R^0, \quad (2.17)$$

and

$$V(x) = \inf_{\tau \in \mathcal{T}_{PM^0}} R(x, \tau).$$

As for the nondiscounted case (Shiryaev [S1]), $V(\cdot)$ also satisfies

$$V(x) = \inf_{\tau \in \mathcal{T}_{R^0}} R(x, \tau).$$

The optimal stopping set is still $B = \{x : V(x) = b(x)\}$, and the function $V(\cdot)$ formally satisfies

$$\mathcal{L}V(x) + k(x) - \lambda(x)V(x) = 0, \qquad x \notin B,$$
$$V(x) \leq b(x), \qquad V(x) = b(x) \quad \text{on} \quad B.$$

There need not be a finite optimal time and B could be empty. For simplicity, suppose that $\Delta t^h(x) \to 0$ as $h \to 0$ uniformly in x. Applying the discretization method of Section 6.5 [which led to (6.5.3)] to the stopping problem yields the equation

$$V^h(x) = \min\{b(x), \exp -\lambda(x)\Delta t^h(x)[E_x V^h(\xi_1^h) + k(x)\Delta t^h(x)]\}. \quad (2.18)$$

For each $m \in \mathscr{T}_{\mathrm{R}}^0(h)$, the family of randomized stopping times for $\{\xi_n^h\}$, define the cost $(A_{-1}^h = 0)$

$$R^h(x, m) = E_x\left[\sum_{i=0}^{m-1} A_i^h k(\xi_i^h)\Delta t_i^h + A_{m-1}^h b(\xi_m^h)\right].$$

Then there is a unique solution to (2.18) which is

$$V^h(x) = \inf_{m \in \mathscr{T}_{\mathrm{PM}}{}^0(h)} R^h(x, m); \qquad \text{also} \qquad V^h(x) = \inf_{m \in \mathscr{T}_{\mathrm{R}}{}^0(h)} R^h(x, m),$$

where $\mathscr{T}_{\mathrm{PM}}^0(h)$ is the class of pure Markov stopping times for the chain $\{\xi_n^h\}$.

It can be proved that $V^h(x) \to V(x)$ as $h \to 0$. We omit the details, which are very similar to those for the undiscounted case, except to mention that the $\{\rho_h\}$ will always (obviously) be tight on $\bar{R}^+$, and that limits of $\{\rho_h\}$ which are infinite with a nonzero probability are acceptable; even if ρ_h or its limit is infinite for some path, the discounting guarantees that the corresponding cost for that path is finite.

Forced Stopping on Leaving a Given Set

Let A denote a given open set, and $\tau = \inf\{t : X(t) \notin A\}$, $\tau' = \inf\{t : X(t) \notin \bar{A}\}$. We can consider the optimal stopping problem, where $X(\cdot)$ is forced to stop at τ, if it has not been stopped before. For $\rho \in \overline{\mathscr{T}}_{\mathrm{R}}$, the appropriate cost is

$$R(x, \rho) = E_x\left[\int_0^{\tau \cap \rho} k(X(s))\, ds + b(X(\tau \cap \rho))\right]$$

and define

$$V(x) = \inf_{\rho \in \overline{\mathscr{T}}\mathrm{PM}} R(x, \rho).$$

The above infimum remains the same if ρ varies over $\overline{\mathscr{T}}_{\mathrm{R}}$. Formally, $V(\cdot)$ satisfies

$$\begin{aligned} \mathscr{L}V(x) + k(x) &= 0, \qquad x \in A - B, \\ B &= \{x : V(x) = b(x)\} \cap A, \\ V(x) &\le b(x) \qquad \text{in } A \\ &= b(x) \qquad \text{on } B \cup \partial A. \end{aligned} \tag{2.19}$$

Define $A_h = A \cap R_h^r$. Then by discretizing (2.19) analogously to the way that (1.1) was discretized, we get the equation

$$\begin{aligned} V^h(x) &= \min\{b(x), E_x V^h(\xi_1^h) + k(x)\,\Delta t^h(x)\}, \qquad x \in A_h, \\ V^h(x) &= b(x) \qquad \text{on} \quad R_h^r - A_h. \end{aligned} \tag{2.20}$$

We will not go into the details of the convergence proof, which involves a combination of the concepts of Chapter 6 and of the previous sections. Only a few comments will be made. For each $m \in \mathcal{T}_R^+(h)$, define the cost

$$R^h(x, m) = E_x\left[\sum_{i=0}^{(N_h \cap m)-1} k(\xi_i^h)\Delta t_i^h + b(\xi_{N_h \cap m}^h)\right],$$

where $N_h = \min\{n : \xi_n^h \notin A_h\}$. If either $k(x) \geq k_0 > 0$ or $E_x N_h < \infty$ or $k(x) \equiv 0$ and $P_x\{N_h < \infty\} = 1$, then there is a unique solution to (2.20). This unique solution is given by

$$V^h(x) = \inf_{m \in \mathcal{T}_{PM}^+(h)} R^h(x, m) = \inf_{m \in \mathcal{T}_R^+(h)} R^h(x, m). \tag{2.21}$$

Assume that $P_x\{\tau = \tau'\} = 1$ and that either $k(x) \geq k_0 > 0$ or $\sup_{y \in \bar{A}} E_y \tau < \infty$, or $k(x) \equiv 0$ and $P_x\{\tau < \infty\} = 1$. In each of these cases, the discretized problem is well formulated for small h and $V^h(x) \to V(x)$ as $h \to 0$.

8.3 Constrained Optimal Stopping Problems

Very little is presently known concerning the solution to the optimal stopping problem (or any stochastic control problem) under side constraints. However, once the problem is discretized, many types of constraints can be readily handled by the linear programming methods of Section 3.9. We continue to use the discretization $\{\xi_n^h\}$ and interpolation $\xi^h(\cdot)$ of Section 6.1 Let M_{y0} and M_{y1} denote the average number of times that $\xi_n^h = y$ and the process does not stop, and does stop, resp. Thus $M_{y1} = P\{\text{stop on reaching state } y\}$. Let $M_y = M_{y0} + M_{y1}$, and fix the initial state to be $x \in R_h^r$; thus $\mu_x = 1$, $\mu_y = 0$, $y \neq x$.

As was done in Section 3.9, we add a *fictitious absorbing state* (to the state space) *which we call the zero or rest state* $\{\mathbf{0}\}$. Define the controlled transition probabilities $\{p_{yz}(i)\}$ by $p_{yz}(0) = p^h(y, z)$, $p_{y\mathbf{0}}(1) = 1$, $p_{\mathbf{00}}(i) = 1$, $y, z \in R_h^r$. Set $b(\mathbf{0}) = k(\mathbf{0}) = 0$, $\Delta t^h(\mathbf{0}) = \infty$. Then the "dynamical" constraints are given by (3.9.2), which we rewrite here:

$$M_y = \mu_y + \sum_{z, j} p_{zy}(j) M_{zj}, \qquad M_{zj} \geq 0. \tag{3.1}$$

The cost to be minimized, namely (1.3), takes the form

$$\sum_y M_{y0} k(y) \Delta t^h(y) + \sum_y M_{y1} b(y) = R^h(x, m), \tag{3.2}$$

where the stopping time m is determined by the $\{M_{il}\}$.

If there are constraints on the path up to the stopping time, besides the constraint (3.1), then the optimal control policy for the chain may not be pure Markov, but will be randomized Markov, as discussed in Section 3.9. If l additional constraints are added to (3.1) and (3.2), then the *simplex procedure yields an optimal policy with which the control (or stopping policy) for at most l states will be randomized; i.e.*, there are at most l values of $y \in R_h^r$ for which both M_{y0} and M_{y1} are positive. If $\xi_n^h = y$, and the state occupancies are given by $\{M_{zj}\}$, then the probability of stopping at time n will be M_{y1}/M_y. Let $\mathscr{T}_{\text{RM}}(h)$ denote the class of such *randomized Markov strategies* for the chain $\{\xi_n^h\}$.

If C denotes a constraint, let $\mathscr{T}_{\text{RM}}(h, C)$, $\mathscr{T}_{\text{R}}(h, C)$, $\mathscr{T}_{\text{PM}}(h, C)$, $\bar{\mathscr{T}}_{\text{PM}}(C)$, $\bar{\mathscr{T}}_{\text{R}}(C)$, $\bar{\mathscr{T}}_{\text{W}}(C)$ denote the appropriate class of strategies for the chain or for $X(\cdot)$, for which C is satisfied. Define

$$V^h(x, C) = \inf_{m \in \mathscr{T}_{\text{RM}}(h, C)} R^h(x, m). \tag{3.3}$$

For the constraints with which we will deal, it will be true that

$$V^h(x, C) = \inf_{m \in \mathscr{T}_{\text{R}}(h, C)} R^h(x, m). \tag{3.4}$$

For notational convenience, let us suppose that h always indexes a sequence for which all the relevant random variables converge to their limits w.p. 1 (Skorokhod imbedding) in the appropriate topology.

We now give several examples of constraints which can be handled, although only Example 1 will be developed in any detail.

Example 1 Let K denote a real number and for stopping times τ, let C denote the constraint $E_x\tau \le K$. In the discrete model, the appropriate constraint is

$$\sum_y M_{y0}\Delta t^h(y) \le K. \tag{3.5}$$

Equation (3.5) assures the existence [whether or not $k(\cdot) > 0$] of an optimal policy and then the minimum of (3.2) under constraints (3.1) and (3.5) is precisely $V^h(x, C)$.

Let m_h denote the optimal constrained stopping time for $\{\xi_n^h\}$, and ρ_h the "interpolation" [the optimal time for $\xi^h(\cdot)$]. Since $E_x\rho_h \le K$ and $\rho_h \to \rho$ w.p. 1, for some random variable ρ, Fatou's lemma implies that

$$E_x\rho \le K.$$

Also, $R^h(x, \rho_h) \to R(x, \rho)$ as $h \to 0$. The probability of stopping at a point $y \ne 0$, given the entire past, is no longer 1 or 0, depending on the value of y as in the unconstrained case, but it is still a function of y since

$$P_x\{m_h = n \mid \xi_j^h, j < n, \xi_n^h = y \ne 0\} = P_x\{m_h = n \mid \xi_n^h = y \ne 0\} = M_{y1}/M_y,$$

where $\{M_{yl}\}$ solve the linear programming problem. Using this property, we can duplicate the proof of Theorem 8.2.3, and show under a Lipschitz condition on $f(\cdot)$ and $\sigma(\cdot)$ that $\rho \in \bar{\mathscr{T}}_{\mathrm{R}}^{+}(x, C)$. We can also show that

$$R(x, \rho) \leq R(x, \tau), \qquad \tau \in \bar{\mathscr{T}}_{\mathrm{W}}(C) \tag{3.6}$$

[still assuming that the space on which $X(\cdot)$ is defined also contains the Wiener process $w(\cdot)$]. The proof is roughly the same as the proof of Theorem 8.2.4 and we will only make a few comments concerning it. Take $\tau \in \bar{\mathscr{T}}_{\mathrm{W}}(C)$, and suppose that $\tau \leq T$ for an arbitrarily large but fixed T. All subsequent terms which are not defined here are defined in the proof of Theorem 8.2.4. Approximate this τ by $\tau(\delta, \Delta)$ and the latter by $\tau_n(\delta, \Delta)$. We have

$$\lim_{\delta, \Delta \to 0} E_x \tau_n^h(\delta, \Delta) \leq K.$$

Owing to the method of approximating τ, it is possible that $E_x \tau_n^h(\delta, \Delta) > K$ for each nonzero δ, Δ. [In that case $R^h(x, \rho_h)$ would not necessarily be no greater than $R^h(x, \tau_n^h(\delta, \Delta))$.] However, by a slight modification of the approximate stop rule, this problem can be avoided. For each $\varepsilon > 0$, there are $\varepsilon_0 > 0$, $\delta_0 > 0$, $\Delta_0 > 0$, and measurable sets $\hat{A}_i(\delta, \Delta)$, $i = 1, 2, \ldots, Q$, which are slight modifications of the $A_i(\delta, \Delta)$ and such that the following properties hold. The sets $\{\hat{A}_i(\delta, \Delta) \times R^{q(Q-i)r}, i = 1, \ldots, Q\}$ are disjoint. If $\hat{\tau}(\delta, \Delta)$ denotes the stopping rule obtained by replacing $A_i(\delta, \Delta)$ by $\hat{A}_i(\delta, \Delta)$ and if $\hat{\tau}_n(\delta, \Delta)$ is obtained from $\hat{\tau}(\delta, \Delta)$ exactly as $\tau_n(\delta, \Delta)$ was obtained from $\tau(\delta, \Delta)$, then

$$\begin{aligned} R(x, \hat{\tau}(\delta, \Delta)) &\leq R(x, \tau(\delta, \Delta)) + \varepsilon, \\ R(x, \hat{\tau}_n(\delta, \Delta)) &\leq R(x, \hat{\tau}(\delta, \Delta)) + \varepsilon, \end{aligned} \tag{3.7}$$

$$E_x \hat{\tau}(\delta, \Delta) \leq K - \varepsilon_0, \qquad E_x \hat{\tau}_n(\delta, \Delta) \leq K - \varepsilon_0,$$
$$\text{for large} \quad n, \ \delta < \delta_0, \ \Delta < \Delta_0. \tag{3.8}$$

Thus it is enough to prove (3.6) if $\hat{\tau}_n(\delta, \Delta)$ replaces τ there and

$$E_x \hat{\tau}(\delta, \Delta) \leq K - \varepsilon_0$$

is assumed. Under this assumption, for small h, δ, Δ and large n, $\tau_n^h(\delta, \Delta)$ satisfies the constraint

$$E_x \tau_n^h(\delta, \Delta) \leq K. \tag{3.9}$$

By (3.9) and the optimality of ρ_h in $\mathscr{T}_{\mathrm{R}}(h, C)$,

$$R^h(x, \rho_h) \leq R^h(x, \tau_n^h(\delta, \Delta)),$$

from which (3.6) follows.

Not much more can be said since little is known concerning the nature of the optimal policy for $X(\cdot)$. The optimal strategy for the discretized problem involves at most one state in which the control is randomized. This suggests (but does not prove) that the optimal strategy for $X(\cdot)$ will not be random. If, in fact, there is a pure Markov optimal strategy for $X(\cdot)$ with stopping set B and ε-optimal stopping set B_ε, and if $B_\varepsilon^0 \supset \bar{B}$ for all $\varepsilon > 0$, we can proceed as in the proof of Theorem 8.2.5 and its Corollary, to show that

$$R(x, \rho) = V(x, C).$$

Example 2 Let A denote an open set and τ a stopping time. The discretization of the constraint

$$E_x \int_0^\tau I_{\{X(s) \notin \bar{A}\}}\, ds \le \varepsilon \tag{3.10}$$

is

$$\sum_{y \in R_h^r - \bar{A}} M_{y0}\, \Delta t^h(y) \le \varepsilon.$$

The limit ρ of the optimal times ρ_h for the discretized constrained problem satisfies the constraint.

Example 3 Let A and τ be as in the previous example. The handling of some constraints requires an augmentation of the state space. Consider the constraint

$$P_x\{X(s) \notin \bar{A}, \text{ some } s \le \tau\} \le \varepsilon, \qquad X(0) = x \in A. \tag{3.11}$$

Let $X_0(\cdot)$ denote the process with initial value zero and which is constant, except for a jump of unity at the first time that $X(\cdot)$ leaves $\bar{A}$. We define the new Markov process $(X_0(\cdot), X(\cdot))$, with discretization $\{\xi_{0,n}^h, \xi_n^h\}$, with the obvious transition function for the first component. Let M_{0y0}, M_{1y0}, M_{0y1}, and M_{1y1} denote the average number of times that ($\xi_{0,n}^h = 0$, $\xi_n^h = y$, do not stop at $(0, y)$), ..., average number of times that ($\xi_{0,n}^h = 1$, $\xi_n^h = y$, stop at $(1, y)$), resp. The discrete form of the constraint is

$$\sum_{y \in R_h^r} (M_{1y0} + M_{1y1}) \le \varepsilon.$$

The objective function (3.2) remains the same and the system (3.1) must be augmented to account for the new component $\{\xi_{0,n}^h\}$. The constraint (3.11) will be satisfied by the limit ρ of the optimal ρ_h. The first component $X_0(\cdot)$ is actually a jump process and, as such, is somewhat outside the main scope of the book. However, much the same techniques can be applied to the jump case. See Kushner and Yu [K12] and the forthcoming work of Kushner and DiMasi [K16].

8.4 Discretization of the Impulsive Control Problem

The treatment of the impulsive control problem of Section 4.7.5 has much in common with the treatment of the optimal stopping problem, and we will use the results of Sections 8.1 and 8.2 wherever possible. The approximating process will still be a controlled chain $\{\xi_n^h\}$ on the state space R_h^r. If there is no action at time n and $\xi_n^h = x$ then the transition probabilities $p^h(x, y)$ are used. If there is an action of value v at time n, then the transition probabilities are $p_{xy}^h(v)$ with $p_{x, x+v}^h(v) = 1$. We need to alter $\mathscr{V}$ slightly since $x + v$ is not always in R_h^r for $x \in R_h^r$ and $v \in \mathscr{V}$. A loss of the property that all $\xi_n^h \in R_h^r$ would greatly complicate the numerical computations—although it does not affect the theoretical development. We will alter $\mathscr{V}$ for each h, x as follows. To each $x \in R_h^r$ and $v \in \mathscr{V}$ for which $x + v \notin R_h^r$, associate a unique point in R_h^r by increasing each coordinate component of $x + v$ which is not equal to an integral multiple of h (positive, negative or zero), until it equals an integral multiple of h. This procedure alters the value of the action by at most h in each direction. The modified action space does not depend on x and will be referred to simply by the symbol $\mathscr{V}^h$. With no loss of generality, we can suppose that $p(x, \cdot)$ is defined on $\mathscr{V}^h$ and is $\geq p_0$ there for each $x \in R_h^r$.

Throughout the discussion on impulsive controls we assume A6.1.1 and the following A8.4.1.

A8.4.1 *$k(\cdot)$ and $p(\cdot, \cdot)$ are continuous real-valued bounded functions defined on R^r and $R^r \times \mathscr{V}$, resp. There is a real $p_0 > 0$ such that $p(x, v) \geq p_0$ for all $x \in R^r$ and $v \in \mathscr{V}$. Let λ denote a given positive number.*

For the sake of notational simplicity only, let us suppose that $\sup_x \Delta t^h(x) \to 0$ as $h \to 0$; the general case is treated in precisely the same way but will use the approximation (6.5.4), rather than (6.5.3) for (4.7.26).

We now use the notation of Section 4.7.5, and let $(\Omega, \mathscr{B}, \mathscr{B}_t, P_x, x \in R^r)$ denote the probability space on which the controlled process $X(\cdot)$ and control times and actions $\{\tau_i, u_i\}$ are defined. Let us use the finite difference approximations (6.2.1)–(6.2.3) and approximate the discounted equation (4.7.26) by the form (6.5.3). Then, using the facts that

$$V(x) \leq \min_{v \in \mathscr{V}}[p(x, v) + V(x + v)] \tag{4.1}$$

and that when there is equality in (4.1) (which equality defines the action set B), we act and use the act which minimizes in (4.1), we get the finite difference equation (see also Section 3.7)

$$\begin{aligned} V^h(x) = \min\{&\exp -\lambda\Delta t^h(x)[k(x)\Delta t^h(x) + E_x V^h(\xi_1^h)], \\ &\min_{v \in \mathscr{V}^h}[p(x, v) + \exp -\lambda\Delta t^h(x)[k(x)\Delta t^h(x) + V^h(x + v)]]\}. \end{aligned} \tag{4.2}$$

Let $\mathscr{C}_R(h)$ and $\mathscr{C}_{PM}(h)$ denote the *classes* of *randomized and pure Markov*, resp., *control laws* for $\{\xi_n^h\}$. For an impulsively controlled chain with action times and values $\{m_i, v_i\} \in \mathscr{C}_R(h)$, define the cost

$$R^h(x, \{m_i, v_i\}) = E_x\left[\sum_{i=1}^{\infty} (\exp -\lambda t_{mi}^h) p(\xi_{mi}^h, v_i) + \sum_{i=0}^{\infty} (\exp -\lambda t_{i+1}^h) k(\xi_i^h) \Delta t_i^h\right], \tag{4.3}$$

where $t_n^h = \sum_{i=0}^{n-1} \Delta t_i^h$. With the interpolated process, we may use the notation $\rho_i = t_{mi}^h$ and write

$$R^h(x, \{m_i, v_i\}) \equiv R^h(x, \{\rho_i, v_i\})$$
$$= E_x\left[\sum_{i=1}^{\infty} (\exp -\lambda\rho_i) p(\xi^h(\rho_i), v_i) + \int_0^{\infty} e^{-\lambda s} k(\xi^h(s))\, ds\right] + \varepsilon_h .$$

The term ε_h tends to zero as $h \to 0$; it is an error due to approximating the piecewise constant function which takes the value $\exp -\lambda t_{i+1}^h$ on $[t_i^h, t_{i+1}^h)$ by the exponential $e^{-\lambda s}$.

Equation (4.2) is the dynamic programming equation for the minimal value of $R^h(x, \{m_i, v_i\})$. At each time n, there is a choice of no action or of action. In the former case, the (discounted) cost of $(\exp -\lambda\Delta t^h(x))k(x)\Delta t^h(x)$ is incurred. In the latter case, if the action is $v \in \mathscr{V}^h$, then the next state is $x + v$, and an immediate cost of $p(x, v) + (\exp -\lambda\Delta t^h(x))k(x)\Delta t^h(x)$ is incurred. Note that if $v_i \neq 0$, then the actual jump occurs at time $m_i + 1$, for the chain, and at $\rho_i + \Delta t_i^h$ for $\xi^h(\cdot)$.

There is a unique solution to (4.2) and it is the right-hand side of (4.4)

$$V^h(x) = \inf_{\{mi,\, vi\} \in \mathscr{C}_{PM}(x)} R^h(x, \{m_i, v_i\}). \tag{4.4}$$

The class $\mathscr{C}_{PM}(h)$ can be replaced by $\mathscr{C}_R(h)$ in (4.4), with no change in the infimum. Let $\{m_i^h, v_i^h\}$ (or $\{\rho_i^h, v_i^h\}$ for the interpolated process) denote the optimum Markov policy for the chain $\{\xi_n^h\}$.

Some Useful Bounds

The following bounds will be derived for the continuous parameter system but they also hold and will be used for the Markov chain. Define $k_1 = \sup_x |k(x)|$. An upper bound K_1 to the optimal cost is obtained by *never acting and setting* $k(x) \equiv k_1$:

$$k_1 \int_0^{\infty} e^{-\lambda s}\, ds = k_1/\lambda \equiv K_1. \tag{4.5}$$

Suppose that there are N_T jumps on $[0, T]$ for the optimal process. Then

$$E_x p_0 N_T(\exp -\lambda T) - K_1 \leq \text{optimal cost} \leq K_1.$$

We would obtain the left-hand side if $k(x) \equiv -k_1$, all acts occurred at $t = T$, and $p(x, v) \equiv p_0$, the minimum. Thus, for the optimal policy

$$E_x N_T \leq 2K_1 \exp \lambda T/p_0. \tag{4.6}$$

In fact, (4.6) gives an upper bound for the (optimal) number of acts on any interval of width T. For each $\varepsilon > 0$, there is a real $T_\varepsilon < \infty$, such that confining all acts to $[0, T_\varepsilon]$ increases the minimum cost by no more than $\varepsilon/2$. To see this, we note that the minimal cost attainable over $[T, \infty)$ is in $[K_1 e^{-\lambda T}, -K_1 e^{-\lambda T}]$, and that not acting on $[T, \infty]$ achieves a cost in that interval.

For each $\varepsilon > 0$, there is an integer $N_\varepsilon < \infty$, such that by restricting control actions to no more than N_ε acts on the interval $[0, T_\varepsilon]$ and no acts on (T_ε, ∞), we increase the minimum cost by no more than ε. To see this, note that any set of acts on $[0, T_\varepsilon]$ can reduce the average cost by no more than $2K_1$ (over no acts) and that

$$2K_1 P_x\{\geq N \text{ acts required on } [0, T_\varepsilon] \text{ by the optimal policy}\}$$
$$\leq 4K_1^2 \exp \lambda T_\varepsilon / p_0 N.$$

Choose N so that the right-hand side is less than $\varepsilon/2$.

Tightness and Weak Convergence

Throughout the discussion on weak convergence, the initial condition $X(0) = \xi_0^h = x$ is fixed. If there is *no action*, then

$$E_x[\xi_{n+1}^h - \xi_n^h \mid \xi_n^h = y] = f(y)\Delta t^h(y),$$
$$\operatorname{cov}_x[\xi_{n+1}^h - \xi_n^h \mid \xi_n^h = y] = \Sigma_h(y)\Delta t^h(y).$$

exactly as in Chapter 6. If there is no action at time n, define β_n^h and δW_n^h as in Chapter 6. For notational convenience, we define $\beta_n^h = f(\xi_n^h) = 0$ if there is an action at time n. We can write

$$\xi_{n+1}^h = \xi_n^h + f(\xi_n^h)\Delta t_n^h + \beta_n^h + \sum_{i=1}^{\infty} v_i^h I_{\{m_i^h = n\}}. \tag{4.7}$$

Define $F^h(\cdot)$, $B^h(\cdot)$, $W^h(\cdot)$ as in Sections 6.3 and 6.6. The proofs of Theorems 6.3.1 and 6.6.1 imply that the sequence $\{F^h(\cdot), B^h(\cdot), W^h(\cdot)\}$ is tight on $D^{3r}[0, \infty)$ and that, if $F(\cdot)$, $B(\cdot)$, $W(\cdot)$ is a limit of any convergent subsequence, then $W(\cdot)$ is a Wiener process, $B(\cdot)$ is a continuous martingale, and $F(\cdot)$ is an absolutely continuous process.

Since each v_i^h takes values in a set which is at most an rh neighborhood of the compact set $\mathscr{V}$, there is some $\varepsilon > 0$ such that the sequence (in h) $\{v_i^h, i = 1, \ldots; h \to 0\}$ is tight on $(\bar{\mathscr{V}}_\varepsilon)^\infty$, where $\mathscr{V}_\varepsilon$ is an ε-neighborhood of $\mathscr{V}$. If $\{v_i\}$ is a limit of a convergent subsequence of $\{v_i^h, i = 1, \ldots; h \to 0\}$, then each v_i has support on $\mathscr{V}$. Obviously $\{\rho_i^h, i = 1, \ldots; h \to 0\}$ is tight on $(\bar{R}^+)^\infty$. Define the jump process $J^h(\cdot)$ by $J^h(t) = \sum_{t_{i+1}^h \le t} v_i^h$. Then

$$\xi^h(t) = x + J^h(t) + F^h(t) + B^h(t). \tag{4.8}$$

The sequence $\{J^h(\cdot)\}$ is not necessarily tight on $D^r[0, \infty)$. The bound (4.6) implies that there can be no accumulation points of jumps of $J^h(\cdot)$ as $h \to 0$. However, it is possible that $\rho_{n+1}^h - \rho_n^h \to 0$ (use Skorokhod imbedding) on a set A_n of probability > 0 for one or more values of n for which $\{\rho_n^h\}$ converges to a finite ρ_n on A_n. This implies that the limit $J(\cdot)$ (if it exists) would have *multiple simultaneous jumps*, at some random time [and so would the limit of $\xi^h(\cdot)$]. Actually, this possibility causes no real difficulty. It only complicates the notation a bit. First, we will proceed under the assumption that the limits of $\{\rho_n^h\}$ for different values of n do not coincide (except perhaps at infinity), and then we will describe the necessary modifications, if some limits do coincide w.p. > 0. Thus, by our temporary assumption, the bound (4.6) implies that $\{J^h(\cdot)\}$ is tight† on $D^r[0, \infty)$.

Since each term on the right-hand side of (4.8) is an element of a tight sequence, $\{\xi^h(\cdot)\}$ is tight on $D^r[0, \infty)$. Henceforth, let *h index a convergent subsequence of* $\{F^h(\cdot), B^h(\cdot), W^h(\cdot), \{\rho_i^h, v_i^h\}\}$ with limit $(F(\cdot), B(\cdot), W(\cdot), \{\rho_i, v_i\})$. Then $J^h(\cdot) \to J(\cdot)$, a pure jump process with (distinct) jump times $\{\rho_i\}$ and jump values $\{v_i\}$. The tightness of $\{J^h(\cdot)\}$ guarantees that there is a separable process $J(\cdot)$ with only discontinuities of the first kind and whose multivariate distributions (except at times at which $P\{\text{jump}\} > 0$) are the limits of those of $\{J^h(\cdot)\}$, but it does not guarantee that $J(\cdot)$ is right continuous (Billingsley [B3, p. 136]). However, we can assume that both $J(\cdot)$ and $\xi(\cdot)$ are right continuous w.p. 1. Thus the results of Section 6.3 imply that [even though the limit $\xi(\cdot)$ may be discontinuous here]

$$\xi(t) = x + J(t) + \int_0^t f(\xi(s))\, ds + \int_0^t \sigma(\xi(s))\, dW(s). \tag{4.9}$$

Furthermore,

$$\xi^h(\rho_i^h) = \xi_{m_i^h}^h \to \xi(\rho_i^-), \qquad \text{as} \quad h \to 0.$$

† A slight problem arises if $\rho_1^h \to 0$ (w.p. > 0) as $h \to 0$. Then, owing to the behavior of $\{J^h(\cdot)\}$ at $t = 0$, we would not have tightness; but the cost functionals converge in any case. We can define the processes on $D^r[-1, \infty)$, and hold them constant on $[-1, 0)$ if we wish to avoid the problem.

Thus,

$$R^h(x, \{\rho_i^h, v_i^h\}) \to E_x\left[\sum_{i=1}^{\infty}(\exp -\lambda\rho_i)p(\xi(\rho_i^-), v_i) + \int_0^{\infty}(\exp -\lambda s)k(\xi(s))\,ds\right] = R(x, \{\rho_i, v_i\}), \tag{4.10}$$

and†

$$\begin{aligned}\xi(t) &= \xi(\rho_i) + \int_{\rho_i}^t f(\xi(s))\,ds + \int_{\rho_i}^t \sigma(\xi(s))\,dW(s), \qquad \rho_i \le t < \rho_{i+1},\\ \xi(\rho_i) &= \xi(\rho_i^-) + v_i, \qquad \xi(0) = x.\end{aligned} \tag{4.11}$$

Let us now consider the case where there are multiple simultaneous jumps.

Let h index a convergent subsequence (*use the Skorokhod imbedding again*) of $\{F^h(\cdot), B^h(\cdot), W^h(\cdot), \{\rho_i^h, v_i^h\}\}$. By the construction (4.8), $\xi^h(t)$ converges for each t which is not a limit point of $\{\rho_i^h\}$ for some i. On each interval $[0, T]$, there are (w.p. 1) only a finite number of such "jump" points. More generally, there is a function $\xi(\cdot)$ which is continuous on each interval $\{t : \rho_i < t < \rho_{i+1}\}$, such that $\xi^h(\cdot)$ converges to $\xi(\cdot)$ on each $\{t : \rho_i < t < \rho_{i+1}\}$. Of course, if $\rho_i = \rho_{i+1}$, then the interval is empty.

We also have the following convergences. If $\rho_i < \rho_{i+1}$, then as $h \to 0$,

$$\xi^h_{m_i^h+1} = \xi^h_{m_i^h} + v_i^h = \xi^h(\rho_i^h) + v_i^h \to \xi(\rho_i^+).$$

If $\rho_{i-1} < \rho_i$, then

$$\xi^h_{m_i^h} \to \xi(\rho_i^-) \qquad \text{as} \quad h \to 0.$$

Now, let α be an integer such that

$$\rho_{i-1} < \rho_i = \rho_{i+1} = \cdots = \rho_{i+\alpha} < \rho_{i+\alpha+1}.$$

Then

$$\xi^h_{m_i^h} \to \xi(\rho_i^-), \qquad \xi^h_{m_i+1^h} \to \xi(\rho_i^-) + v_i,$$

$$\vdots$$

$$\xi^h_{m_{i+\alpha}^h} \to \xi(\rho_i^-) + v_i + \cdots + v_{i+\alpha-1},$$

$$\xi^h_{m_{i+\alpha}^h+1} \to \xi(\rho_{i+a}^+) = \xi(\rho_i^-) + v_i + \cdots + v_{i+\alpha}.$$

Also $(0 \le \beta \le \alpha)$,

$$p(\xi^h_{m_{i+\beta}^h}, v_{i+\beta}^h) \to p(\xi(\rho_i^-) + v_i + \cdots + v_{i+\beta-1}, v_{i+\beta}).$$

† Strictly speaking, $\xi(0) = x$ should be replaced by $\xi(0^-) = x$, in case there is a jump at $t = 0$. This possibility should be kept in mind, but we will ignore it in the notation henceforth.

We make the following conventions. Suppose that $\xi(\cdot)$ is right continuous. But, if there is a multiple jump (say, of multiplicity $\alpha + 1$) at some ρ_i, define (for $\alpha \geq \beta \geq 1$)

$$\xi(\rho_{i+\beta}^-) \equiv \xi(\rho_i^-) + v_i + \cdots + v_{i+\beta-1}$$

or, equivalently,

$$\xi(\rho_{i+\beta}^-) = \xi(\rho_{i+\beta-1}^-) + v_{i+\beta-1}.$$

Then (4.10) [and the obvious form of (4.11)] continue to hold.

8.5 Optimality of the Limits $\{\rho_i, v_i\}$ and $R(x, \{\rho_i, v_i\})$

We must show, as we did in Section 8.2 for the optimal stopping problem, that $\{\rho_i, v_i\}$ have certain optimality properties. We will prove the results when $f(\cdot)$ and $\sigma(\cdot)$ satisfy a Lipschitz condition. However, the techniques of Theorem 8.2.5 and its Corollary (and of Chapter 9) can be used to show that $V^h(x) \to V(x)$ in general.

Theorem 8.5.1 *Let h index a weakly convergent subsequence of* $\{F^h(\cdot), B^h(\cdot), W^h(\cdot), \{\rho_i^h, v_i^h\}\}$ *with limit* $\{F(\cdot), B(\cdot), W(\cdot), \{\rho_i, v_i\}\}$. *Then* $\{v_i I_{\{\rho_i \leq t\}}, I_{\{\rho_i \leq t\}}, i = 1, 2, \ldots\}$ *is independent of* $\{W(s) - W(t), s \geq t\}$, *for each* $t \geq 0$. *Also, for each i,* $\{v_j, \rho_j, j \leq i\}$ *is independent of* $\{W(t + \rho_i) - W(\rho_i), t \geq 0\}$.

PROOF The proof is almost identical to that of Corollary 8.2.1, and will be omitted. Q.E.D.

Theorem 8.5.2 *Let* $f(\cdot)$ *and* $\sigma(\cdot)$ *satisfy a uniform Lipschitz condition. Then* $\{\rho_i, v_i\} \in \bar{\mathscr{A}}_R(x)$.

PROOF By Theorem 8.5.1 and the Lipschitz condition, the uniqueness of the solution to (4.11) is obvious, in that if $\{\{\tilde{\rho}_i, \tilde{v}_i\}, W(\cdot)\}$ is any sequence with the same distributions as $\{\{\rho_i, v_i\}, W(\cdot)\}$, then the distributions of the corresponding sets $(\tilde{\xi}(\cdot), \{\tilde{\rho}_i, \tilde{v}_i\})$ and $(\xi(\cdot), \{\rho_i, v_i\})$, resp., are the same.

Let $(\Omega, \mathscr{B}, P)$ denote the space used in the Skorokhod imbedding, and let $\mathscr{B}_t$ denote the minimal σ-algebra which measures $\{W(s), \xi(s), s < t, I_{\{\rho_i \leq s\}}, v_i I_{\{\rho_i \leq s\}}$, all i and $s \leq t\}$. Clearly each ρ_i is a stopping time with respect to $\{\mathscr{B}_t\}$ and each v_i is $\mathscr{B}_{\tau_i}$ measurable. The uniqueness, Theorem 8.5.1, and the last two sentences imply that $\{\rho_i, v_i\} \in \bar{\mathscr{A}}_R(x)$. Q.E.D.

Theorem 8.5.3 *Under the conditions of Theorem 8.5.2,* $R(x, \{\rho_i, v_i\}) = V(x)$ *for each x. Hence* $V^h(x)$ *converges to the minimal cost as* $h \to 0$ *for each x.*

Also, $R(x, \{\rho_i, v_i\}) \leq R(x, \{\tau_i, u_i\})$ *for any* $\{\tau_i, u_i\} \in \bar{\mathscr{A}}_W(x)$, *even in the absence of a Lipschitz condition.*

PROOF The second sentence is implied by the first, since the limit $V(x)$ is independent of the particular convergent subsequence. Many of the details are similar to those of the proof of Theorem 8.2.4 and we will only outline the ideas. For each $\Delta > 0$, define the discrete parameter impulse controlled process $\{X_n^\Delta\}$ by

$$X_{n+1}^\Delta = X_n^\Delta + f(X_n^\Delta)\Delta + \sigma(X_n^\Delta)[w(n\Delta + \Delta) - w(n\Delta)],$$
$$\text{if there is no action at time } n \qquad (5.1a)$$

and

$$X_{n+1}^\Delta = \tilde{X}_n^\Delta + f(\tilde{X}_n^\Delta)\Delta + \sigma(\tilde{X}_n^\Delta)[w(n\Delta + \Delta) - w(n\Delta)],$$
$$\text{if there is an action } Q_n \text{ at time } n, \qquad (5.1b)$$

where

$$\tilde{X}_n^\Delta = X_n^\Delta + Q_n, \qquad Q_n \in \mathscr{V},$$

and the control policy with realizations $\{Q_n\}$ is in $\mathscr{C}_R$. ($\mathscr{C}_R$ = randomized control policies for the discrete parameter process; each Q_n must be $\mathscr{V}$-valued.) Let the cost corresponding to the sequence of realizations $\{Q_n\}$ be defined by

$$R^\Delta(x, \{Q_n\}) = E_x\left[\sum_{n=0}^{\infty} (\exp -\lambda n\Delta) p(X_n^\Delta, Q_n) I_{\{Q_n \neq 0\}} + \sum_{n=0}^{\infty} (\exp -\lambda n\Delta) k(X_n^\Delta)\Delta\right]. \qquad (5.2)$$

If each Q_n is a functional of $w(s)$, $s \leq n\Delta$, then we say that $\{Q_n\} \in \mathscr{C}_W$. We will abuse notation and refer to $\{Q_n\}$ also as the control policy.

It can be shown that (under the Lipschitz condition)

$$\inf_{\{Q_n\} \in \mathscr{C}_R} R^\Delta(x, \{Q_n\}) = \inf_{\{Q_n\} \in \mathscr{C}_W} R^\Delta(x, \{Q_n\}) = \inf_{\{Q_n\} \in \mathscr{C}_{PM}} R^\Delta(x, \{Q_n\}) = V^\Delta(x). \qquad (5.3)$$

Under the Lipschitz condition we have: Let $X(\cdot)$, $w(\cdot)$, $\{\tau_i, u_i\}$ satisfying (4.7.23) be defined on some probability space, where $\{\tau_i, u_i\} \in \bar{\mathscr{A}}_R(x)$. Then, for each $\varepsilon > 0$, there is a $\Delta_1 > 0$ such that, for each $\Delta < \Delta_1$, there are sequences $\{X_n^\Delta\}$ and $\{Q_n^\Delta\} \in \mathscr{C}_R$, defined on the same probability space that $X(\cdot)$, $w(\cdot)$, $\{\tau_i, u_i\}$ is defined on, and such that

$$|R(x, \{\tau_i, u_i\}) - R^\Delta(x, \{Q_i^\Delta\})| \leq \varepsilon. \qquad (5.4)$$

Now suppose that $\{X_n^\Delta, Q_n\}$, $w(\cdot)$ satisfy (5.1), with $\{Q_n\} \in \mathscr{C}_W$. For each $\varepsilon > 0$, there is a $\Delta_0 > 0$ such that $\Delta < \Delta_0$ implies that there are $X(\cdot)$, $w(\cdot)$,

$\{\tau_i, u_i\}$ satisfying (4.7.23), and defined in the same probability space that $\{X_n^\Delta, Q_n\}$, $w(\cdot)$ are defined on, $\{\tau_i, u_i\} \in \bar{\mathscr{A}}_R(x)$, and such that

$$|R(x, \{\tau_i, u_i\}) - R^\Delta(x, \{Q_i\})| \leq \varepsilon. \tag{5.5}$$

Equations (5.3)–(5.5) imply that

$$\lim_{\Delta \to 0} \inf_{\{Q_n\} \in \mathscr{C}_R} R^\Delta(x, \{Q_i\}) = V(x),$$

and that

$$\inf_{\{\tau_i, u_i\} \in \bar{\mathscr{A}}_W(x)} R(x, \{\tau_i, u_i\}) = V(x). \tag{5.6}$$

Thus if

$$R(x, \{\rho_i, v_i\}) \leq R(x, \{\tau_i, u_i\}), \qquad \text{for all } \{\tau_i, u_i\} \in \bar{\mathscr{A}}_W(x), \tag{5.7}$$

then the theorem holds. For each integer N and real $T < \infty$, define the class of strategies $\bar{\mathscr{A}}_W^{T, N}$ by

$$\bar{\mathscr{A}}_W^{T, N} = \{\{\tau_i, u_i, i \leq N\} \in \bar{\mathscr{A}}_W(x)\text{: either } \tau_i = \infty \text{ or } \tau_i \leq T\}.$$

By the bounds in Section 8.4, Eq. (5.7) holds if it holds with $\bar{\mathscr{A}}_W^{T, N}$ replacing $\bar{\mathscr{A}}_W(x)$ for each N and T.

For each $\Delta > 0$, integers N and s, real $T > 0$, and values $\bar{v}_1, \ldots, \bar{v}_s$ in $\mathscr{V}$, let $\bar{\mathscr{A}}_W^{T, N}(\Delta, \bar{v}_1, \ldots, \bar{v}_s)$ denote the subset of sequences in $\bar{\mathscr{A}}_W^{T, N}$, in which each control action can take only the values $\bar{v}_1, \ldots, \bar{v}_s$, and each action time must be an integral multiple of Δ and $N\Delta = T$. It is enough to show (5.7) for each such subset.

Let $\{\tau_i, u_i\} \in \bar{\mathscr{A}}_W^{T, N}(\Delta, \bar{v}_1, \ldots, \bar{v}_s)$. We can approximate τ_i by a $\tau_i(\delta, \Delta)$ of the type dealt with in Theorem 8.2.4, where $\tau_i(\delta, \Delta)$ depends only on $\{w(j\delta), j\delta \leq i\Delta\}$. We thus assume that each τ_i is in that class. There are measurable sets $A_{m, i}$, $m = 1, \ldots, N$, $i = 1, \ldots, s$, such that the $\{\tau_i, u_i\}$ strategy implies an action of value $\bar{v}_i$ at time $m\Delta$ if and only if

$$\{w(j\delta), j\delta \leq m\Delta\} \in A_{m, i}.$$

We now proceed as in the proof of Theorem 8.2.4 by smoothing the indicator functions $I_{A_{m,i}}(\cdot)$, obtaining a family of open sets $\{A_{m, i}^n\}$ whose boundaries, $\partial A_{m, i}^n$, have zero measure and where $I_{A_{m, i}^n} \to I_{A_{m, i}}(\cdot)$ almost everywhere, as $n \to \infty$. Next, define the approximation to $\{\tau_i, u_i\}$, which uses action $\bar{v}_i$ at time $m\Delta$ if and only if $\{w(j\delta), j\delta \leq m\} \in A_{m, i}^n$. The cost corresponding to this approximation converges to the cost for $\{\tau_i, u_i\}$ as $n \to \infty$. We then use the sets $\{A_{m, i}^n\}$ to derive a policy for the chain $\{\xi_n^h\}$, by simply replacing each $w(t)$ by $W^h(t)$ (and each $\bar{v}_i$ by $\bar{v}_i^h$, the value in $\mathscr{V}^h$ which corresponds to it) as done in Theorem 8.2.4. Finally a weak convergence argument yields the assertion. We omit the details. The last sentence of the theorem is implied by the method of proof of the second sentence.

8.6 Numerical Data for the Optimal Stopping Problem

Results of some numerical calculations are illustrated in Figs. 8.1–8.4 for the system

$$dX_1 = X_2\, dt,$$

$$dX_2 = (-X_1 - 3X_2)\, dt + \sigma\, dw,$$

$b(x) = x_1^2 + x_2^2$, $k(x) \equiv 1$, with forced stopping on first leaving the set $G = \{x : |x_1| < 3,\ |x_2| < 3\}$. Only the "left half" of the data is plotted since the results were antisymmetric about the x_2 axis.

Figure 8.1 plots the stopping sets for $h = 0.3$, 0.15, and 0.075. The curves for the latter two values are very close and from them we can obtain a "reasonable" idea of the appearance of the optimal stopping set for the diffusion. At first glance, the stopping set for $h = 0.3$ does not seem to be too close to the other stopping sets but it is better than it may appear. We can see this from Fig. 8.2, where the $\varepsilon = 0.2$ stopping set is plotted for $h = 0.15$, along with the values of $V^h(x)$ at some selected points on the boundary.

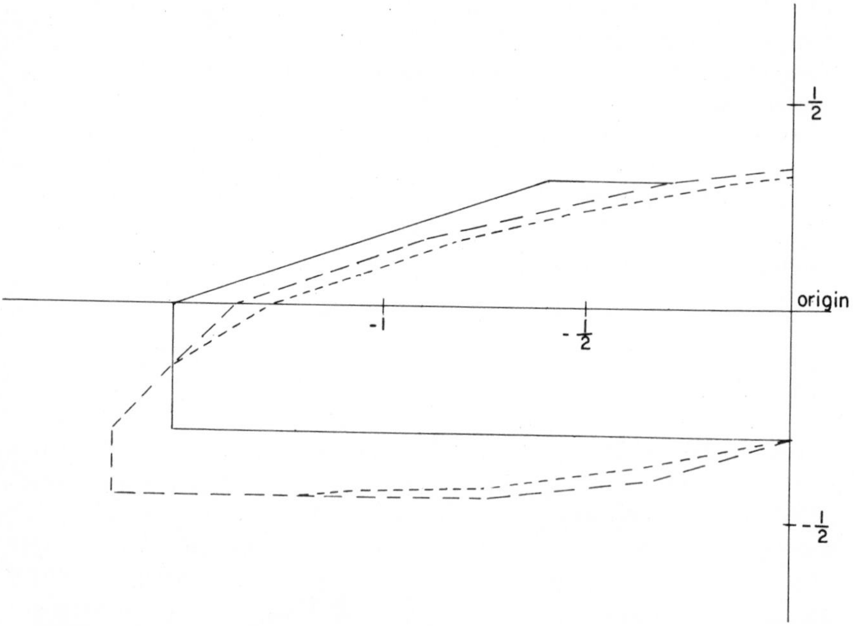

FIG. 8.1 Optimal stopping sets for the chain ξ_n^h. (——) $h = 0.3$, (– – –) $h = 0.15$, (- - -) $h = 0.075$.

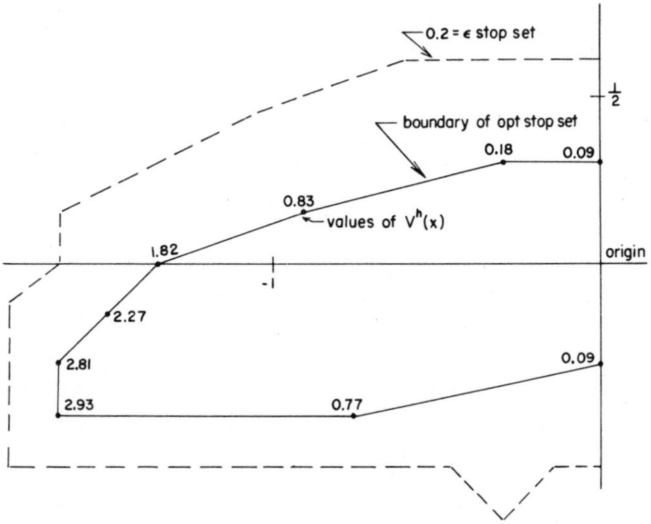

FIG. 8.2 $\varepsilon = 0.2$. Optimal stopping set, $h = 0.15$.

Generally, we need a stopping set for which the cost is "reasonably close" to the minimum cost. Here, even the set for $h = 0.3$ gives this to us, if "reasonably close" is 0.2.

Several types of iterative procedures were used to solve (1.2). Let us order the points as in Fig. 8.5. We selected that ordering, since it is preferable to order "against the flow" whenever possible for the various versions of the

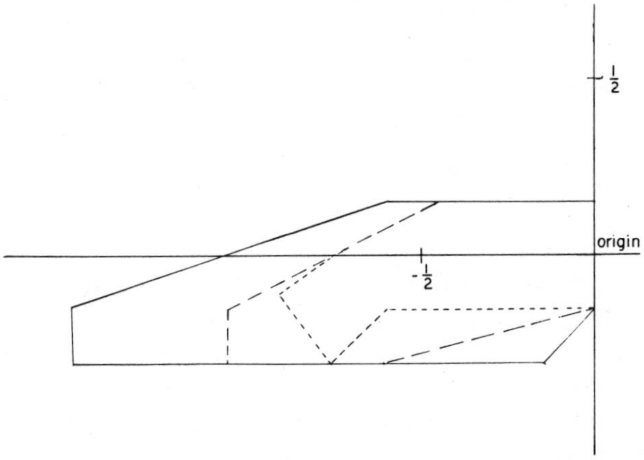

FIG. 8.3 Convergence for the Gauss–Seidel iteration, $h = 0.3$. (——) at 5 iterations, (- - -) at 10 iterations, (- - -) at 20 iterations.

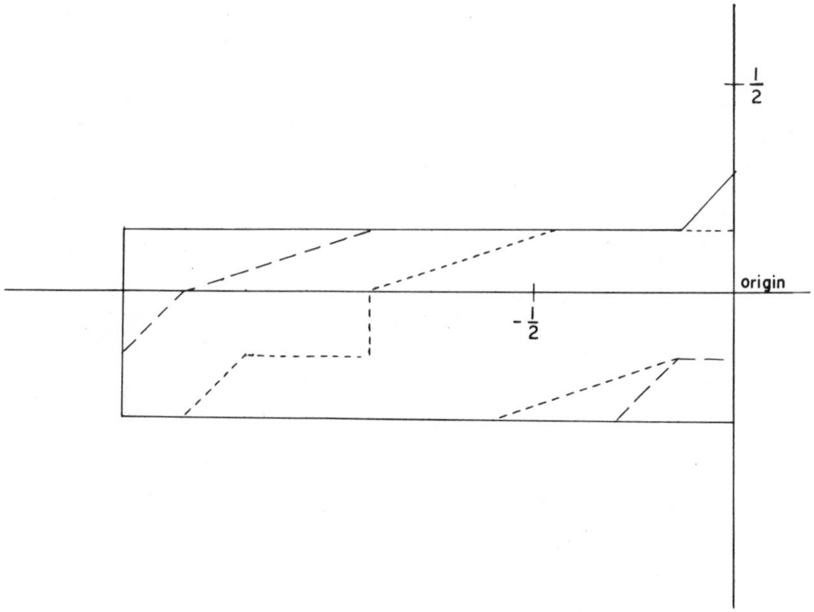

FIG. 8.4 Convergence for the Jacobi iteration, $h = 0.3$. (——) at 5 iterations, (– – –) at 10 iterations, (- - -) at 20 iterations.

Gauss–Seidel procedure. The points on the boundary are absorbing. The Jacobi iteration is defined by $V_0^h(x) \equiv b(x)$ and

$$V_{n+1}^h(i) = \min\left[b(i), \sum_{j=1}^{N2} p^h(i,j)V_n^h(j) + \Delta t^h(i)\right], \qquad i \notin \partial G.$$

The Gauss–Seidel procedure is defined by the same initial value and boundary condition and

$$V_{n+1}^h(i) = \min\left[b(i), \sum_{j=1}^{i-1} p^h(i,j)V_{n+1}^h(j) + \sum_{j=i}^{N2} p^h(i,j)V_n^h(j) + \Delta t^h(i)\right], \qquad i \notin \partial G.$$

The progress of the convergence of the stopping sets is illustrated in Fig. 8.3 for the Gauss–Seidel and in Fig. 8.4 for the Jacobi. With $h = 0.3$, convergence of the stopping sets occurred at 30 iterations for the Gauss–Seidel and at 60 for the Jacobi. Convergence of the $V_n^h(\cdot)$ (as $n \to \infty$) itself was slower, depending on the exact convergence criterion used. But in any case, the Gauss–Seidel was always preferable. An alternating direction Gauss–Seidel was somewhat (but not significantly) faster (alternating the ordering in Fig. 8.5 with the alternative ordering in that figure). The accelerated procedures were even faster.

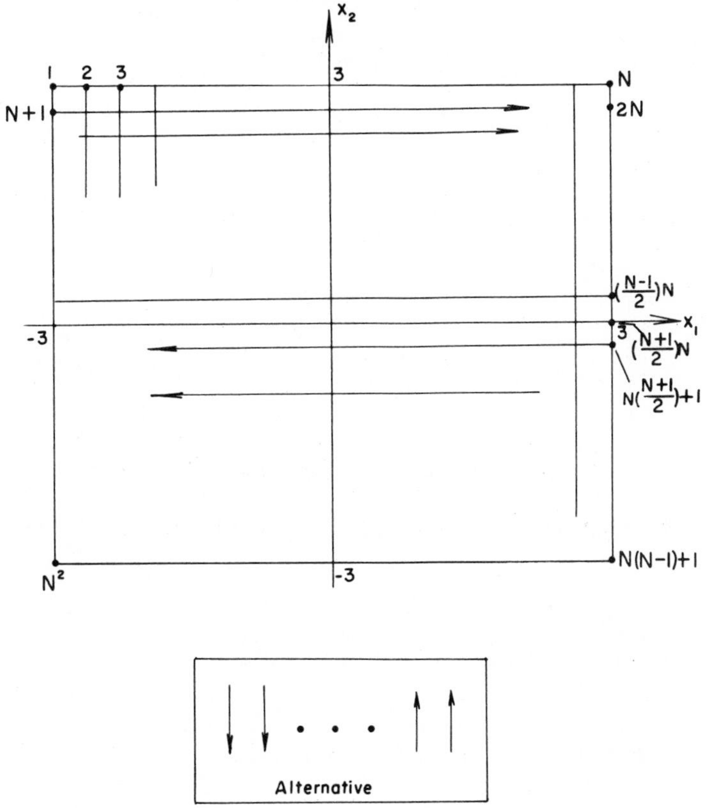

FIG. 8.5 The ordering of the grid points.

CHAPTER 9

Approximations to Optimal Controls and Nonlinear Partial Differential Equations

In Chapter 8, we dealt with the case where there was either one control action or at most a countable set of "impulsive" actions. In this chapter, we deal with the case where there may also be a control which acts continuously in time. In Section 9.1 we formulate the optimal stopping problem and the relevant discretizations when there is also a continuously acting control. Several methods for getting an approximating chain are discussed. The discretizations actually lead to optimal control problems for the chain.

In Section 9.2, it is shown, via weak convergence arguments and an implicit function theorem, that the sequence of interpolations for the optimal approximating chains actually converges to a controlled and stopped diffusion. In Section 9.3, we turn our attention to an examination of the optimality of the limiting cost. In order to show that the limit is optimal, it is necessary to compare the limiting cost to the cost corresponding to arbitrary controls and stopping times. To do this (in Section 9.3), we first give several ways of approximating the arbitrary controls by simpler ones. Then we adapt the simpler ones for use in controlling the approximating chain and compare the corresponding cost with the cost for control for the chain. The types of approximations that are developed should be of wider use in stochastic control theory in proving approximation, representation, or existence results.

Some of the approximation theorems in Section 9.3 are rather technical, but they seem to be necessary to show that "general" control policies can be approximated by somewhat simpler control policies. A basic problem causing the complexity is that we do not usually assume a Lipschitz condition. Then in order to approximate a control and still have a corresponding solution to the controlled stochastic differential equation, it is often necessary to emphasize measures rather than paths. Also, we assume that to certain types of very simple controls, there correspond solutions to the controlled Itô equation. The controls are given, basically, in terms of their joint measures with a Wiener process. Thus we frequently have to make statements such as: given a probability space on which is defined a pair $(u(\cdot), w(\cdot))$, there is another probability space on which processes $(\tilde{X}(\cdot), \tilde{u}(\cdot), \tilde{w}(\cdot))$ are defined, where the triple solves the controlled stochastic differential equation and $(\tilde{u}(\cdot), \tilde{w}(\cdot))$ has the same probability law as $(u(\cdot), w(\cdot))$. The assumption of a Lipschitz condition would considerably simplify the the development.

In Section 9.4, the results are extended to the discounted case. Sections 9.5 and 9.6 treat the case where the control stops when a target set is first reached. The discretized problem is an optimal control problem for a Markov chain, which stops on reaching a target set. As in Section 9.3, in order to show that the limiting discrete costs are optimal, we must approximate a class of controls by simpler ones and then adapt the simpler ones for use on the chain. Section 9.7 gives some results for the impulsive control problem, where there is also a continuously acting control. Some numerical results and comments appear in Section 9.8.

Owing to the relationship between the optimal control problem and the nonlinear partial differential equations of Section 4.7, the discretization methods of this chapter are also approximation methods for weak solutions of these equations.

9.1 Optimal Stopping and Continuously Acting Control: Formulation and Approximations

An interesting and very useful generalization of the stopping problem of Chapter 8 arises when we are allowed to control the process up to the stopping time. For example, we may be given a control problem with a "running" cost $k(x, u)$ and a stopping cost $b(x)$, where $b(\cdot)$ denotes a penalty for stopping at a point x which is not in a given desired stopping set $\tilde{G}$ (in $\tilde{G}$, we would set $b(x) = 0$). If $x = X(0) \notin \tilde{G}$, then this example yields a mechanism for controlling the process up to "almost" when it hits $\partial\tilde{G}$, the boundary of the desired stopping set, without explicitly introducing hard boundaries. The generalization also serves as an introduction to the problem, considered in later sections, where we control up to the actual time at which we reach a desired target set.

We will use the following assumptions.

A9.1.1 *$f(\cdot, \cdot)$ and $\sigma(\cdot)$ are bounded continuous R^r and $r \times r$ matrix-valued functions on $R^r \times \mathscr{U}$ and R^r, resp., where $\mathscr{U}$ is a compact convex set in some Euclidean space, say R^m.*

A9.1.2 *$k(\cdot, \cdot)$ and $b(\cdot)$ are bounded continuous real-valued functions on $R^r \times \mathscr{U}$ and R^r, resp. Up to Section 9.3, assume that* $\inf_{x, \alpha} k(x, \alpha) \geq k_0 > 0$.

A9.1.3 *Define $g(\cdot, \cdot) = (f(\cdot, \cdot), k(\cdot, \cdot))$. The sets $g(x, \mathscr{U}) = \{g(x, \alpha): \alpha \in \mathscr{U}\}$, $x \in R^r$, are convex and compact.*

The next assumption says only that certain very simple controls are meaningful, in that there are controlled processes [solutions to (1.1)] corresponding to such controls. The assumption is used for certain technical arguments in the approximation theorems. It seems to be rather unrestrictive. Of course, under a Lipschitz condition on $f(\cdot, \cdot)$ and $\sigma(\cdot)$, the condition always holds.

A9.1.4 *Suppose that there is a probability space on which a Wiener process $\tilde{w}(\cdot)$ and a pair $(\tilde{u}(\cdot), \tilde{\tau})$ are defined, where the latter pair is nonanticipative† with respect to $\tilde{w}(\cdot)$, where $\tilde{\tau}$ takes only finitely many values, and where $\tilde{u}(\cdot)$ takes only finitely many values in $\mathscr{U}$ and there is a real $\Delta > 0$ such that $\tilde{u}(\cdot)$ is constant on each interval $[i\Delta, i\Delta + \Delta)$. Then, for the initial condition x of interest, there is a probability space $(\Omega, \mathscr{B}, \mathscr{B}_t, P)$ supporting $X(\cdot)$, $u(\cdot)$, $w(\cdot)$, τ, where $(w(\cdot), u(\cdot), \tau)$ has the same probability law as does $(\tilde{w}(\cdot), \tilde{u}(\cdot), \tilde{\tau})$, τ is a stopping time relative to $\{\mathscr{B}_t\}$, the triple $X(\cdot)$, $u(\cdot)$, $w(\cdot)$ solves (1.1), and for each $t > 0$ the functions $X(s)$, $u(s)$, $w(s)$, $s \leq t$, are $\mathscr{B}_t$-measurable and $w(t + s) - w(t)$, $s \geq 0$, is independent of $\mathscr{B}_t$. The solution is unique in the sense that $u \in \mathscr{C}_R(x)$, $\tau \in \bar{\mathscr{T}}_R^0(u, x)$.*

In Section 8.5 on, where we refer to A9.1.4 but do not treat the stopping problem, we delete the reference to $\bar{\mathscr{T}}_R$ in A9.1.4.

† When we say that $\tilde{\tau}$ is nonanticipative, we always mean that the function $I(\cdot)$ with values $I(t) = I_{\{\tilde{\tau} \leq t\}}$ is nonanticipative.

The terminology of Section 4.7 (in particular, that of Section 4.7.3) will be used, and we will deal with the *controlled diffusion model*

$$X(t) = x + \int_0^t f(X(s), u(s))\, ds + \int_0^t \sigma(X(s))\, dw(s). \tag{1.1}$$

If $u(\cdot)$ is a control in any of the classes $\bar{\mathscr{C}}_{\mathrm{R}}(x)$, etc, it will be considered as either a random element of $L_{2,l}^m$ or as a sequence $\{u(t), t \geq 0\}$ of $\mathscr{U}$-valued random variables (which forms a separable process). As pointed out at the end of Chapter 2, the two views are mutually consistent.

Some additional terminology will now be introduced.

NOTE When considering the controlled stopping problem, the random stopping time τ should properly be considered as part of the control. So we modify the definition of *uniqueness* to read that if $(w(\cdot), u(\cdot), \tau)$have the same law as $(\tilde{w}(\cdot), \tilde{u}(\cdot), \tilde{\tau})$ and if $X(\cdot)$ and $\tilde{X}(\cdot)$ are two nonanticipative solutions to (1.1) corresponding to the two triples above, then $(X(\cdot), u(\cdot), \tau)$ and $(\tilde{X}(\cdot), \tilde{u}(\cdot), \tilde{\tau})$ have the same probability law. To be consistent with this definition, we say that $u \in \bar{\mathscr{C}}_{\mathrm{R}}(x)$, $\tau \in \bar{\mathscr{T}}_{\mathrm{R}}^0(u, x)$, if there is a probability space $(\Omega, \mathscr{B}, \mathscr{B}_t, P)$ supporting $X(\cdot)$, $u(\cdot)$, $w(\cdot)$, τ, where τ is a stopping time relative to $\{\mathscr{B}_t\}$ and $X(\cdot)$, $u(\cdot)$, $w(\cdot)$ solve (1.1); $X(s)$, $u(s)$, $w(s)$, $s \leq t$, are $\mathscr{B}_t$-measurable, and $X(\cdot)$, $u(\cdot)$, τ are nonanticipative with respect to $w(\cdot)$ and $w(t+s) - w(t)$, $s \geq 0$, is independent of $\mathscr{B}_t$; if $(\tilde{\Omega}, \tilde{\mathscr{B}}, \tilde{\mathscr{B}}_t, \tilde{P})$ is another such probability space supporting $\tilde{X}(\cdot)$, $\tilde{u}(\cdot)$, $\tilde{w}(\cdot)$, $\tilde{\tau}$, where the last three elements have the same distribution as does $(u(\cdot), w(\cdot), \tau)$, then the laws of $(X(\cdot), u(\cdot), \tau)$ and of $(\tilde{X}(\cdot), \tilde{u}(\cdot), \tilde{\tau})$ are the same. A similar definition is used when $\bar{\mathscr{C}}$ or $\bar{\mathscr{T}}$ have affixes W, G, 0, or +, etc.

TERMINOLOGY The symbol $\mathscr{U}_D$ is used to denote an arbitrary but *finite set* $\{\bar{u}_i\}$, where $\bar{u}_i \in \mathscr{U}$. The set will be fixed in each usage, but may vary from usage to usage. For each $\Delta > 0$, let $\bar{\mathscr{C}}_{\mathrm{R}}^{\Delta}(x)$ denote the subclass of controls in $\bar{\mathscr{C}}_{\mathrm{R}}(x)$ which take *values in some* $\mathscr{U}_D$ and which are *constant on each interval* $[i\Delta, i\Delta + \Delta)$, $i = 0, 1, \ldots$. Let $u(\cdot) \in \bar{\mathscr{C}}_{\mathrm{R}}^{\Delta}(x)$. Suppose that there are nonnegative numbers $\{t_j\}$ and Borel sets (of the appropriate dimension, say d_i) A_{il} such that $\{A_{il}, l = 1, 2, \ldots\}$ are disjoint for each i and

$$P\{(w(t_j), t_j \leq i\Delta) \in \partial A_{il}\} = 0, \qquad \bigcup_l A_{il} = R^{d_i},$$

and suppose that $u(i\Delta)$ is a function of $w(t_j)$, $t_j \leq i\Delta$, and is constant on each set A_{il}. Then we say that $u \in \bar{\mathscr{C}}_{\mathrm{W}}^{\Delta}(x)$.

The reason for the requirement (whenever such a condition occurs) that the probability be zero that $(w(t_j), t_j \leq i\Delta)$ be on the boundary of the decision set A_{il} is that the indicator function $I_{\tilde{A}_{il}}(x(\cdot))$ of the set $\tilde{A}_{il} = \{x(\cdot) \in C^r[0, \infty) : (x(t_j), t_j \leq i\Delta) \in A_{il}\}$ is then a continuous function on $C^r[0, \infty)$ w.p. 1 (Wiener measure). This latter fact is very useful in the weak convergence analysis since we will have $W^h(\cdot) \to W(\cdot)$, a Wiener process w.p. 1 (assuming Skorokhod imbedding) and the condition then implies that the control, as a function of $W^h(\cdot)$ instead of $w(\cdot)$, converges w.p. 1 to the same control, but where the Wiener process is the limit $W(\cdot)$.

If $\rho \in \bar{\mathcal{T}}_R^0(u, x)$ and ρ takes values $i\Delta$, $i = 0, 1, \ldots$, we say that $\rho \in \bar{\mathcal{T}}_R^\Delta(u, x)$. If $\rho \in \bar{\mathcal{T}}_R^\Delta(u, x)$ and there are real $t_j \geq 0$ and Borel-measurable A_i (of, say, dimension c_i) which satisfy

$$P\{(w(t_j), t_j \leq i\Delta) \in \partial A_i\} = 0,$$

and if $\rho = i\Delta$ if and only if $\{w(t_j), t_j \leq i\Delta\} \in A_i$, then we say that $\rho \in \bar{\mathcal{T}}_W^\Delta(u, x)$.

The Cost Functional and Finite Difference Assumptions

For each $u \in \bar{\mathscr{C}}_R(x)$ and $\rho \in \bar{\mathcal{T}}_R(u, x)$, define the cost functional

$$R(x, \rho, u) = E_x^u \int_0^\rho k(X(s), u(s))\, ds + E_x^u b(X(\rho)). \tag{1.2}$$

Equation (1.2) is a special case of the functional defined in Section 4.7.3, when we put $G = R^r$. We will take the liberty of defining $u(\cdot)$ for $t > \rho$ in any convenient manner. Let $V(\cdot)$ denote the function defined by

$$V(x) = \inf_{\rho, u} R(x, \rho, u),$$

where the infimum is taken over the defining class above (1.2) Define the stopping set $B = \{x : V(x) = b(x)\}$ and suppose that $V(\cdot)$ is "smooth." Then (see Section 4.7) $V(\cdot)$ satisfies the differential equation

$$\begin{aligned} &\inf_{\alpha \in \mathscr{U}} [\mathscr{L}^\alpha V(x) + k(x, \alpha)] = 0, \qquad x \notin B \\ &V(x) \leq b(x), \qquad V(x) = b(x) \qquad \text{on } B, \end{aligned} \tag{1.3}$$

where the operator $\mathscr{L}^\alpha$ is defined by

$$\mathscr{L}^\alpha = \sum_{i,j} a_{ij}(x) \frac{\partial^2}{\partial x_i\, \partial x_j} + \sum_i f_i(x, \alpha) \frac{\partial}{\partial x_i}$$

and is just the differential generator of the controlled process when the control takes the value α.

We will approximate (1.3) by finite differences. If the finite difference approximations to the derivatives are carefully chosen, then for each finite difference interval, the finite difference approximation to (1.3) will be the dynamic programming equation for the minimal cost of a certain controlled Markov chain. The limit (as the finite difference interval goes to zero) of any convergent subsequence of the interpolations of the chains will be a controlled and optimally stopped diffusion. The costs (finite difference solutions) will converge to $V(x)$ as the finite difference interval goes to zero. It is immaterial whether (1.3) holds in the strong sense or not. The finite difference method of getting an approximating chain is used for convenience, but any sequence of approximating chains which is consistent with the controlled diffusion can be used.

The general technique of approximation is that used in the previous chapters, and we will use (6.2.2) and (6.2.3) for the second derivative terms. There are several possibilities for the approximations of the first derivatives $V_{x_i}(\cdot)$, and we will describe three of them. The methods differ according to whether or not we are willing to let $\Delta t^h(x)$ depend on the control value. Other types of approximations are also possible. In fact, although the techniques to be discussed are useful and natural, the entire subject of the technique of the discretization needs further work.

Define

$$Q_h(x, \alpha) = 2 \sum_i a_{ii}(x) - \sum_{i, j, i \neq j} |a_{ij}(x)| + h \sum_i |f_i(x, \alpha)|,$$

$$\bar{Q}_h(x) = \max_{\alpha \in \mathscr{U}} Q_h(x, \alpha)$$

$$\Delta t^h(x) = h^2/\bar{Q}_h(x), \qquad \Delta t^h(x, \alpha) = h^2/Q_h(x, \alpha), \qquad \alpha \in \mathscr{U}.$$

Approximation 1 For each x, let I_x denote the set of indices i for which $f_i(x, \alpha)$ actually depends on the argument α. If $i \notin I_x$, then use the finite difference approximation (6.2.1) for $V_{x_i}(x)$. If $i \in I_x$, then use the central difference approximation

$$V_{x_i}(x) \to [V(x + e_i h) - V(x - e_i h)]/2h. \tag{1.4}$$

An advantage of (1.4) is that the values of y for which $V^h(y)$ actually enters into the finite difference equations do not depend on the actual value of the chosen control. But in order to guarantee that the coefficients of the $V^h(y)$ terms in the finite difference equations are ≥ 0 and hence can serve as transition probabilities for a controlled Markov chain, we need to assume that

$$a_{ii}(x) - \sum_{i \neq j, j} |a_{ij}(x)| \geq \begin{cases} h|f_i(x, \alpha)|, & \text{all } \alpha \in \mathscr{U}; \quad i \in I_x \\ 0, & i \notin I_x. \end{cases} \tag{1.5}$$

The requirement (1.5) can be rather stringent in applications. It may force us to use (numerically) a very small value of h. A technique for weakening (1.5) will be discussed below.

Define $\tilde{Q}_h(\cdot)$ and $\Delta\tilde{t}^h(\cdot)$ by

$$\tilde{Q}_h(x) = 2\sum_i a_{ii}(x) - \sum_{i\neq j,\, i,\, j} |a_{ij}(x)| + h\sum_{i\notin I_x} |f_i(x, \alpha)|,$$

$$\Delta\tilde{t}^h(x) = h^2/\tilde{Q}_h(x).$$

Define the *controlled transition function* $\tilde{p}^h(\cdot, \cdot\,|\,\cdot)$ on R_h^r, with *control parameter* $\alpha \in \mathscr{U}$, by

$$\begin{aligned}
\tilde{p}^h(x, x \pm e_i h\,|\,\alpha) &= \left[a_{ii}(x) - \sum_{j\neq i,\, j} |a_{ij}(x)| \pm h\frac{f_i(x, \alpha)}{2}\right]\Big/\tilde{Q}_h(x), \quad i \in I_x \\
&= \left[a_{ii}(x) - \sum_{j\neq i,\, j} |a_{ij}(x)| + hf_i^{\pm}(x, \alpha)\right]\Big/\tilde{Q}_h(x), \quad i \notin I_x \\
\left.\begin{aligned}\tilde{p}^h(x, x + e_i h \pm e_j h\,|\,\alpha) &= a_{ij}^{\pm}(x)/\tilde{Q}_h(x) \\ \tilde{p}^h(x, x - e_i h \pm e_j h\,|\,\alpha) &= a_{ij}^{\mp}(x)/\tilde{Q}_h(x)\end{aligned}\right\} & \quad i \neq j, \\
\tilde{p}^h(x, y\,|\,\alpha) = 0, &\qquad \text{all other } x, y \in R_h^r.
\end{aligned} \tag{1.6}$$

Let $V^h(\cdot)$ denote the solution to the finite difference equation. Substituting (6.2.2), (6.2.3) and either (1.4) or (6.2.1) (depending on whether $i \in I_x$ or $I \notin I_x$) into (1.3) and collecting terms yields the finite difference equation

$$0 = \min_{\alpha\in\mathscr{U}}\left[-V^h(x)\tilde{Q}_h(x) + \sum_y \tilde{Q}_h(x)V^h(y)\tilde{p}^h(x, y\,|\,\alpha) + h^2k(x, \alpha)\right], \qquad \text{if} \quad V^h(x) < b(x),$$

$$V^h(x) = b(x) \qquad \text{otherwise.}$$

Equivalently,

$$V^h(x) = \min\left[b(x), \min_{\alpha\in\mathscr{U}}\left\{\sum_y \tilde{p}^h(x, y\,|\,\alpha)V^h(y) + k(x, \alpha)\Delta\tilde{t}^h(x)\right\}\right]. \tag{1.7}$$

The $\{\tilde{p}^h(x, y\,|\,\alpha)\}$ are the *one-step controlled transition probabilities for a Markov chain* whose random variables we denote by $\{\tilde{\xi}_n^h\}$ and (1.7) is just the dynamic programming equation for an optimal control and stopping problem on this chain. For each α, we have

$$\begin{aligned}
E_x[\tilde{\xi}_{n+1}^h - \tilde{\xi}_n^h\,|\,\tilde{\xi}_n^h = y, \alpha \text{ used}] &= f(y, \alpha)\Delta\tilde{t}^h(y) \\
\operatorname{Cov}_x[\tilde{\xi}_{n+1}^h - \tilde{\xi}_n^h\,|\,\tilde{\xi}_n^h = y, \alpha \text{ used}] &= 2a(y)\Delta\tilde{t}^h(y) + O(h\Delta\tilde{t}^h(y)),
\end{aligned} \tag{1.8}$$

which is consistent with the infinitesimal coefficients of the controlled diffusion $X(\cdot)$.

We will now describe a slightly different method of approximation, which allows a weakening of (1.5). Let $\bar{u}$ denote a fixed element of $\mathscr{U}$, and define

$\bar{f}_i(x) = f_i(x, \bar{u})$ and $\tilde{f}_i(x, \alpha) = f_i(x, \alpha) - f_i(x, \bar{u})$. Rewrite $\mathscr{L}^\alpha$ in the form of $\tilde{\mathscr{L}}^\alpha$, where

$$\tilde{\mathscr{L}}^\alpha = \sum_{i,j} a_{ij}(x) \frac{\partial^2}{\partial x_i \, \partial x_j} + \sum_i \tilde{f}_i(x, \alpha) \frac{\partial}{\partial x_i} + \sum_i \bar{f}_i(x) \frac{\partial}{\partial x_i}.$$

Now, use the approximation (1.4) for the $\tilde{f}_i(x, \alpha)(\partial/\partial x_i)$ term and (6.2.1) for the $\bar{f}_i(x)(\partial/\partial x_i)$ term. The finite difference equation takes the form (1.7), where $\tilde{p}^h(\cdot, \cdot \,|\, \cdot)$ is replaced by a slightly different set of transition probabilities and we can replace (1.5) by the weaker requirement

$$a_{ii}(x) - \sum_{i \neq j, j} |a_{ij}(x)| \geq h |\tilde{f}_i(x, \alpha)|, \qquad \alpha \in \mathscr{U}, \quad \text{all } i, x.$$

The denominator of the equation for the new transition probabilities is (replacing $\tilde{Q}_h(x)$)

$$2 \sum_i a_{ii}(x) - \sum_{i \neq j, i, j} |a_{ij}(x)| + h \sum_i |\bar{f}_i(x)|.$$

Our limited experience with numerical calculations indicates that the alternative approximation is preferable, in that larger values of h can be used with the same degree of numerical accuracy. Note that $\tilde{f}_i(x, \alpha) = 0$ if $i \notin I_x$.

Approximation 2 A second type of approximating chain is obtained by letting the finite difference approximation for $V_{x_i}(\cdot)$ depend on the unknown control function. Let us use

$$V_{x_i}(x) \to \begin{cases} [V(x + e_i h) - V(x)]/h & \text{if } f_i(x, \alpha) \geq 0, \\ [V(x) - V(x - e_i h)]/h & \text{if } f_i(x, \alpha) < 0. \end{cases} \tag{1.9}$$

Substituting (1.9), (6.2.2), and (6.2.3) into (1.3) yields the finite difference equation

$$0 = \min_{\alpha \in \mathscr{U}} \left[-V^h(x) Q_h(x, \alpha) + \sum_y Q_h(x, \alpha) \hat{p}^h(x, y \,|\, \alpha) V^h(y) + k(x, \alpha) h^2 \right]$$
$$\text{if} \quad V^h(x) < b(x), \tag{1.10}$$
$$V^h(x) = b(x) \qquad \text{otherwise},$$

where we define $\hat{p}(\cdot, \cdot \,|\, \cdot)$ by

$$\hat{p}(x, x \pm e_i h \,|\, \alpha) = \left[a_{ii}(x) - \sum_{j \neq i, j} |a_{ij}(x)| + h f_i^{\pm}(x, \alpha) \right] / Q_h(x, \alpha), \tag{1.11a}$$

$$\left.\begin{aligned} \hat{p}(x, x + e_i h \pm e_j h \,|\, \alpha) &= a_{ij}^{\pm}(x)/Q_h(x, \alpha) \\ \hat{p}(x, x - e_i h \pm e_j h \,|\, \alpha) &= a_{ij}^{\mp}(x)/Q_h(x, \alpha) \end{aligned}\right\} \quad i \neq j,$$

$$\hat{p}(x, y \,|\, \alpha) = 0, \qquad \text{all other } x, y \in R_h^r, \quad \alpha \in \mathscr{U}. \tag{1.11b}$$

The $\hat{p}(\cdot, \cdot|\cdot)$ is a *transition function for a controlled Markov chain* whose random variables we denote by $\{\hat{\xi}_n^h\}$. Since the coefficient $Q_h(\cdot, \cdot)$ depends on the control, it is not immediately clear whether we can simply divide each term inside the brackets of (1.10) by $Q_h(x, \alpha)$ and obtain the dynamic programming equation (1.12) for the optimally controlled and stopped chain.

$$V^h(x) = \min\left[b(x), \min_{\alpha \in \mathscr{U}}\left\{\sum_y \hat{p}^h(x, y|\alpha)V^h(y) + k(x, \alpha)\Delta t^h(x, \alpha)\right\}\right]. \tag{1.12}$$

In fact, doing this yields an approximation $V^h(x)$ which will converge to $V(x)$ under the same conditions as do approximations 1 and 3. In particular, we can verify that

$$\begin{aligned} E_x[\hat{\xi}_{n+1}^h - \hat{\xi}_n^h \mid \hat{\xi}_n^h = y, \alpha \text{ used}] &= f(y, \alpha)\Delta t^h(y, \alpha), \\ \operatorname{cov}_x[\hat{\xi}_{n+1}^h - \hat{\xi}_n^h \mid \hat{\xi}_n^h = y, \alpha \text{ used}] &= \Sigma_h(y)\Delta t^h(y, \alpha), \end{aligned} \tag{1.13}$$

where $\sum_h(\cdot)$ is given by (6.2.11).

Consistency of (1.12) with a finite difference equation for (1.3) is not important. It is important that we obtain an approximating controlled process with the correct weak convergence properties and that is all that needs to be ascertained when deriving the approximation. With $\{\hat{\xi}_n^h\}$, the "current" interpolation interval depends explicitly on the "current" control (they always depend implicitly on past controls). This dependence is natural, for if a particular control action implies a faster drift (bigger $|f_i(x, \alpha)|$) or a greater diffusion, then the interpolation intervals should be smaller. The interpolation times $\Delta t^h(x, \alpha)$ have this property. A drawback to Approximation 2 is that the $\{\}$ term in (1.12) is more difficult to (numerically) minimize than the equivalent term for the other approximations.

Approximation 3 A third method, which is derived from Approximation 2 and in which the interpolation intervals do not explicitly depend on the current control, is sometimes simpler for computation. In particular, the inner minimization in (1.12) may become simpler since the replacement for the $\hat{p}(x, x \pm e_i h|\alpha)$ will have control-dependent terms only in its numerator, and the replacements for the $\hat{p}(x, x \pm e_i h \pm e_j h|\alpha)$ will not depend on α at all. To get this approximation, define $p^h(\cdot, \cdot|\cdot)$ by

$$p^h(x, x|\alpha) = [\bar{Q}_h(x) - Q_h(x, \alpha)]/\bar{Q}_h(x) \qquad \text{for } x = y. \tag{1.14}$$

For all other x, y, we define $p^h(x, y|\alpha)$ by the right-hand sides of (1.11) but with $Q_h(x, \alpha)$ replaced by $\bar{Q}_h(x)$.

The $p^h(\cdot, \cdot|\cdot)$ are a *one-step transition function of a controlled Markov chain* $\{\xi_n^h\}$, but now each state x may possibly communicate with itself. To motivate the use of (1.14), add $\bar{Q}_h(x)V^h(x)$ to each side of (1.10), divide each side by $\bar{Q}_h(x)$ and rewrite the resulting equation to yield

$$V^h(x) = \min\left[b(x), \min_{\alpha \in \mathscr{U}}\{E_x^\alpha V^h(\xi_1^h) + k(x, \alpha)\Delta t^h(x)\}\right]. \tag{1.15}$$

The symbol E_x^α denotes the expectation, using $\{p^h(x, y \mid \alpha)\}$. The inner minimization in (1.15) is usually easier to do than the inner minimization in (1.12).

In the sequel only the controlled chain $\{\xi_n^h\}$ will be dealt with. We do this purely for definiteness; the techniques are the same for all the approximations 1–3. Note that

$$\begin{aligned} E_x[\xi_{n+1}^h - \xi_n^h \mid \xi_n^h = y, \alpha \text{ used}] &= f(y, \alpha)\Delta t^h(y), \\ \operatorname{cov}_x[\xi_{n+1}^h - \xi_n^h \mid \xi_n^h = y, \alpha \text{ used}] &= \Sigma_h(y)\Delta t^h(y). \end{aligned} \tag{1.16}$$

A variant of method 3, which is analogous to the variant of method 1, is often *preferable* for numerical calculations. Define $\tilde{\mathscr{L}}^\alpha$, $\tilde{f}_i(x, \alpha)$ and $\bar{f}_i(x)$ as in method 1. Define

$$\hat{Q}_h(x) = 2\sum_i a_{ii} - \sum_{i \neq j;\, i, j} |a_{ij}| + h\sum_i |\bar{f}_i(x)| + \max_{\alpha \in \mathscr{U}} h \sum_i |\tilde{f}_i(x, \alpha)|.$$

The transition probabilities and Δt^h for the variant are obtained by simply replacing $\bar{Q}_h(x)$ by $\hat{Q}_h(x)$ and using (6.2.1) separately for $\bar{f}_i(x)\partial/\partial x_i$ and $\tilde{f}_i(x, \alpha)\partial/\partial x_i$. In particular, we have the new values

$$p^h(x, x \pm e_i h \mid \alpha) = \left[a_{ii}(x) - \sum_{j \neq i;\, j} |a_{ij}| + h\tilde{f}^\pm(x, \alpha) + h\bar{f}^\pm(x)\right] / \hat{Q}_h(x),$$

$$p^h(x, x \mid \alpha) = h\left[\max_{\alpha \in \mathscr{U}} \sum_i |\tilde{f}_i(x, \alpha)| - \sum_i |\tilde{f}_i(x, \alpha)|\right] / \hat{Q}_h(x).$$

The other $p^h(x, y \mid \alpha)$ and $\Delta t^h(x)$ change only in the denominator.

Optimal Policy for $\{\xi_n^h\}$

Let $\mathscr{C}(h)$ (with subscripts PM or R for pure Markov or randomized, resp.) denote the class of control policies for $\{\xi_n^h\}$. For each policy $\pi \in \mathscr{C}(h)$, let $\mathscr{T}(h, \pi)$ (with additional superscripts 0 or +, subscripts PM or R, and additional argument x where appropriate) denote the class of stopping times m for which $P_x^\pi\{m < \infty\} = 1$, each x. For each $\pi \in \mathscr{C}_R(h)$ and $m \in \mathscr{T}_R(h, \pi)$ and corresponding chain $\{\xi_n^h\}$, define the cost

$$R^h(x, m, \pi) = E_x^\pi\left[b(\xi_m^h) + \sum_{i=0}^{m-1} k(\xi_i^h, U_i)\Delta t_i^h\right], \tag{1.17}$$

where $\Delta t_i^h = \Delta t^h(\xi_i^h)$, and U_i is the *actual sample control action* used at time i. Define $V^h(\cdot)$ by

$$\inf_{m \in \mathcal{T}_{PM}(h,\pi);\ \pi \in \mathcal{C}_{PM}(h)} R^h(x, m, \pi) = V^h(x).$$

The infimum over the class of randomized stopping times and control policies is also $V^h(x)$. If a policy π is pure Markov and stationary, then there is a function $u(\cdot)$ such that the control action at time i is $u(\xi_i^h)$. We then write u or $u(\cdot)$ for π. *There is a stationary pure Markov policy $u^h(\cdot)$ and a pure Markov stopping time which is optimal for $\{\xi_i^h\}$.* The policy, and the minimal cost $V^h(\cdot)$, are given by the appropriate dynamic programming equation, say by (1.15) or its variant, if approximation 3 is used. Let $U_i^h = u^h(\xi_i^h)$ denote the *optimal action* at time i.

Since

$$f(y, \alpha)\Delta t^h(y) = E_x[\xi_{n+1}^h - \xi_n^h / \xi_n^h = y, U_n = \alpha],$$

we can write

$$\xi_{n+1}^h = \xi_n^h + f(\xi_n^h, U_n^h)\Delta t_n^h + \beta_n^h$$

as in earlier chapters, where

$$\beta_n^h = \xi_{n+1}^h - \xi_n^h - E_x[\xi_{n+1}^h - \xi_n^h \mid \xi_n^h, U_n^h].$$

When δW_n^h or $W^h(\cdot)$ is referred to, it is constructed from the $\{\beta_n^h\}$ and an independent Wiener process $\psi(\cdot)$ as in Chapter 6. Sometimes, we will apply a nonoptimal control to ξ_n^h. Then the δW_n^h and $W^h(\cdot)$ will be computed from the corresponding $\{\beta_n^h\}$ and $\psi(\cdot)$ but the actual process used should be clear from the context.

Interpolations and Tightness

Henceforth, the *initial state will be fixed and denoted by* x. Let $\{\xi_n^h\}$ denote the chain with the *optimal* control applied, and let m_h denote the *optimal* stopping time for the chain. Let $\rho_h = \sum_{i=0}^{m_h-1} \Delta t_i^h$ be the interpolated optimal stopping time, where $\Delta t_i^h = \Delta t^h(\xi_i^h)$, and define the interpolations $\Phi^h(\cdot) = (\xi^h(\cdot), F^h(\cdot), K^h(\cdot), B^h(\cdot), W^h(\cdot))$ exactly as was done in Chapter 6. The process $\Phi^h(\cdot)$ only needs to be defined on the time interval $[0, \rho_h)$. However, it is convenient to define it on $[0, \infty)$. To do this, we choose a $\bar{u} \in \mathcal{U}$, set $U_n^h = \bar{u}$ for $n \geq m_h$, and define the chain $\{\xi_n^h\}$ accordingly for $n > m_h$. The extension of the optimal control policy to $[0, \infty)$ will still be denoted by $u^h(\cdot)$.

The bound (8.1.5) holds for $\{\rho_h\}$ and we need only consider stopping times (for either the diffusion or the interpolated chain) which satisfy (8.1.5). Hence $\{\rho_h\}$ is tight on $[0, \infty)$.

The tightness of $\{\Phi^h(\cdot)\}$ on $D^{4r+1}[0, \infty)$ as well as the continuity w.p. 1 of the limit of any convergent subsequence follow from the results of Chapter 6. Henceforth let h index a subsequence which converges in distribution (or w.p. 1 by use of the Skorokhod imbedding) and denote the limit by $\Phi(\cdot) = (\xi(\cdot), F(\cdot), K(\cdot), B(\cdot), W(\cdot))$, and let $\rho_h \to \rho$ as $h \to 0$. The arguments of Chapters 6 and 8 yield that $W(\cdot)$ is a Wiener process, $B(\cdot)$ is a continuous martingale with representation $\int_0^t \sigma(\xi(s))\, dW(s)$,

$$\xi(t) = x + F(t) + B(t), \tag{1.18}$$

and $\xi(\cdot)$, $F(\cdot)$, $K(\cdot)$, $B(\cdot)$, and the function $I(\cdot)$ with value $I(t) = I_{\{\rho \le t\}}$ are all nonanticipative with respect to $W(\cdot)$. Also, if we write $u^h(s) = u^h(\xi^h(s))$, then $V^h(x)$ has the representation

$$V^h(x) = E_x\left[b(\xi^h(\rho_h)) + \int_0^{\rho_h} k(\xi^h(s), u^h(s))\, ds\right] \tag{1.19}$$

and

$$V^h(x) \to E_x[b(\xi(\rho)) + K(\rho)], \qquad \text{as } h \to 0. \tag{1.20}$$

So far, we have said nothing about the nature of the limits; i.e., whether $\xi(\cdot)$ is a controlled diffusion, ρ a stopping time, etc. This is the subject of the next two sections.

9.2 The Limit Is a Controlled, Stopped Diffusion

We must show that $\xi(\cdot)$ is indeed a controlled and stopped process. Since $\{\rho_h\}$ satisfy (8.1.5), so does ρ by Fatou's Lemma. Also, there is no problem in showing that ρ is a stopping time with respect to an appropriate sequence of σ-algebras. Generally, we can show only that $\xi(\cdot)$ is a controlled process, with a control $\bar{u}(\cdot, \cdot)$ which is a $\mathscr{U}$-valued function of ω, t, with values $\bar{u}(\omega, t)$, and which is nonanticipative with respect to $W(\cdot)$. However, under a Lipschitz condition, $\bar{u} \in \bar{\mathscr{C}}_R(x)$, $\rho \in \bar{\mathscr{T}}_R^+(u, x)$, and $V^h(x) \to V(x)$ as $h \to 0$. The problem is, simply, that we do not know in general whether the control $\bar{u}(\cdot)$ yields a *unique* solution, in the sense that the law of the triple $(\xi(\cdot), \bar{u}(\cdot), \rho)$ is uniquely determined by that of $(\bar{u}(\cdot), \rho, W(\cdot))$. In any case, $\bar{u}(\cdot)$ will turn out to be as good as any control which does yield a unique solution and it is optimal in that sense. The optimality is dealt with in the next section.

Several preparatory results are needed.

Theorem 9.2.1 *Under* (A9.1.1) *to* (A9.1.3), *there are measurable* ω, t *functions* $\bar{f}(\cdot, \cdot)$, $\bar{k}(\cdot, \cdot)$, *which are nonanticipative with respect to* $W(\cdot)$ *and for which*

$$(\bar{f}(\omega, t), \bar{k}(\omega, t)) \in (f(\xi(t), \mathscr{U}), k(\xi(t), \mathscr{U})), \tag{2.1}$$

$$F(t) = \int_0^t \bar{f}(\omega, s)\, ds, \qquad K(t) = \int_0^t \bar{k}(\omega, s)\, ds, \qquad \text{all } t \geq 0. \tag{2.2}$$

PROOF We suppose that the Skorokhod imbedding is used, so that all the processes are defined on the same probability space. Define $g(\cdot, \cdot) = \{f(\cdot, \cdot), k(\cdot, \cdot)\}$, $G^h(t) = \{F^h(t), K^h(t)\}$ and $G(t) = \{F(t), K(t)\}$. Since $g(\cdot, \cdot)$ is bounded, there is a constant K such that (w.p. 1) $\lim_{h \to 0} |G^h(t+s) - G^h(t)| \leq Ks$. Hence (w.p. 1) $|G(t+s) - G(t)| \leq Ks$ and $G(\cdot)$ is absolutely continuous w.p. 1. Thus there are measurable ω, t functions $\bar{g}(\cdot, \cdot) = \{\bar{f}(\cdot, \cdot), \bar{k}(\cdot, \cdot)\}$ which satisfy (2.2) (w.p. 1), and we can suppose that $|\bar{g}(\omega, t)| \leq K$ for all ω, t and that $\bar{g}(\cdot, \cdot)$ is nonanticipative with respect to $W(\cdot)$, because $G(\cdot)$ is.

The sequence $\{g^h(\cdot, \cdot)\}$ defined by $g^h(\omega, t) = g(\xi^h(\omega, t), u^h(\omega, t))$ converges to $\bar{g}(\cdot, \cdot)$ weakly in $L_1(\Omega \times [0, T])$ for each $T < \infty$. To see this, first note that (by the w.p. 1 convergence of $G^h(\cdot)$ to $G(\cdot)$) if (2.3) holds for a monotone increasing sequence of measurable (ω, t) sets A_n,

$$E \int I_{A_n} g^h(\omega, s)\, ds \to E \int I_{A_n} \bar{g}(\omega, s)\, ds \qquad (\text{as } h \to 0), \tag{2.3}$$

then it holds for the limit $A = \bigcup_n A_n$. The relation (2.3) holds for any set A_n which is of the form $A_n = \sum_i B_i \times C_i$, where B_i are disjoint intervals in $[0, T]$ and C_i are measurable ω sets. These facts imply that (2.3) holds for all measurable (ω, t) sets A_n. Hence $\bar{g}^h(\cdot, \cdot)$ converges weakly in $L_1(\Omega \times [0, T])$ as stated.

Let l denote an arbitrary vector in R^{r+1}. Then (using an argument like that used in Roxin [R1]), it follows from the L_1-weak convergence that [for almost all (ω, s)]

$$\overline{\lim_h}\, l'g(\xi(s), u^h(s)) = \overline{\lim_h}\, l'g(\xi^h(s), u^h(s)) \geq l'\bar{g}(\omega, s), \tag{2.4}$$

with the reverse inequalities holding for $\underline{\lim}$ replacing $\overline{\lim}$. The left-hand equality in (2.4) is obvious. To prove the right-hand inequality of (2.4), we first suppose that it does not hold on a measurable (ω, s) set A_l, of nonzero measure. Then Fatou's lemma yields

$$\overline{\lim_h}\, E \int_{A_l} l'g^h(\omega, s)\, ds \leq E \int_{A_l} \left(\overline{\lim_h}\, l'g^h(\omega, s) \right) ds < E \int_{A_l} l'\bar{g}(\omega, s)\, ds,$$

which contradicts the convergence in (2.3) for $A_n = A_l$. Thus the measure of A_l is zero. Now, l is arbitrary, and $g(\xi(t), \mathscr{U})$ is convex and the $\overline{\lim}_h\, l'g^h(\omega, t)$ and $\underline{\lim}_h\, l'g^h(\omega, t)$ lie in $l'g(\xi(t), \mathscr{U})$ for $(\omega, t) \notin A_l$. Define $A = \bigcup_l A_l$, where l ranges over a countable dense set of vectors in R^{r+1}. Since the measure of A is zero, the previous assertions of the paragraph imply that (2.1) holds for almost all (ω, t). We can define $\bar{g}(\cdot, \cdot)$ on the exceptional set so that (2.1) holds everywhere. Q.E.D.

We will need the following implicit function theorem of McShane and Warfield [M1]. The theorem will help answer the question: given the inclusion relation (2.1) can we actually choose a nonanticipative $\mathscr{U}$-valued measurable function $\bar{u}(\cdot, \cdot)$ for which

$$\begin{aligned} \bar{f}(\omega, t) &= f(\xi(t), \bar{u}(\omega, t)), \\ \bar{k}(\omega, t) &= k(\xi(t), \bar{u}(\omega, t)), \qquad \text{for all} \quad \omega, t. \end{aligned} \tag{2.5}$$

It is clearly possible to select a vector $u(\omega, t) \in \mathscr{U}$ for each ω, t such that (2.5) holds for $\bar{u} = u$. But the measurability is not so obvious.

Theorem 9.2.2 *Let $(M, \mathscr{M})$ be a measure space, A a separable metric space, and B a compact metric space. Let $q : M \times B \to A$ be continuous in the second argument for each value of the first, and $\mathscr{M}$ measurable in the first for each value of the second. Let $y : M \to A$ be $\mathscr{M}$-measurable with*

$$y(z) \in q(z, B) \qquad \text{all} \quad z \in M.$$

Then there is an $\mathscr{M}$-measurable function $u : M \to B$ such that

$$y(z) = q(z, u(z)) \qquad \text{all} \quad z \in M.$$

Theorems 9.2.1 and 9.2.2 combine to yield the next theorem, which states that $\xi(\cdot)$ is a controlled process of the form (1.1).

Theorem 9.2.3 *Under the conditions of Theorem 9.2.1, there is a nonanticipative $\mathscr{U}$-valued measurable function $\bar{u}(\cdot, \cdot)$ for which (2.5) holds and*

$$\lim_{h \to 0} V^h(x) \to R(x, \rho, \bar{u}) \equiv \bar{V}(x) \qquad \text{as} \quad h \to 0. \tag{2.6}$$

PROOF The relation (2.6) follows from (2.5), (1.19), and (1.20). Denote the probability space by $(\hat{\Omega}, \hat{\mathscr{B}}, \hat{P})$, and let $\hat{\mathscr{B}}_t$ be the minimal σ-algebra over which $\bar{g}(\cdot, s)$, $\xi(\cdot, s)$, $s \le t$, are measurable. Let $\mathscr{B}^0$ be the minimal sub-σ-algebra of† $\hat{\mathscr{B}} \times \mathscr{B}[0, \infty)$ over which $\bar{\xi}(\cdot, \cdot)$ and $\bar{g}(\cdot, \cdot)$ are measurable. Let $M = \hat{\Omega} \times [0, \infty)$, $\mathscr{M} = \mathscr{B}^0$, $A = R^{r+1}$, $B = \mathscr{U}$, $y(\cdot, \cdot) = \bar{g}(\cdot, \cdot)$, and $q(\cdot, \cdot) = g(\cdot, \cdot)$. Then an application of Theorem 9.2.2. yields that there is a

† $\mathscr{B}[0, \infty)$ is the Lebesgue field over $[0, \infty)$.

$\mathscr{B}^0$-measurable $\mathscr{U}$-valued function $\bar{u}(\cdot, \cdot)$ such that (2.5) holds. Since the fixed s section of $\mathscr{B}^0$ is contained in $\hat{\mathscr{B}}_t$, for each $0 \le s \le t$, any $\mathscr{B}^0$-measurable function is nonanticipative with respect to $W(\cdot)$ since any $\hat{\mathscr{B}}_t$,measurable random variable is independent of $W(t+s) - W(t)$, $s \ge 0$. Q.E.D.

9.3 Optimality of the Limit

In Theorems 9.3.1 and 9.3.2, we show that we can approximate the classes $\bar{\mathscr{T}}_R^+(u, x)$ and $\bar{\mathscr{C}}_R(x)$ by much simpler classes of stopping times and controls. Theorem 9.3.3 uses these approximations to get the optimality properties of $\bar{u}(\cdot, \cdot)$ and ρ and the minimality of $\bar{V}(x)$. The proofs of both Theorems 9.3.1 and 9.3.2 involve a number of detailed estimates and approximations but the results are useful.

Theorem 9.3.1 *Assume* A9.1.1, A9.1.2, *and* A9.1.4. *Suppose that* $u \in \bar{\mathscr{C}}_R(x)$ *and that* $\tau \in \bar{\mathscr{T}}_R(u, x)$ *and* $\tau \le T$. *For each* $\varepsilon > 0$, *there is a* $\Delta > 0$ *and a set* $\mathscr{U}_D$ *and a* $u_\Delta \in \bar{\mathscr{C}}_R^\Delta(x)$ *and a* $\tau_\Delta \in \bar{\mathscr{T}}_R^\Delta(u_\Delta, x)$ *with* $\tau_\Delta \le T$, *such that*

$$R(x, \tau_\Delta, u_\Delta) \le R(x, \tau, u) + \varepsilon. \tag{3.1}$$

The above statement holds with W *replacing* R *in* $\bar{\mathscr{C}}_R^\Delta(x)$ *and* $\bar{\mathscr{T}}_R^\Delta(u_\Delta, x)$.

REMARK Statements such as that of Theorem 9.3.1 will be made frequently in the rest of this chapter. We start with a Wiener process $w(\cdot)$ and a control $u(\cdot)$ and stopping time τ, all defined on the same probability space. We then, for example, approximate $u(\cdot)$ by some $u_\Delta(\cdot)$ and claim that $u_\Delta(\cdot) \in \bar{\mathscr{C}}_R^\Delta(x)$, without explicitly mentioning which Wiener process is involved. The notation can become awkward, because sometimes the probability space on which u_Δ is defined is the original one, and the Wiener process is $w(\cdot)$, and sometimes it is the space used in the Skorokhod imbedding, and the Wiener process may be different.

All that we are usually after is the construction of a controlled diffusion whose corresponding *cost of control* is close to the *cost* for some other process [say, for $\xi(\cdot)$, $\bar{u}(\cdot)$, ρ].

In the present theorem, $u_\Delta \in \bar{\mathscr{C}}_R^\Delta(x)$, $\tau_\Delta \in \bar{\mathscr{T}}_R^\Delta(u, x)$ means the following. Given $(u(\cdot), \tau, w(\cdot))$, we construct $u_\Delta(\cdot)$, τ_Δ on the same probability space. The values of $u_\Delta(\cdot)$, τ_Δ satisfy the restrictions put on values of members of $\bar{\mathscr{C}}_R^\Delta(x)$ and $\bar{\mathscr{T}}_R^\Delta(u_\Delta, x)$. There is some probability space, which supports a Wiener process $\tilde{w}(\cdot)$, control $\tilde{u}_\Delta(\cdot)$, stopping time $\tilde{\tau}_\Delta$, and $\tilde{X}(\cdot)$, where $\tilde{X}(\cdot)$, $\tilde{u}_\Delta(\cdot)$, $\tilde{\tau}_\Delta$ have the appropriate nonanticipativeness properties, $\tilde{u}_\Delta(\cdot)$ and $\tilde{\tau}_\Delta$ take values in the appropriate sets, $(\tilde{X}(\cdot), \tilde{u}_\Delta(\cdot), \tilde{w}(\cdot))$ solve (1.1), and $(\tilde{u}_\Delta(\cdot),$

$\tilde{\tau}_\Delta$, $\tilde{w}(\cdot)$) has the same law as ($u_\Delta(\cdot)$, τ_Δ, $w(\cdot)$). Also, there is the "usual" uniqueness. We can (and will) write $R(x, \tilde{\tau}_\Delta, \tilde{u}_\Delta)$ as $R(x, \tau_\Delta, u_\Delta)$, without ambiguity.

We do things in this fashion partly because we are concerned primarily with *costs* and not paths. Sometimes the Wiener process will not be mentioned explicitly.

PROOF By hypothesis, we can suppose that there is a Wiener process $w(\cdot)$ and a process $X(\cdot)$ such that $(X(\cdot), u(\cdot), w(\cdot))$ solve (1.1), and that $u \in \mathscr{C}_R(x)$, $\tau \in \bar{\mathscr{T}}_R(u, x)$, $\tau \le T$. Clearly, for each $\varepsilon > 0$, there is a $\Delta > 0$, such that the stopping time $\tau_\Delta = \min\{i\Delta : i\Delta \ge \tau\}$ satisfies $\tau_\Delta \le T$ and

$$R(x, \tau_\Delta, u) \le R(x, \tau, u) + \varepsilon/2.$$

Thus to prove (3.1), we can (and, henceforth, will) assume that $\tau \in \bar{\mathscr{T}}_R^\Delta(u, x)$ whatever u is.

Define a sequence $\{u_n(\cdot)\}$ as follows. If $u(\omega, \cdot)$ is continuous (for almost all ω) on $[0, \infty)$, set $u_n(\cdot) = u(\cdot)$, all n. Otherwise, select an arbitrary $\bar{u}_0 \in \mathscr{U}$, set $u(t) = \bar{u}_0$ for $t < 0$, and define

$$u_n(t) = n \int_0^\infty e^{-ns} u(t - s)\, ds = \int_0^\infty e^{-s} u(t - s/n)\, ds.$$

For each t and n, $u_n(t)$ is $\mathscr{U}$-valued since $\mathscr{U}$ is convex. For each n and $\delta > 0$, define the function $u_{n\delta}(\cdot)$ by

$$u_{n\delta}(t) = u_n(i\delta) \qquad \text{on} \quad [i\delta, i\delta + \delta).$$

Let $\mathscr{U}^0 = \{\bar{u}_1, \ldots\}$ be dense in $\mathscr{U}$ and define $\mathscr{U}^c = \{\bar{u}_1, \ldots, \bar{u}_c\}$ for each integer c. Let $\varepsilon_c > 0$ denote a real number such that† $\bigcup_{i=1}^c \bar{N}_{\varepsilon_c}(\bar{u}_i) \supset \mathscr{U}$ and $\varepsilon_c \to 0$ as $c \to \infty$. For each n, $\delta > 0$, and integer c, define the function $u_{n\delta c}(\cdot)$ by

$$u_{n\delta c}(t) = \begin{cases} \bar{u}_1 & \text{on} \quad [i\delta, i\delta + \delta) \quad \text{if} \quad |u_{n\delta}(i\delta) - \bar{u}_1| \le \varepsilon_c, \\ \bar{u}_l & \text{on} \quad [i\delta, i\delta + \delta) \quad \text{if} \quad |u_{n\delta}(i\delta) - \bar{u}_j| > \varepsilon_c, \\ & j = 1, \ldots, l-1 \quad \text{and} \quad |u_{n\delta}(i\delta) - \bar{u}_l| \le \varepsilon_c. \end{cases}$$

By A9.1.4, each $u_{n\delta c}(\cdot)$ is in‡ $\bar{\mathscr{C}}_R(x)$. There is a subsequence of $\{u_{n\delta c}(\cdot)\}$ which converges to $u(\cdot)$ for almost all (ω, t). We can select a subsequence, denoted by $\{v^p(\cdot),\ p = 1, 2, \ldots\}$, such that

$$E \int_0^p |v^p(t) - u(t)|^2\, dt \le 2^{-p}.$$

† $\bar{N}_\varepsilon(\bar{u}_i)$ denotes the closed sphere with center $\bar{u}_i$ and radius ε.

‡ See remark before the proof.

By A9.1.4, for each p, there is a probability space which supports processes $\tilde{X}^p(\cdot)$, $\tilde{v}^p(\cdot)$, $\tilde{w}^p(\cdot)$, $\tilde{\tau}^p$, where $(\tilde{v}^p(\cdot), \tilde{w}^p(\cdot), \tilde{\tau}^p)$ has the same law as $(v^p(\cdot), w(\cdot), \tau)$, and where $\tilde{X}^p(\cdot)$ is a nonanticipative solution to

$$\tilde{X}^p(t) = x + \int_0^t f(\tilde{X}^p(s), \tilde{v}^p(s))\, ds + \int_0^t \sigma(\tilde{X}^p(s))\, d\tilde{w}^p(s).$$

We can augment the probability space by adding† a separable and measurable process $\tilde{u}^p(\cdot)$, where $(\tilde{u}^p(\cdot), \tilde{w}^p(\cdot), \tilde{\tau}^p, \tilde{v}^p(\cdot))$ has the same law as $(u(\cdot), w(\cdot), \tau, v^p(\cdot))$. Then, of course, we can suppose that $\tilde{v}^p(\cdot)$ is obtained from $\tilde{u}^p(\cdot)$ in the same way that $v^p(\cdot)$ was obtained from $u(\cdot)$. Each $\tilde{u}^p(\cdot)$ has the *same measure*. Thus $\{\tilde{u}^p(\cdot)\}$ is tight on $L^m_{2,l}$. The sequence $\{\tilde{X}^p(\cdot), \tilde{w}^p(\cdot), \tilde{u}^p(\cdot), \tilde{\tau}^p\}$ is tight on $C^{2r}[0, \infty) \times L^m_{2,l} \times [0, \infty)$.

Let p index a convergent subsequence. Using the *Skorokhod imbedding* we can suppose that all the processes are defined on the same probability space, and that convergence to the limit $\tilde{X}(\cdot)$, $\tilde{w}(\cdot)$, $\tilde{u}(\cdot)$, $\tilde{\tau}$ is w.p. 1, in the appropriate topology. The triple $(\tilde{u}(\cdot), \tilde{w}(\cdot), \tilde{\tau})$ has the same law as either $(\tilde{u}^p(\cdot), \tilde{w}^p(\cdot), \tilde{\tau}^p)$ or $(u(\cdot), w(\cdot), \tau)$. Thus $\tilde{u}(\omega, t) \in \mathscr{U}$ for almost all ω, t. Furthermore,

$$E\int_0^p |\tilde{v}^p(t) - \tilde{u}^p(t)|^2\, dt \le 2^{-p}.$$

Thus $\tilde{v}^p(\omega, t) \to \tilde{u}(\omega, t)$ for almost all ω, t.

By either construction or definition, $(\tilde{v}^p(\cdot), \tilde{X}^p(\cdot), \tilde{\tau}^p)$ and $(\tilde{v}^p(\cdot), \tilde{u}^p(\cdot), \tilde{\tau}^p)$ are nonanticipative with respect to $\tilde{w}^p(\cdot)$. By weak convergence methods, such as those used in Section 8.2, it can be shown that $(\tilde{X}(\cdot), \tilde{u}(\cdot), \tilde{\tau})$ is nonanticipative with respect to $\tilde{w}(\cdot)$, and that $\tilde{X}(\cdot)$ satisfies

$$\tilde{X}(t) = x + \int_0^t f(\tilde{X}(s), \tilde{u}(s))\, ds + \int_0^t \sigma(\tilde{X}(s))\, d\tilde{w}(s).$$

Since $u \in \mathscr{C}_R(x)$, $\tau \in \mathscr{T}_R(u, x)$, and $(\tilde{u}(\cdot), \tilde{w}(\cdot), \tilde{\tau})$ has the same law as $(u(\cdot), w(\cdot), \tau)$, the triple $(X(\cdot), u(\cdot), \tau)$ has the same law as $(\tilde{X}(\cdot), \tilde{u}(\cdot), \tilde{\tau})$ and hence

$$R(x, \tau, u) = R(x, \tilde{\tau}, \tilde{u}).$$

By the weak convergence,

$$R(x, \tilde{\tau}^p, \tilde{v}^p) \to R(x, \tilde{\tau}, \tilde{u}).$$

Thus for each $\varepsilon > 0$, we can find a p for which the left- and right-hand sides of the above equation differ by no more than ε, from which (3.1) follows.

† To construct the probability measure on the enlarged space, use the formula

$$P(d\tilde{X}^p \cdot d\tilde{v}^p \cdot d\tilde{u}^p \cdot d\tilde{w}^p \cdot d\tilde{\tau}^p) = P_{\tilde{v}^p, \tilde{w}^p, \tilde{\tau}^p}(d\tilde{X}^p) P_{\tilde{v}^p, \tilde{w}^p, \tilde{\tau}^p}(d\tilde{u}^p) P(d\tilde{v}^p \cdot d\tilde{w}^p \cdot d\tilde{\tau}^p),$$

where the subscripted terms are regular conditional probabilities.

Next, we shall prove the last assertion of the theorem. Let $u \in \bar{\mathscr{C}}_W(x)$, $\tau \in \bar{\mathscr{T}}_W(u, x)$, $\tau \leq T$. According to the first part of the proof, we can suppose that $u(\cdot)$ takes values $\{\bar{u}_1, \ldots, \bar{u}_c\}$, where $\bar{u}_i \in \mathscr{U}$, and that there is a $\Delta > 0$ such that $u(\cdot)$ is constant on the intervals $[i\Delta, i\Delta + \Delta)$, $i = 0, 1, \ldots$. If $\tau \in \bar{\mathscr{T}}_W(u, x)$, $\tau \leq T$ (whatever $u(\cdot)$ is), then by the method of the proof of Theorem 8.2.4, for each $\varepsilon > 0$, we can find a $\delta > 0$ and a $\tau_\delta \in \bar{\mathscr{T}}_W^\Delta(u, x)$ such that

$$R(x, \tau_\delta, u) \leq R(x, \tau, u) + \varepsilon/2, \qquad \tau_\delta \leq T.$$

Thus we can suppose that $\tau \in \bar{\mathscr{T}}_W^\Delta(u, x)$, and turn to the approximation of $u(\cdot)$.

Let $x(\cdot)$ denote the generic element of $C^r[0, \infty)$ and P the Wiener measure on $C^r[0, \infty)$, and let $\mathscr{B}_t$ and $\mathscr{B}_{i\Delta}^\delta$ $(0 < \delta < \Delta < T)$ denote the minimal sub-σ-algebras of (the Borel algebra over) $C^r[0, \infty)$ over which $\{x(s), s \leq t\}$ and $\{x(j\delta), j\delta \leq i\Delta\}$, resp., are measurable. Since $u(\cdot)$ is a $w(\cdot)$ function and for each ω there is an $x(\cdot)$ such that $w(\omega, \cdot) = x(\cdot)$, let us write the value $u(\omega, t)$ in the form $u(x(\cdot), t)$. For each $i = 0, 1, \ldots$ and $l = 1, \ldots, c$ define the subset of $C^r[0, \infty) : \tilde{B}_{il} = \{x(\cdot) : u(x(\cdot), i\Delta) = \bar{u}_l\}$. For each $v > 0$, there is a $\delta > 0$ and sets $\tilde{C}_{il}^v \in \mathscr{B}_{i\Delta}^\delta$ such that

$$\sum_{i, l} P\{\tilde{C}_{il}^v \, \Delta \tilde{B}_{il}\} \leq v.$$

There are finite-dimensional sets C_{il}^v such that $x(\cdot) \in \tilde{C}_{il}^v$ is equivalent to $\{x(j\delta), j\delta \leq i\Delta\} \in C_{il}^v$, and there are open sets $D_{il}^v \supset C_{il}^v$ such that

$$P\{(w(j\delta), j\delta \leq i\Delta) \in D_{il}^v \, \Delta C_{il}^v\} < v.$$

For each i and $v > 0$, define the control function $v^v(\cdot)$ as follows: $v^v(t)$ is constant on each interval $[i\Delta, i\Delta + \Delta)$ and

$$\begin{aligned} v^v(i\Delta) &= \bar{u}_1 \qquad \text{if} \quad \{w(j\delta), j\delta \leq i\Delta\} \in D_{i1}^v \\ &= \bar{u}_l, \qquad l > 1, \quad \text{if} \quad \{w(j\delta), j\delta \leq i\Delta\} \in D_{i1}^v - \bigcup_{j=1}^{l-1} D_{ij}^v. \end{aligned}$$

Each $v^v(\cdot)$ is in $\bar{\mathscr{C}}_W^\Delta(x)$ and $\tau \in \bar{\mathscr{T}}_W^\Delta(v^v, x)$ by A9.1.4. The sequence $\{v^v(\cdot)\}$ converges to $u(\cdot)$ for almost all $w(\cdot)$, t if $v \to 0$ fast enough. Also, a repetition of the embedding argument of the first part of the proof yields that

$$R(x, \tau, v^v) \to R(x, \tau, u) \qquad \text{as } v \to 0.$$

Hence for each $\varepsilon > 0$, there is a $v > 0$ such that

$$R(x, \tau, v^v) \leq R(x, \tau, u) + \varepsilon,$$

which proves the assertion. Q.E.D.

Theorem 9.3.2 *Assume* A9.1.1, A9.1.2, *and* A9.1.4. *Let* $u \in \mathscr{C}_{\mathrm{R}}(x)$, $\tau \in \bar{\mathscr{T}}_{\mathrm{R}}(u, x)$, $\tau \le T$. *Then for each* $\varepsilon > 0$, *there is a set* $\mathscr{U}_{\mathrm{D}}$, *a* $\Delta > 0$, *and a* $v_\Delta \in \mathscr{C}_{\mathrm{W}}^{\Delta}(x)$ *and a* $\tau_\Delta \in \bar{\mathscr{T}}_{\mathrm{W}}^{\Delta}(v_\Delta, x)$, $\tau_\Delta \le T$, *such that*

$$R(x, \tau_\Delta, v_\Delta) \le R(x, \tau, u) + \varepsilon. \tag{3.2}$$

Proof By Theorem 9.3.1, (3.2) will hold if it holds for all $u \,\varepsilon\, \bar{\mathscr{C}}_{\mathrm{R}}^{\Delta}(x)$ and $\tau \in \bar{\mathscr{T}}_{\mathrm{R}}^{\Delta}(u, x)$, $\tau \le T$, and we will assume that $u(\cdot)$ and τ are in this class for some fixed Δ and $\mathscr{U}_D$. Thus, we can suppose that there is a probability space supporting the Wiener process $w(\cdot)$ and the nonanticipative $X(\cdot)$, $u(\cdot)$ and stopping time τ, where $(X(\cdot), u(\cdot), w(\cdot))$ solve (1.1).

Let $\delta < \Delta$, where $\Delta/\delta =$ integer. Define the processes $\{X_n^\delta\}$ and $X^\delta(\cdot)$ by $X_0^\delta = X^\delta(0) = x$ and

$$X_{n+1}^\delta = X_n^\delta + f(X_n^\delta, u(n\delta))\delta + \sigma(X_n^\delta)[w(n\delta + \delta) - w(n\delta)], \tag{3.3}$$

$$X^\delta(t) = X_n^\delta, \qquad t \in [n\delta, n\delta + \delta). \tag{3.4}$$

Suppose that [say, where $w(\cdot)$ is the Wiener process] $\hat{u}$ is an arbitrary control in $\bar{\mathscr{C}}_{\mathrm{R}}^{\Delta}(x)$, where the set $\mathscr{U}_D$ is the same one used for $u(\cdot)$ and $\hat{\tau}$ is an arbitrary stopping time in $\bar{\mathscr{T}}_{\mathrm{R}}^{\Delta}(\hat{u}, x)$. Let $N_\delta = \hat{\tau}/\Delta\delta$ and let $\{\hat{X}_i^\delta\}$ be the solution to (3.3) using $(\hat{u}(\cdot), w(\cdot))$. Then, for each such $\hat{u}$, $\hat{\tau}$, *define the cost*

$$\begin{aligned} R^\delta(x, \hat{\tau}, \hat{u}) &= E_x^{\hat{u}}\left[\int_0^{\hat{\tau}} k(\hat{X}^\delta(s), \hat{u}(s))\, ds + b(\hat{X}^\delta(\hat{\tau}))\right] \\ &= E_x^{\hat{u}}\left[\sum_{i=0}^{N_\delta - 1} k(\hat{X}_i^\delta, \hat{u}(i\delta))\delta + b(\hat{X}_{N_\delta}^\delta)\right]. \end{aligned} \tag{3.5}$$

The sequence $\{X^\delta(\cdot), w(\cdot), u(\cdot), \tau; \delta > 0\}$ is tight on $D^r[0, T] \times C^r[0, T] \times L_2^m[0, T] \times [0, T]$ [note that the measures of $(w(\cdot), u(\cdot), \tau)$ do not depend on δ]. Let δ index a convergent subsequence. Let us suppose that the Skorokhod imbedding is used, and that the limit is denoted by $(\tilde{X}(\cdot), \tilde{w}(\cdot), \tilde{u}(\cdot), \tilde{\tau}(\cdot))$. Then $(\tilde{X}(\cdot), \tilde{u}(\cdot), \tilde{\tau})$ is nonanticipative with respect to $\tilde{w}(\cdot)$ and (w.p. 1)

$$\tilde{X}(t) = x + \int_0^t f(\tilde{X}(s), \tilde{u}(s))\, ds + \int_0^t \sigma(\tilde{X}(s))\, d\tilde{w}(s),$$

where $(\tilde{w}(\cdot), \tilde{u}(\cdot), \tilde{\tau})$ has the same probability law as does $(w(\cdot), u(\cdot), \tau)$. Thus by A9.1.4 the law of $(\tilde{X}(\cdot), \tilde{u}(\cdot), \tilde{\tau})$ does not depend on the subsequence. Hence

$$R^\delta(x, \tau, u) \to R(x, \tau, u) \qquad \text{as} \quad \delta \to 0. \tag{3.6}$$

The cost (3.5) is a well-defined functional of the control $\hat{u}(\cdot)$, the stopping time $\hat{\tau}$ and corresponding process $\{\hat{X}_n^\delta\}$ for *each* $\delta > 0$, $\hat{u} \in \bar{\mathscr{C}}_{\mathrm{R}}^{\Delta}(x)$ and $\hat{\tau} \in \bar{\mathscr{T}}_{\mathrm{R}}^{\Delta}(u, x)$, $\hat{\tau} \le T$. Let us *optimize* (3.5) over this class, for each $\delta > 0$. Since $\hat{u}(\cdot)$ can change values only at (and $\hat{\tau}$ take values) $0, \Delta, 2\Delta, \ldots,$ the problem is actually an optimization problem on a Markov chain. Let $u^\delta(i\Delta)$, $i = 0, 1, \ldots$, and τ^δ denote the *optimal* pair, and $\hat{X}^\delta(\cdot)$ the optimal interpolated [as in (3.4)] process. Then the value of $u^\delta(i\Delta)$, and whether or not $\tau^\delta = i\Delta$ (if $\tau^\delta \geq i\Delta$), depends only on the value of $\hat{X}^\delta(i\Delta)$. Thus, $u^\delta(i\Delta)$ and $I_{\{\tau^\delta = i\Delta\}}$ are functions of $w(j\delta)$, $j\delta \leq i\Delta$.

By optimality,

$$R^\delta(x, \tau^\delta, u^\delta) \leq R^\delta(x, \tau, u). \tag{3.7}$$

Define $u^\delta(\cdot)$ by $u^\delta(t) = u^\delta(i\Delta)$ on $[i\Delta, i\Delta + \Delta)$ and suppose (as we may) that $u^\delta(\cdot)$ is a $w(\cdot)$ function.

Owing to the properties of $u^\delta(\cdot)$ and τ^δ, by A9.1.4 there is a probability space supporting a Wiener process $\bar{w}^\delta(\cdot)$ and nonanticipative $\bar{X}^\delta(\cdot)$, $\bar{u}^\delta(\cdot)$, $\bar{\tau}^\delta$, where $(\bar{u}^\delta(\cdot), \bar{w}^\delta(\cdot), \bar{\tau}^\delta)$ has the same law as $(u^\delta(\cdot), w(\cdot), \tau^\delta)$ and $(\bar{X}^\delta(\cdot), \bar{u}^\delta(\cdot), \bar{w}^\delta(\cdot))$ solve (1.1). The function $\bar{u}^\delta(i\Delta)$ has (w.p. 1) the same dependence on $(\bar{w}^\delta(j\delta), j\delta \leq i\Delta)$ that $u^\delta(i\Delta)$ has on $(w(j\delta), j\delta \leq i\Delta)$.

The sequences $\{\hat{X}^\delta(\cdot), w(\cdot), \tau^\delta; u^\delta(n\Delta), n = 0, 1, \ldots\}$ and $\{\bar{X}^\delta(\cdot), \bar{w}^\delta(\cdot), \bar{\tau}^\delta; \bar{u}^\delta(n\Delta), n = 0, 1, \ldots\}$ are each tight on $D^r[0, T] \times C^r[0, T] \times [0, T] \times \mathscr{U}_D^{T/\Delta}$. Let δ index a convergent subsequence of both sequences, use the Skorokhod embedding, and denote the limits by $(\hat{X}(\cdot), \hat{w}(\cdot), \hat{\tau}; \hat{u}(n\Delta), n = 0, \ldots)$ and $(\bar{X}(\cdot), \bar{w}(\cdot), \bar{\tau}; \bar{u}(n\delta), n = 0, \ldots)$. The laws of $(\hat{w}(\cdot), \hat{\tau}, \hat{u}(\cdot))$ and of $(\bar{w}(\cdot), \bar{\tau}, \bar{u}(\cdot))$ are the same, and $(\hat{X}(\cdot), \hat{\tau}, \hat{u}(\cdot))$ is nonanticipative with respect to $\hat{w}(\cdot)$, and similarly for the second set of limits. Each set satisfies (1.1). By the weak convergence,

$$\begin{aligned} &R^\delta(x, \tau^\delta, u^\delta) \to R(x, \hat{\tau}, \hat{u}) \\ &R(x, \bar{\tau}^\delta, \bar{u}^\delta) \to R(x, \bar{\tau}, \bar{u}) = R(x, \hat{\tau}, \hat{u}) \qquad \text{as} \quad \delta \to 0. \end{aligned} \tag{3.8}$$

Equations (3.6), (3.7), and (3.8) imply that

$$R(x, \tau, u) \geq R(x, \hat{\tau}, \hat{u}). \tag{3.9}$$

and that, for each $\varepsilon > 0$, there is a $\delta_0 > 0$ such that, for $\delta < \delta_0$,

$$R(x, \bar{\tau}^\delta, \bar{u}^\delta) \leq R(x, \tau, u) + \varepsilon/2. \tag{3.10}$$

Fix $\delta \leq \delta_0$. Let A_{il} and B_i denote the decision sets† for $\bar{u}^\delta(\cdot)$ and $\bar{\tau}^\delta$ [i.e., $\bar{u}^\delta(i\Delta) = \bar{u}_l$ if and only if $\{w(j\delta), j\delta \leq i\Delta\} \in A_{il}$]. If the boundaries of the A_{il} and B_i have zero probability, then $\bar{u}^\delta \in \mathscr{C}_W^\Delta(x)$, $\bar{\tau}^\delta \in \mathscr{T}_W^\Delta(\bar{u}^\delta, x)$, and we are done. In general, we need to show that there are $u_\Delta \in \mathscr{C}_W^\Delta(x)$, $\tau_\Delta \in \mathscr{T}_W^\Delta(u_\Delta, x)$ such that

$$R(x, \tau_\Delta, u_\Delta) \leq R(x, \bar{\tau}^\delta, \bar{u}^\delta) + \varepsilon/2. \tag{3.11}$$

† The decision sets are the same as those for $u^\delta(\cdot)$ and τ^δ.

For each $\nu > 0$, approximate the decision sets A_{il} and B_i by sets A_{il}^ν and B_i^ν, as was done in the proof of Theorem 9.3.1, and then *define the corresponding control* $u^\nu(\cdot)$ *and stopping time* τ^ν, *as in that theorem*. Then, a proof similar to the one used in Theorem 9.3.1 yields that $R(x, \tau^\nu, u^\nu) \to R(x, \bar{\tau}^\delta, \bar{u}^\delta)$ as $\nu \to 0$. Thus, for sufficiently small ν, set $u_\Delta = u^\nu$, $\tau_\Delta = \tau^\nu$, and (3.11) will hold. Q.E.D.

We will assume, henceforth, in the proofs, that $\sup_x \Delta t^h(x) \to 0$ as $h \to 0$. Otherwise, with minor modifications in the definition of the $v^h(\cdot)$, the proofs still hold.

Theorem 9.3.3 *Assume* A9.1.1 *through* A9.1.4, *and let* $u \in \bar{\mathscr{C}}_R(x)$, $\tau \in \bar{\mathscr{T}}_R(u, x)$. *Then*

$$\bar{V}(x) = R(x, \rho, \bar{u}) \le R(x, \tau, u). \tag{3.12}$$

If $f(\cdot, \cdot)$ *and* $\sigma(\cdot)$ *satisfy a uniform Lipschitz condition in* x, *then* $\bar{u} \in \bar{\mathscr{C}}_R(x)$ *and* $\rho \in \bar{\mathscr{T}}_R(\bar{u}, x)$ *and* $\bar{V}(x) = V(x)$ (*i.e., we have the uniqueness property*).

PROOF By Theorem 9.3.2, and the fact that we need only consider τ satisfying $E_x^u \tau \le 2 \max_x |b(x)|/k_0$, (3.12) will hold as stated if, for each $T < \infty$, $\mathscr{U}_D$, and $\Delta > 0$, it holds for all $u \in \bar{\mathscr{C}}_W^\Delta(x)$ and $\tau \in \bar{\mathscr{T}}_W^\Delta(u, x)$, $\tau \le T$. Henceforth let $(u(\cdot), \tau)$ be a fixed element in this class, with decision sets $\{A_{il}\}$, $\{B_i\}$, and where $w(\cdot)$ denotes the Wiener process. We can suppose that there is a $\delta > 0$ such that $u(\cdot)$ and τ depend only on the values of the Wiener process at time $\{j\delta, j\delta \le T\}$. Let $X(\cdot)$ denote the solution to (1.1) corresponding to $w(\cdot)$, $u(\cdot)$, and τ.

Define a control $v^h(\cdot)$ and stopping time τ^h for the *interpolated chain* $\xi^h(\cdot)$ as follows. If $\{W^h(j\delta), j\delta \le i\Delta\} \in A_{il}$, set†

$$v^h(s) = \bar{u}_l \qquad \text{on} \quad [i\Delta, i\Delta + \Delta).$$

If $\{W^h(j\delta), j\delta \le i\Delta\} \in B_i$, set $\tau^h = i\Delta$. Let $\xi^h(\cdot)$ denote the process under control $v^h(\cdot)$. Of course, the $W^h(\cdot)$ used to calculate $\xi^h(\cdot)$ is obtained (nonanticipatively) from $\xi^h(\cdot)$ and an independent Wiener process $\psi(\cdot)$. By the optimality of $u^h(\cdot)$ and ρ_h,

$$R^h(x, \rho_h, u^h) \le R^h(x, \tau^h, v^h). \tag{3.13}$$

The sequence $\{\xi^h(\cdot), W^h(\cdot), \tau^h; v^h(i\Delta), i\Delta < T\}$ is tight on $D^{2r}[0, T] \times [0, T] \times \mathscr{U}_D^{T/\Delta}$. Let h index a convergent subsequence with limit $(\bar{X}(\cdot), \bar{W}(\cdot), \bar{\tau}; \bar{v}(i\Delta), i\Delta < T)$, and suppose that the convergence is w.p. 1 in the appropriate topology. Then $\bar{X}(\cdot)$, $\bar{\tau}$ and $\bar{v}(\cdot)$ are nonanticipative with respect

† Actually, to be consistent with the definition of ξ_n^h or $\xi^h(\cdot)$, the control may change at one of the t_n^h only. When we write that the control changes at $i\Delta$, we in fact suppose that it changes at the t_n^h for which $n = \min\{i : t_i^h \ge i\Delta\}$.

to $\overline{W}(\cdot)$, where we define $\bar{v}(\cdot)$ by $\bar{v}(s) = \bar{v}(i\Delta)$ on $[i\Delta, i\Delta + \Delta)$, and the limits satisfy

$$\overline{X}(t) = x + \int_0^t f(\overline{X}(s), \bar{v}(s))\, ds + \int_0^t \sigma(\overline{X}(s))\, d\overline{W}(s). \tag{3.14}$$

Since

$$P\{(\overline{W}(j\delta), j\delta \leq i\Delta) \in \partial A_{il} \cup \partial B_i\} = 0,$$

for all i, l, the weak convergence implies that the probability law of $\{\bar{v}(\cdot), \overline{W}(\cdot), \bar{\tau}\}$ is the same as that of $\{u(\cdot), w(\cdot), \tau\}$. Hence, by A9.1.4, the distributions of $\{\overline{X}(\cdot), \bar{v}(\cdot), \tau\}$ are the same as those of $\{X(\cdot), u(\cdot), \tau\}$. Thus

$$R^h(x, \tau^h, v^h) \to R(x, \bar{\tau}, \bar{v}) = R(x, \tau, u).$$

This and (3.13) prove (3.12).

The last assertion of the theorem is proved by showing that, if $u(\cdot)$ is any $\mathscr{U}$-valued measurable random function which is nonanticipative with respect to $w(\cdot)$, then the solution to

$$X(t) = x + \int_0^t f(X(s), u(s))\, ds + \int_0^t \sigma(X(s))\, dw(s) \tag{3.15}$$

is pathwise unique under the Lipschitz condition.

Suppose that $\xi_1(\cdot)$ and $\xi_2(\cdot)$ are both nonanticipative solutions. Define $\delta(\cdot) = |\xi_1(\cdot) - \xi_2(\cdot)|$. Then for any $T_1 \leq T < \infty$,

$$E \max_{0 \leq t \leq T_1} |\delta(s)|^2 \leq 2\, E \max_{0 \leq t \leq T_1} \left| \int_0^t [f(\xi_1(s), u(s)) - f(\xi_2(s), u(s))]\, ds \right|^2$$
$$+ 2\, E \max_{0 \leq t \leq T} \left| \int_0^t [\sigma(\xi_1(s)) - \sigma(\xi_2(s))]\, dw(s) \right|^2$$

Let K_i denote real numbers whose value depends on T. Using the Lipschitz condition and Schwarz's inequality, the first term in the right-hand side is bounded above by

$$K_1 E \int_0^{T_1} |\delta(s)|^2\, ds.$$

By the martingale estimate (1.2.3), the second term is bounded above by

$$K_2 E \int_0^{T_1} |\sigma(\xi_1(s)) - \sigma(\xi_2(s))|^2\, ds$$

hence by

$$K_3 E \int_0^{T_1} |\delta(s)|^2\, ds.$$

Putting these estimates together and defining $\Delta(t) = E \max_{0 \le s \le t} |\delta(s)|^2$ yields

$$\Delta(t) \le K_4 \int_0^t \Delta(s)\, ds.$$

Then, since $0 \le \Delta(t) < \infty$, we have $\Delta(t) \equiv 0$. Let τ denote a nonnegative random variable which is nonanticipative with respect to $w(\cdot)$. By using the Picard iteration technique of the proof of Theorem 1.5.1, we can construct a solution to (3.15). By the method of construction, there is a $C^r[0, T]$-valued measurable function $F(\cdot, \cdot)$ on $C^r[0, T] \times L_2^m[0, T]$, such that $X(\cdot) = F(w(\cdot), u(\cdot))$ solves (3.15) w.p. 1, and $X(\cdot)$ is nonanticipative with respect to $w(\cdot)$. By path uniqueness F does not (w.p. 1) depend on the probability space. Thus, the distributions of $(w(\cdot), u(\cdot), \tau)$ imply the distributions of $(X(\cdot), u(\cdot), \tau)$. Hence, $\bar{u} \in \bar{\mathscr{C}}_{\mathrm{R}}(x)$ and $\rho \in \bar{\mathscr{T}}_{\mathrm{R}}(\bar{u}, x)$. Q.E.D.

9.4 Discounted Cost

In this section, we will briefly outline the changes required (from the development in Sections 9.1–9.3) when the cost function is of the discounted type.

Assume A9.1.1 through A9.1.4, but drop the assumption concerning the positivity of $k(\cdot, \cdot)$. For each $u \in \bar{\mathscr{C}}_{\mathrm{R}}(x)$, $\tau \in \bar{\mathscr{T}}_{\mathrm{R}}^0(u, x)$, and corresponding controlled process $X(\cdot)$, define the cost

$$R(x, \tau, u) = E_x^u \left[A(\tau) b(X(\tau)) + \int_0^\tau A(s) k(X(s), u(s))\, ds \right], \tag{4.1}$$

where $\lambda(\cdot)$ is a bounded continuous real-valued function on R^r satisfying $\lambda(x) \ge \lambda_0 > 0$ and $A(t) = \exp - \int_0^t \lambda(X(s))\, ds$. Define

$$V(x) = \inf_{u \in \bar{\mathscr{C}}_{\mathrm{R}}(x),\, \tau \in \bar{\mathscr{T}}_{\mathrm{R}}(u, x)} R(x, \tau, u). \tag{4.2}$$

Suppose that $\Delta t^h(x) \to 0$ as $h \to 0$, uniformly in x. Then one type of discretization for the discounted problem leads to (see Chapters 6.5 and 8.2)

$$V^h(x) = \min\left\{ b(x), \min_{\alpha \in \mathscr{U}} [(\exp - \Delta t^h(x)\lambda(x))(E_x^\alpha V^h(\xi_1^h) + k(x, \alpha)\Delta t^h(x))] \right\}, \tag{4.3}$$

which is just the dynamic programming equation for the minimum cost for a discounted controlled optimally stopped Markov chain.

For each $\pi \in \mathscr{C}_R(h)$, $m \in \mathscr{T}_R^0(h, \pi)$, and corresponding controlled chain $\{\xi_n^h\}$, define the cost

$$R^h(x, m, \pi) = E_x^\pi\left[A_{m-1}^h b(\xi_m^h) + \sum_{i=0}^{m-1} A_i^h k(\xi_i^h, U_i)\Delta t_i^h\right], \tag{4.4}$$

where A_i^h is defined below (6.5.5), and U_i is the actual realization of the control at time i, and the infimum

$$V^h(x) = \inf_{\pi \in \mathscr{C}_R(h),\, m \in \mathscr{T}_R^0(h, \pi)} R^h(x, m, \pi). \tag{4.5}$$

Then (4.3) has a unique solution which is just the right-hand side of (4.5). The *optimal policy* π is *stationary pure Markov* and the *optimal stopping time* is *pure Markov*. Let $u^h(\cdot)$, m_h (with sample values $\{U_i^h\}$ and interpolation ρ_h resp.) denote the *optimizing pair* in (4.5). In this discounted case, define $K^h(\cdot)$ to be the interpolation of $\{\sum_{i=0}^{n-1} A_i^h k(\xi_i^h, U_i^h)\Delta t_i^h\}$. The (optimal) sequence $\{\xi^h(\cdot), W^h(\cdot), F^h(\cdot), K^h(\cdot), \rho_h\}$ is tight on $D^{3r+1}[0, \infty) \times \bar{R}^+$. Let h index a convergent subsequence, with limit $(\xi(\cdot), W(\cdot), F(\cdot), K(\cdot), \rho)$. Then $\xi(\cdot)$ satisfies

$$\xi(t) = x + F(t) + \int_0^t \sigma(\xi(s))\, dW(s), \qquad t \geq 0,$$

and via an argument exactly like that in Section 9.2, there is a nonanticipative $\mathscr{U}$-valued control $\bar{u}(\cdot)$ and a stopping time ρ, such that

$$F(t) = \int_0^t f(\xi(s), \bar{u}(s))\, ds$$

$$K(t) = \int_0^t \left(\exp - \int_0^s \lambda(\xi(v))\, dv\right) k(\xi(s), \bar{u}(s))\, ds,$$

and

$$R^h(x, \rho_h, u^h) \to R(x, \rho, \bar{u}) \qquad \text{as} \quad h \to 0.$$

Furthermore, Theorem 9.3.3 continues to hold, with $\overline{\mathscr{T}}_R(u, x)$ replaced by $\overline{\mathscr{T}}_R^0(u, x)$.

If $\Delta t^h(x) \not\to 0$ as $h \to 0$, uniformly in x, but one makes the replacements indicated in connection with (6.5.4), then the above convergence results remain valid.

9.5 Control until a Boundary Is Reached: Discounted Case

In this section, optimization of the cost (4.7.14) will be treated. The development is similar to that in the previous sections—but with the added complication that the nature of the process at the time that it first reaches the boundary must be considered. The infimum of the costs, if sufficiently smooth, satisfies (4.7.15). Any one of the approximations 1, 2, or 3 of Section 9.1 can be used, but we will stay with approximation 3, for definiteness. Let $\lambda(x) \geq \lambda_0 > 0$, and assume A9.1.1 through A9.1.4, except for the positivity condition on $k(\cdot, \cdot)$, and the assumptions concerning the stopping time. (We now stop only on first hitting the boundary ∂G.)

Let $\{\xi_n^h\}$ denote the approximating controlled chain. Assume, for definiteness, that $\sup_x \Delta t^h(x) \to 0$ as $h \to 0$; the assumption can be dropped if we make the minor alterations in the discretized cost which were indicated in Chapter 6. For each $\pi \in \mathscr{C}_R(h)$, define the corresponding escape time from G_h by $m = \min\{n : \xi_n^h \notin G_h\}$, and the cost

$$R^h(x, \pi) = E_x^\pi\left[A_{m-1}^h b(\xi_m^h) + \sum_{i=0}^{m-1} A_i^h k(\xi_i^h, U_i)\Delta t_i^h\right],$$

and let $V^h(x)$ denote the *infimum* of the costs over $\mathscr{C}_R(h)$. The dynamic programming equation for $V^h(x)$ is

$$\begin{aligned} V^h(x) &= \min_{\alpha \in \mathscr{U}}[(\exp - \lambda(x)\Delta t^h(x))(E_x^\alpha V^h(\xi_1^h) + k(x, \alpha)\Delta t^h(x))], \qquad x \in G_h, \\ &= b(x), \qquad x \notin G_h. \end{aligned} \tag{5.1}$$

The equation (5.1) has a unique solution and there is a *pure Markov minimizing control* $u^h(\cdot)$. *For the chain under control* $u^h(\cdot)$, let m_h denote the hitting time of ∂G_h (it may be infinite). Select an arbitary $\bar{u}_0 \in \mathscr{U}$ and, for $n \geq m_h$, set $u^h(\xi_n^h) = U_n^h = \bar{u}_0$. Define the usual interpolations $\xi^h(\cdot)$, $W^h(\cdot)$, $F^h(\cdot)$, $K^h(\cdot)$, and let ρ_h be the interpolation of m_h, namely $t_{m_h}^h$. The interpolation $K^h(\cdot)$ is defined as in the last section.

The sequence $\{\xi^h(\cdot), W^h(\cdot), F^h(\cdot), K^h(\cdot), \rho_h\}$ is tight on $D^{3r+1}[0, \infty) \times \bar{R}^+$. Let h index a convergent subsequence, with limit $(\xi(\cdot), W(\cdot), F(\cdot), K(\cdot), \rho)$. Suppose that the Skorokhod imbedding is used. The result of Theorem 9.2.3 remains valid and there is a measurable $\mathscr{U}$-valued function $\bar{u}(\cdot)$, which is nonanticipative with respect to the Wiener process, such that

$$F(t) = \int_0^t f(\xi(s), \bar{u}(s))\, ds$$

$$K(t) = \int_0^t \left(\exp - \int_0^s \lambda(\xi(v))\, dv\right) k(\xi(s), \bar{u}(s))\, ds,$$

$$\xi(t) = x + F(t) + \int_0^t \sigma(\xi(s))\, dW(s),$$

and the random variable ρ is also nonanticipative with respect to $W(\cdot)$. On the set $\{\rho < \infty\}$, $\xi(\rho) \in \partial G$ since $\rho_h \to \rho$ w.p. 1 and $\xi^h(\cdot) \to \xi(\cdot)$ w.p. 1, uniformly on finite intervals. *But* ρ *is not necessarily the first time that* $\xi(t) \in \partial G$. If it were, then the w.p. 1 convergence would immediately yield that

$$R^h(x, u^h) \to R(x, \bar{u}) \qquad \text{as} \quad h \to 0. \tag{5.2}$$

For each control $v(\cdot)$, let $\tau(v)$ and $\tau'(v)$ denote the escape times (of the corresponding controlled process) from G and $\bar{G}$, resp., when the initial condition is x. The initial condition will be fixed at x throughout this section. In general, there seems to be little that we can do at present about the problem of whether (w.p. 1) $\rho = \tau(\bar{u}) \equiv \inf\{t : \xi(t) \in \partial G\}$. In fact, one of the more serious deficiencies of the method concerns the lack of adequate information on $\bar{u}(\cdot)$. There are two main problems. The first concerns the uniqueness of solutions under the control $\bar{u}(\cdot)$. The second concerns the escape times. If $\bar{u} \in \hat{\mathscr{C}}_{\mathrm{R}}^0(x, G)$, we will be able to show that $R^h(x, u^h) \to V(x)$ as $h \to 0$. To do this, some more results are needed on the approximations of controls. First, let us *assume* A9.5.1.

A9.5.1 $\rho = \tau(\bar{u})$ w.p. 1

There are many cases where A9.5.1 holds. For example, it holds in both the controlled versions of the example in Section 4.4 and in the case where $a(\cdot)$ is continuous and positive definite at all $x \in \partial G$ and ∂G satisfies a cone condition.

In the absence of general information concerning $\tau(\bar{u})$ and ρ, it seems that we must consider special cases. Perhaps a detailed investigation of the nature of the discrete and continuous optimization near the boundary will shed more light on the problem. But it is clear that A9.5.1 holds in very many cases of interest. The exceptional cases are harder to find. *Under* A9.5.1, *Eq.* (5.2) *holds.* If $k(y, \alpha) \geq 0$ and $b(\cdot) \equiv 0$, then Theorems 9.5.1 and 9.5.2 hold with $\bar{V}(x)$ replacing $R(x, \bar{u})$, where $\bar{V}(x) = \lim_{h \to 0} V^h(x)$ and is no less than $R(x, \bar{u})$ and A9.5.1 is not needed.

We will now prove an optimality theorem under A9.5.1. Let $u \in \hat{\mathscr{C}}_{\mathrm{W}}^0(x, G)$ and let $\{v^\nu(\cdot)\}$ denote the sequence constructed from $u(\cdot)$ and $w(\cdot)$ in the proof of Theorem 9.3.1. In particular, $v^\nu(\cdot) \in \mathscr{C}_{\mathrm{W}}^{\Delta}(x)$ and (let $p = 1/\nu$)

$$E \int_0^p |v^\nu(\omega) - u(\omega)|^2 \, ds \leq 2^{-p}.$$

Since $v^\nu(\cdot) \in \mathscr{C}_{\mathrm{W}}^{\Delta}(x)$, there is a probability space supporting a Wiener process $\tilde{w}^\nu(\cdot)$ and nonanticipative processes $\tilde{X}^\nu(\cdot)$, $\tilde{v}^\nu(\cdot)$ such that

$$\tilde{X}^\nu(t) = x + \int_0^t f(\tilde{X}^\nu(s), \tilde{v}^\nu(s)) \, ds + \int_0^t \sigma(\tilde{X}^\nu(s)) \, d\tilde{w}^\nu(s),$$

and $(\tilde{v}^\nu(\cdot), \tilde{w}^\nu(\cdot))$ has the same law as $(v^\nu(\cdot), w(\cdot))$. Let us enlarge the latter probability space to add a nonanticipative process $\tilde{u}^\nu(\cdot)$, where $(\tilde{u}^\nu(\cdot), \tilde{v}^\nu(\cdot),$

$\tilde{w}^\nu(\cdot)$) has the same law as $(u(\cdot), v^\nu(\cdot), w(\cdot))$. Thus, we can suppose that $\tilde{v}^\nu(\cdot)$ is obtained from $\tilde{u}^\nu(\cdot)$ in the same way that $v^\nu(\cdot)$ is obtained from $u(\cdot)$.

The sequence $\{\tilde{X}^\nu(\cdot), \tilde{w}^\nu(\cdot), \tilde{u}^\nu(\cdot)\}$ is tight on $C^{2r}[0, \infty) \times L^m_{2,l}$. Let ν index a convergent subsequence, with limit $(\tilde{X}(\cdot), \tilde{w}(\cdot), \tilde{u}(\cdot))$, and suppose that the Skorokhod imbedding is used. Then $(\tilde{w}(\cdot), \tilde{u}(\cdot))$ has the same law as $(w(\cdot), u(\cdot))$, and $(\tilde{X}(\cdot), \tilde{w}(\cdot), \tilde{u}(\cdot))$ satisfy (1.1). Furthermore, by uniqueness $(u \in \hat{\mathscr{C}}^0_W(x, G))$, $(\tilde{X}(\cdot), \tilde{u}(\cdot))$ has the same law as $(X(\cdot), u(\cdot))$. Thus $\tau(\tilde{u}) = \tau'(\tilde{u})$ w.p. 1 and

$$R(x, \tilde{u}) = R(x, u).$$

Furthermore, since $\tilde{X}^\nu(\cdot) \to \tilde{X}(\cdot)$ as $v \to 0$, and $\tau(\tilde{u}) = \tau'(\tilde{u})$ w.p. 1,

$$\begin{gathered} \tau(\tilde{v}^\nu) \to \tau(\tilde{u}), \qquad \text{w.p. } 1, \\ P\{\tau(\tilde{v}^\nu) \neq \tau'(\tilde{v}^\nu) \mid \tau(\tilde{u}) < \infty\} \to 0 \qquad \text{as} \quad \nu \to 0. \end{gathered} \tag{5.3}$$

Thus, for each $\varepsilon > 0$, there is a $v(\cdot)$ in some $\bar{\mathscr{C}}^\Delta_W(x)$ such that

$$R(x, v) \le R(x, u) + \varepsilon, \qquad P\{\tau(v) \neq \tau'(v) \mid \tau(v) < \infty\} \le \varepsilon. \tag{5.4}$$

Let $v^h(\cdot)$ (a control for the chain $\{\xi^h_n\}$ and interpolated process $\xi^h(\cdot)$) be obtained from $v(\cdot)$ in exactly the same way as $v^h(\cdot)$ was obtained from $u(\cdot)$ in the proof of Theorem 9.3.3. Then there is a $\delta > 0$, which goes to zero as $\varepsilon \to 0$, such that

$$R(x, v) - \delta \le \varliminf_{h\to 0} R^h(x, v^h) \le \varlimsup_{h\to 0} R^h(x, v^h) \le R(x, v) + \delta. \tag{5.5}$$

We use $\delta > 0$ and $\varliminf$ and $\varlimsup$, since $\tau(v) \neq \tau'(v)$ w.p. 1, although (5.4) holds. Thus, the sample path costs (for $\xi^h(\cdot)$) may not converge to the sample path costs for the limit on the set $\{\tau(v) \neq \tau'(v)\}$ as $h \to 0$. But the probability of this set can be made as small as desired. By the optimality of $u^h(\cdot)$.

$$R^h(x, v^h) \ge R^h(x, u^h). \tag{5.6}$$

Thus we have proved

Theorem 9.5.1 *Assume* A9.1.1 *to* A9.1.4 *and* A9.5.1. *Then*

$$\lim_{h\to 0} R^h(x, u^h) = R(x, \bar{u}) \le R(x, u), \qquad \text{for all } u \in \hat{\mathscr{C}}^0_W(x, G). \tag{5.7}$$

If $k(y, \alpha) \ge 0$ *and* $b(\cdot) \equiv 0$, *then the right-hand inequality of* (5.7) *holds without* A9.5.1.

If $u \in \hat{\mathscr{C}}^0_R(x, G)$ but not in $\hat{\mathscr{C}}^0_W(x, G)$, then, owing to the presence of the boundary ∂G, we cannot use the method of approximation of Theorem 9.3.2, where a $w(\cdot)$-dependent control (which was at least as good as $u(\cdot)$) was found. A problem arises, since the degree to which $\tau = \tau'$ for the new "approximating" control is not clear. Another, more direct, type of approximation will now be described. It will allow us to circumvent the above difficulty.

An Approximation to Controls in $\hat{\mathscr{C}}_{\mathrm{R}}^{0}(x, G)$

Let $u \in \hat{\mathscr{C}}_{\mathrm{R}}^{0}(x, G)$. Then, we can suppose that there is a probability space supporting a Wiener process $w(\cdot)$ and nonanticipative process $X(\cdot)$, $u(\cdot)$, where $(X(\cdot), u(\cdot), w(\cdot))$ solve (1.1). Let $\{v^p(\cdot)\}$ denote the sequence of functions which were constructed from $u(\cdot)$ in the first part of the proof of Theorem 9.3.1. Then $v^p(\cdot) \in \bar{\mathscr{C}}_{\mathrm{R}}^{\Delta}(x)$, and we can suppose that

$$E \int_0^p |v^p(s) - u(s)|^2 \, ds \leq 2^{-p}.$$

Since $v^p \in \bar{\mathscr{C}}_{\mathrm{R}}^{\Delta}(x)$, for each p there is a probability space supporting a Wiener process $\tilde{w}^p(\cdot)$ and nonanticipative $(\tilde{X}^p(\cdot), \tilde{v}^p(\cdot))$, where $(\tilde{v}^p(\cdot), \tilde{w}^p(\cdot))$ has the same law as $(v^p(\cdot), w(\cdot))$, and $(\tilde{X}^p(\cdot), \tilde{v}^p(\cdot), \tilde{w}^p(\cdot))$ solves (1.1). Let us enlarge the probability space by adding $\tilde{u}^p(\cdot)$, where $(\tilde{u}^p(\cdot), \tilde{v}^p(\cdot), \tilde{w}^p(\cdot))$ has the same law as $(u(\cdot), v^p(\cdot), w(\cdot))$. We can (and will) suppose that $\tilde{v}^p(\cdot)$ is obtained from $\tilde{u}^p(\cdot)$ in the same way that $v^p(\cdot)$ is obtained from $u(\cdot)$. The sequence $\{\tilde{X}^p(\cdot), \tilde{w}^p(\cdot), \tilde{u}^p(\cdot)\}$ is tight on $C^{2r}[0, \infty) \times L_{2,l}^m$. Let p index a convergent subsequence, use Skorokhod imbedding, and denote the limit by $(\tilde{X}(\cdot), \tilde{w}(\cdot), \tilde{u}(\cdot))$. By uniqueness, the limit has the same law as $(X(\cdot), w(\cdot), u(\cdot))$. *Hence* (as in the proof of Theorem 9.5.1) *for each* $\varepsilon > 0$, *there is a* $\mathscr{U}_D$, *a* $\Delta > 0$, *and a* $v \in \bar{\mathscr{C}}_{\mathrm{R}}^{\Delta}(x)$ *such that* (5.4) *holds.*

Fix ε *and the corresponding* $v(\cdot)$. Let $w(\cdot)$ and $X(\cdot)$ denote the associated Wiener process and controlled diffusion, resp. We will approximate $v(\cdot)$ by a "simpler" control in a way that allows us to define a "similar" control $v^h(\cdot)$ for use on $\xi^h(\cdot)$.

Let $\Delta/\delta = q$, an integer. Let $A_1^q, \ldots,$ denote a collection of disjoint sets in R^r, each of diameter $\leq 1/q$, *whose boundaries have Lebesgue measure zero*, and such that $R^r = \bigcup_j A_j^q$. Let $\mathscr{B}_{i\Delta}^q$ denote the minimal σ-algebra which contained all the sets (for all $j\delta \leq i\Delta$, $n < i$, $\bar{u}_l \in \mathscr{U}_D$, and k)

$$\{\omega : w(j\delta) \in A_{kj}^q\}, \{\omega : v(n\Delta) = \bar{u}_l\},$$

and let $P_{w,v}^q\{v(i\Delta) = \bar{u}_l\}$ denote the conditional probability, given $\mathscr{B}_{i\Delta}^q$.

For each $\bar{u}_l$ and i, $P_{w,v}^q\{v(i\Delta) = \bar{u}_l\}$ is a function of $w(j\delta)$, $j\delta \leq i\Delta$, and of $v(n\Delta)$, $n < i$. For each set $\{k_1, \ldots, k_{iq}; m_0, \ldots, m_{i-1}\}$, it is constant on the product set

$$A_{k_1} \times \cdots \times A_{k_{iq}} \times \{v(n\Delta) = \bar{u}_{m_n}, n < i\}.$$

As $q \to \infty$, the function converges (w.p. 1) to the conditional probability, given $w(s)$, $s \leq i$, and $v(s)$, $s \leq i\Delta - \Delta$.

For each q, there is a probability space on which are defined a Wiener process $w^q(\cdot)$ and a nonanticipative $\mathcal{U}_D$-valued function $v^q(\cdot)$, where $v^q(\cdot)$ is constant on each interval $[i\Delta, i\Delta + \Delta)$, and where $P^q_{w,v}\{v^q(i\Delta) = \bar{u}_l\}$ is the conditional probability that $v^q(i\Delta) = \bar{u}_l$, given $w^q(s)$, $s < \infty$, $v^q(n\Delta)$, $n < i$. In fact, by the definition, $P\{v^q(i\Delta) = \bar{u}_l | w^q(s), \; s < \infty, \; v^q(n\Delta), \; n < i\} = P\{v^q(i\Delta) = \bar{u}_l \,|\, w^q(j\delta), \, j\delta \le i\Delta, \, v^q(n\Delta), \, n < i\}$. Since $v^q \in \mathscr{C}^\Delta_{\mathrm{R}}(x)$, we can suppose by A9.1.4 that there is a nonanticipative $X^q(\cdot)$ also defined on the space, such that $(X^q(\cdot), v^q(\cdot), w^q(\cdot))$ solve (1.1).

The sequence $\{X^q(\cdot), w^q(\cdot); v^q(i\Delta), i = 0, 1, \ldots\}$ is tight on $C^{2r}[0, \infty) \times \mathcal{U}_D^\infty$. Let q index a convergent subsequence with limit $(\tilde{X}(\cdot), \tilde{w}(\cdot), \tilde{v}(i\Delta), i = 0, 1, \ldots)$ and suppose that Skorokhod imbedding is used. The limit has the same law as $(X(\cdot), w(\cdot); v(i\Delta), i = 0, 1, \ldots)$. Thus we can assert that

$$\begin{aligned} R(x, \tilde{v}) &= R(x, v), & \lim_q P\{\tau(v^q) \nrightarrow \tau(\tilde{v})\} &\le \varepsilon, \\ \lim_q P\{\tau(v^q) &\neq \tau'(v^q)\} \le \varepsilon, & \lim_q R(x, v^q) &\le R(x, v) + \varepsilon. \end{aligned} \tag{5.8}$$

It follows from (5.8) that, for purposes of comparing $v(\cdot)$ to $\bar{u}(\cdot)$, we can suppose that $v(\cdot)$ is of the form of $v^q(\cdot)$ for some finite, but arbitrary, q. The sets A_i^q will be denoted by A_i. Suppose this henceforth. Now, let us define $v^h(\cdot)$, the adaptation of $v(\cdot)$ to the interpolated process $\xi^h(\cdot)$.

Choose $v^h(\cdot)$ as follows. Select $v^h(0)$ randomly according to

$$P\{v^h(0) = \bar{u}_l\} = P\{v(0) = \bar{u}_l\},$$

and let the $\xi^h(\cdot)$ evolve until time Δ. (Actually, we must let the system evolve until the "discrete" time t_n^h, where $n = \min\{i : t_i^h \ge \Delta\}$ before changing controls. Here and in the sequel, it should be understood that the controls should change only at such discrete times.) In general, at $t = i\Delta$, choose $v^h(i\Delta)$ at random according to

$$\begin{aligned} P\{v^h(i\Delta) &= \bar{u}_l \,|\, W^h(s), \, s \le i\Delta, \, \xi^h(s), \, s \le i\Delta, \, v^h(n\Delta), \, n < i\} \\ &= P\{v^h(i\Delta) = \bar{u}_l \,|\, W^h(j\delta), \, j\delta \le i\Delta, \, v^h(n\Delta), \, n < i\} \\ &= P\{v(i\Delta) = \bar{u}_l \,|\, w(j\delta) = W^h(j\delta), \, j\delta \le i\Delta, \, v(n\Delta) = v^h(n\Delta), \, n < i\}, \end{aligned}$$

where, by the last expression, we mean that we simply substitute $W^h(j\delta)$ and $v^h(n\Delta)$ for $w(j\delta)$ and $v(n\Delta)$ in the $P^q_{w,v}\{v(i\Delta) = \bar{u}_l\}$, which we consider as a function of $w(j\delta)$, $j\delta \le i\Delta$ and of $v(n\Delta)$, $n < i$.

The sequence $\{\xi^h(\cdot), W^h(\cdot); v^h(i\Delta), i = 0, 1, \ldots\}$ is tight on $D^{2r}[0, \infty) \times \mathcal{U}_D^\infty$. Let h index a convergent subsequence, and use Skorokhod imbedding. Denote the limit by $(\xi(\cdot), W(\cdot); \bar{v}(i\Delta), i = 0, 1, \ldots)$. For arbitrary n, $t < s$, and $t_i \le t$, $i \le n$, and an arbitrary bounded and continuous function $g(\cdot)$

defined on the appropriate space, we have

$$G_n \equiv Eg(\xi^h(t_i), W^h(t_i), v^h(t_i), i \le n)(W^h(s) - W^h(t)) \to 0,$$

$$G_n \to Eg(\xi(t_i), W(t_i), \bar{v}(t_i), i \le n)(W(s) - W(t)),$$

$$E_n = Eg(\xi^h(t_i), W^h(t_i), v^h(t_i), i \le n)(W^h(s) - W^h(t))(W^h(s) - W^h(t))'$$

$$\to Eg(\xi(t_i), W(t_i), \bar{v}(t_i), i \le n)(s - t),$$

$$E_n \to Eg(\xi(t_i), W(t_i), \bar{v}(t_i), i \le n)\,(W(s) - W(t))(W(s) - W(t))'$$

where $\bar{v}(\cdot)$ is the piecewise constant interpolation of the function with values $\bar{v}(i\Delta)$ at $i\Delta$. Thus, $\xi(\cdot)$ and $\bar{v}(\cdot)$ are *nonanticipative* with respect to the Wiener process $W(\cdot)$ (see, e.g., Theorems 8.2.1 and 6.3.1 and Section 6.6). Also,

$$\xi(t) = x + \int_0^t f(\xi(s), \bar{v}(s))\, ds + \int_0^t \sigma(\xi(s))\, dW(s).$$

Next, let us compute the law of $(\bar{v}(\cdot), W(\cdot))$. Let C_α and B_α denote measurable subsets of the A_k and R^r, resp., each with a *boundary of zero measure*. Let $t_1, \ldots, t_m \le i\Delta$ and $t_k \ne j\delta$, all j, k, and let each $\bar{u}_{j_s} \in \mathscr{U}_D$. Then (assuming that the denominator of the second term is >0)

$$P\{\bar{v}(i\Delta) = \bar{u}_l \mid W(t_q) \in B_q\,,\, q \le m,\, W(j\delta) \in C_j\,,\, j\delta \le i\Delta,\, \bar{v}(p\Delta) = \bar{u}_{j_p}\,,\, p < i\}$$

$$= \frac{P\left\{\begin{array}{l}\bar{v}(i\Delta) = \bar{u}_l\,,\, W(t_q) \in B_q\,,\, q \le m,\\ \qquad W(j\delta) \in C_j\,,\, j\delta \le i\Delta,\, \bar{v}(p\Delta) = \bar{u}_{j_p}\,,\, p < i\end{array}\right\}}{P\left\{\begin{array}{l}W(t_q) \in B_q\,,\, q \le m,\, W(j\delta) \in C_j\,,\\ \qquad j\delta \le i\Delta,\, \bar{v}(p\Delta) = \bar{u}_{j_p}\,,\, p < i\end{array}\right\}}$$

$$= \lim_{h\to 0} \frac{P\left\{\begin{array}{l}v^h(i\Delta) = \bar{u}_l\,,\, W^h(t_q) \in B_q\,,\, q \le m,\\ \qquad W^h(j\delta) \in C_j\,,\, j\delta \le i\Delta,\, v^h(p\Delta) = \bar{u}_{j_p}\,,\, p < i\end{array}\right\}}{P\left\{\begin{array}{l}W^h(t_q) \in B_q\,,\, q \le m,\, W^h(j\delta) \in C_j\,,\\ \qquad j\delta \le i\Delta,\, v^h(p\Delta) = \bar{u}_{j_p}\,,\, p < i\end{array}\right\}}$$

$$= \lim_{h\to 0} \frac{P\{v^h(i\Delta) = \bar{u}_l\,,\, W^h(j\delta) \in C_j\,,\, j\delta \le i\Delta,\, v^h(p\Delta) = \bar{u}_{j_p}\,,\, p < i\}}{P\{W^h(j\delta) \in C_j\,,\, j\delta \le i\Delta,\, v^h(p\Delta) = \bar{u}_{j_p}\,,\, p < i\}}$$

$$= \frac{P\{\bar{v}(i\Delta) = \bar{u}_l\,,\, W(j\delta) \in C_j\,,\, j\delta \le i\Delta,\, \bar{v}(p\Delta) = \bar{u}_{j_p}\,,\, p < i\}}{P\{W(j\delta) \in C_j\,,\, j\delta \le i\Delta,\, \bar{v}(p\Delta) = \bar{u}_{j_p}\,,\, p < i\}}$$

$$= P\{\bar{v}(i\Delta) = \bar{u}_l \mid W(j\delta) \in C_j\,,\, j\delta \le i\Delta,\, \bar{v}(p\Delta) = \bar{u}_{j_p}\,,\, p < i\}. \tag{5.9}$$

By the construction of $v^h(\cdot)$, the last term equals (using an obvious abuse of notation)

$$P\{v(i\Delta) = \bar{u}_l \mid w(j\delta) \in C_j\,,\, j\delta \le i\Delta,\, v(p\Delta) = \bar{u}_{j_p}\,,\, p < i\}. \tag{5.10}$$

The equality between (5.9) and (5.10) and the nonanticipative property of $\bar{v}(\cdot)$ imply that $(\bar{v}(\cdot), W(\cdot))$ has the same probability law as $(v(\cdot), w(\cdot))$. By the uniqueness assumption A9.1.4, the law of $(\xi(\cdot), \bar{v}(\cdot))$ is uniquely determined by the law of $(\bar{v}(\cdot), W(\cdot))$, and $(\xi(\cdot), \bar{v}(\cdot))$ has the same law as $(X(\cdot), v(\cdot))$. Thus, by the foregoing argument, Eqs. (5.5) and (5.6), and the arbitrariness of $\varepsilon > 0$, we can conclude that

Theorem 9.5.2 *Assume* A9.1.1 *to* A9.1.4 *and* A9.5.1. *Then*

$$R(x, \bar{u}) \leq R(x, u) \textit{ for all } u \in \hat{\mathscr{C}}_{\mathrm{R}}^{0}(x, G).$$

REMARK Let $k(y, \alpha) \geq 0$, all y, α, and let $b(\cdot) \equiv 0$. Then, whether or not $\rho = \tau(\bar{u})$, we have $\rho \geq \tau(\bar{u})$ and hence

$$R(x, \bar{u}) \leq \varliminf_{h \to 0} R^h(x, u^h).$$

In this case, we can drop condition A9.5.1 from Theorems 9.5.1 and 9.5.2. Note also that Assumption A9.5.1 is implied by

$$P_x\{\tau(\bar{u}) = \tau'(\bar{u})\} = 1.$$

The Optimality of $\bar{u}$ and $R(x, \bar{u})$

If $f(\cdot, \cdot)$ and $\sigma(\cdot)$ satisfy a uniform Lipschitz condition in x, then according to Theorem 9.3.3, $\bar{u} \in \mathscr{C}_{\mathrm{R}}(x)$; i.e., the solutions corresponding to the law of the limiting $(\bar{u}(\cdot), W(\cdot))$ are unique in probability law. We insist, at least for practical reasons, that the solutions be unique in law. Thus under a Lipschitz condition, $\bar{u}(\cdot)$ is a "legitimate" control and is optimal with respect to each $u \in \hat{\mathscr{C}}_{\mathrm{R}}^{0}(x, G)$. We selected the class $\hat{\mathscr{C}}_{\mathrm{R}}^{0}(x, G)$ because it was possible to "discretize" its members; i.e., for each $u(\cdot)$ in that class, we can define a $v^h(\cdot)$ which we can compare to $u^h(\cdot)$, and prove the optimality of $\bar{u}(\cdot)$ from the optimality of $u^h(\cdot)$ and weak convergence arguments. Any other class of "discretizable" controls would do as well. But it is difficult to discretize or to approximate an arbitrary control $u(\cdot)$ if control stops at the boundary ∂G, unless $\tau'(u) = \tau(u)$ w.p. 1. More work needs to be done on this point.

9.6 Control until a Boundary Is Reached: No Discounting

When the discount factor of Section 9.5 is dropped, a new problem arises. It is not always true that the optimization problem is well formulated for either the chain or for the diffusion. For example, there may not be any control under which the boundary ∂G is attainable, or the structure of the cost may be such that it is "cheaper" never to reach the boundary. We will treat some aspects of this problem. It should be emphasized that the problem does not arise because of the method of approximation, but because the original optimization problem may not be well formulated.

Assume A9.1.1 through A9.1.4 but drop the positivity assumption on $k(\cdot, \cdot)$. We will treat the cost $R(x, u)$ as defined by (4.7.9). The infimum of $R(x, u)$ over controls in $\bar{\mathscr{C}}_R^+(x, G)$, when considered to be a function of x, formally satisfies the partial differential equation (4.7.13). Again, any of the approximation methods 1–3 of Section 9.1 can be used, but for definiteness we will continue to use approximation 3 (although the only place where the actual properties of the particular approximation would appear is in the use of $\Delta t^h(x)$, $\Delta t^h(x, u)$, or $\Delta \tilde{t}^h(x)$). The discretization of (4.7.13) is

$$V^h(x) = \begin{cases} \min_{\alpha \in \mathscr{U}} [E_x^\alpha V^h(\xi_1^h) + k(x, u)\Delta t^h(x)], & x \in G_h, \\ b(x), & x \notin G_h. \end{cases} \tag{6.1}$$

Let $\mathscr{C}_R^+(h, x, G)$ denote the class of randomized policies for the chain $\{\xi_n^h\}$ with a finite average escape time from G_h, when $\xi_0^h = x$. Denote the *escape time* by m. For each $\pi \in \mathscr{C}_R^+(h, x, G)$ and corresponding chain $\{\xi_n^h\}$, define the cost

$$R^h(x, \pi) = E_x^\pi \left[\sum_{i=0}^{m-1} k(\xi_i^h, U_i)\Delta t_i^h + b(\xi_m^h) \right],$$

where U_i is the *actual* sample control realization at time i, and define

$$V^h(x) = \inf_{\pi \in \mathscr{C}_R^+(h, x, G)} R^h(x, \pi). \tag{6.2}$$

Equation (6.1) is only a formal representation for $V^h(\cdot)$. Unlike the situation for the stopping or discounted problems, it is not always guaranteed that (6.1) has a unique solution, which is the right-hand side of (6.2). For example, there may not be a control in $\mathscr{C}_R^+(h, x, G)$ even if there is one in $\bar{\mathscr{C}}_R^+(x, G)$ or in $\hat{\mathscr{C}}_R^+(x, G)$. Furthermore, even if such a control $u(\cdot)$ exists, it is conceivable that a lower cost (than achieved with the use of $u(\cdot)$) can be obtained by use of another control, which always keeps the process in† G_h. The problem is complicated because it involves both the properties of $k(\cdot, \cdot)$ and $b(\cdot)$ and the structure of the transition probabilities. We will show that the discretized problem is well formulated in two particular, but broad, cases. Intermediate cases can certainly be handled also provided that there is some condition which guarantees that the boundary will be reached under the optimal control. The following cases are illustrative of the possibilities and techniques.

† Recall that part of the problem statement is that we must reach the boundary or target in finite time.

CASE 1 There is a $T_0 < \infty$ such that, for each $\mathscr{U}$-valued nonanticipative measurable control $u(\cdot)$,

$$\sup_{y \in \bar{G}} E_y^u \tau' \le T_0. \tag{6.3}$$

[The sup should be taken over all $y \in \bar{G}$ for which there is a solution $X(\cdot)$, unique or not, and over all nonanticipative solutions.]

CASE 2 Fix the initial condition to be x. There is a real $k_0 > 0$ such that $k(y, \alpha) \ge k_0$, all $y \in R^n$, $\alpha \in \mathscr{U}$, and there is at least one $u \in \hat{\mathscr{C}}_R^+(x, G)$.

In case 2, there may not be a minimizing policy for $\{\xi_n^h\}$ for all initial conditions, but (for small h) where will be one when $\xi_0^h = x$. The optimal policy can, of course, be assumed to be *pure Markov and stationary*. Then, the optimal pure Markov and stationary policy $u^h(\cdot)$ will be defined for x and for all states which communicate with x under $u^h(\cdot)$.

Case 1 includes all problems with positive definite $a(\cdot)$ and various generalizations of the example in Section 4.4. In particular, it holds if there is an $a_0 > 0$ and an i such that $a_{ii}(x) \ge a_0$, $x \in \bar{G}$. For $t > \tau(u)$, the control $u(\cdot)$ can be defined arbitrarily. But the cases imply the assumption that, for each relevant $u(\cdot)$, there is an extension of $u(\cdot)$ to $[\tau(u), \infty)$ such that $\tau'(u)$ is defined, and the above properties hold. If $k(\cdot, \cdot) = 0$, then the discretized problem has a solution under weaker conditions, which will be remarked upon at the end of the section.

Theorem 9.6.1 *Assume* A9.1.1–A9.1.3. *Then for each small h, there is an optimal stationary pure Markov control $u^h(\cdot)$ and a unique solution to* (6.1) *which is the right-hand side of* (6.2). *Also*

$$\limsup_{h \to 0} E_x^{u^h} \rho_h < \infty, \tag{6.4}$$

where ρ_h is the interpolation of the hitting time m_h (under u^h).

PROOF CASE 1 Let $\pi^h(\cdot)$ denote an arbitrary policy in $\mathscr{C}_R(h)$ for the chain $\{\xi_n^h\}$ (and interpolated process $\xi^h(\cdot)$). Let q_h denote the corresponding interpolated escape time from G_h. Suppose that there is a $y \in \bar{G}$ and a sequence $x_h \to y$ as $h \to 0$ such that

$$P_{x_h}^{\pi^h}\{q_h \le T\} \to 0 \qquad \text{as} \quad h \to 0, \quad x_h \in G_h, \tag{6.5}$$

for each $T < \infty$.

The sequence $\{\xi^h(\cdot), W^h(\cdot), F^h(\cdot), K^h(\cdot), q_h\}$ is tight on $D^{3r+1}[0, \infty) \times \bar{R}^+$. Let h index a convergent subsequence with limit $(\xi(\cdot), W(\cdot), F(\cdot), K(\cdot), q)$.

Then (by use of the results of Section 9.2) there is a $\mathscr{U}$-valued nonanticipative measurable process $\bar{v}(\cdot)$ such that

$$\xi(t) = y + \int_0^t f(\xi(s), \bar{v}(s))\, ds + \int_0^t \sigma(\xi(s))\, dW(s),$$

$$K(t) = \int_0^t k(\xi(s), \bar{v}(s))\, ds.$$

By (6.3),

$$E_y^{\bar{v}} \tau' \leq T_0 .$$

Suppose that Skorokhod imbedding is used. This implies that there is an $\alpha_1 > 0$ and a $T_1 < \infty$ for which†

$$P_y^{\bar{v}}\{\tau' \leq T_1\} \geq \alpha_1. \tag{6.6}$$

Also, we have $\lim_{h \to 0} q_h \equiv q \leq \tau'(\bar{v})$ w.p. 1, which, together with (6.6), contradicts (6.5). Thus, there is a $T < \infty$ and an $\alpha > 0$ for which

$$P_y^{\pi^h}\{q_h \leq T\} \geq \alpha, \qquad \text{for all small } h,\ y \in G_h \quad \text{and} \quad \pi^h \in \mathscr{C}_R(h). \tag{6.7}$$

Let $\tilde{\xi}^h(\cdot)$ denote the stopped process $\xi^h(t \cap q_h)$. By (6.7) and an induction argument,

$$\begin{aligned} P_x^{\pi^h}\{q_h > nT\} &= E_x^{\pi^h} I_{\{\tilde{\xi}^h(nT - T) \in G\}} I_{\{\tilde{\xi}^h(nT) \in G\}} \\ &\leq E_x^{\pi^h} I_{\{\tilde{\xi}^h(nT - T) \in G\}} P_x^{\pi^h}\{q_h > nT \mid \tilde{\xi}^h(nT - T) \in G\} \\ &\leq E_x^{\pi^h} I_{\{\tilde{\xi}^h(nT - T) \in G\}}(1 - \alpha) \leq (1 - \alpha)^n. \end{aligned} \tag{6.8}$$

[We used the fact that (6.7) is uniform in $\pi^h(\cdot)$ and y.] Thus

$$E_x^{\pi^h} q_h \leq \int_0^\infty P_x^{\pi^h}\{q_h > t\}\, dt \leq T + T/\alpha,$$

which implies the theorem for Case 1.

CASE 2 Let $u(\cdot)$ denote an element of $\hat{\mathscr{C}}_R^+(x, G)$. By the assumption of Case 2, such an element exists. Consider the discounted problem with constant discount factor $\lambda > 0$ (replacing the function $\lambda(\cdot)$) and denote the discrete and continuous *discounted* costs (with policy π and control u, resp.) by $R_\lambda^h(x, \pi)$ and $R_\lambda(x, u)$, resp. Let $u_\lambda^h(\cdot)$ denote the pure Markov stationary control which optimizes $R_\lambda^h(x, \pi)$. By the results of the last section

$$\lim_{h \to 0} R_\lambda(x, u_\lambda^h) \leq R_\lambda(x, u). \tag{6.9}$$

† $\tau'(v)$ is the escape time from $\bar{G}$, under $v(\cdot)$, If $v(\cdot)$ is obvious, say from the operator E_x^v, we drop the argument of τ'.

Also,

$$R_\lambda(x, u) \le R(x, u) < \infty \tag{6.10}$$

for all $\lambda > 0$. The inequalities (6.9) and (6.10), the positivity of $k(\cdot, \cdot)$, and the arbitrariness of $\lambda > 0$, imply that there is a $\lambda > 0$ with corresponding optimal discrete controls $u_\lambda^h(\cdot)$ and (interpolated) hitting times ρ_h^λ and a constant K such that

$$E_x^{u_\lambda^h} \rho_h^\lambda \le K, \qquad \text{all small } h. \tag{6.11}$$

Equation (6.11) implies that the nondiscounted discrete problem is well formulated under Case 2. Q.E.D.

Theorem 9.6.2 *Assume* A9.1.1 *through* A9.1.3 (*except for the positivity condition on* $k(\cdot, \cdot)$). *Fix* x, *and let the optimal* $u^h(\cdot)$ *exist for each small* h. *Then the sequence (corresponding to* $u^h(\cdot)$) $\{\xi^h(\cdot), W^h(\cdot), F^h(\cdot), K^h(\cdot)\}$ *is tight on* $D^{3r+1}[0, \infty)$. *Let* h *index a weakly convergent subsequence with limit* $(\xi(\cdot), W(\cdot), F(\cdot), K(\cdot))$. *Then* $\xi(t) = x + F(t) + \int_0^t \sigma(\xi(s))\, dW(s)$. *There is a* $\mathscr{U}$*-valued, nonanticipative and* (ω, t) *measurable function* $\bar{u}(\cdot)$ *for which* (w.p. 1)

$$F(t) = \int_0^t f(\xi(s), \bar{u}(s))\, ds, \qquad K(t) = \int_0^t k(\xi(s), \bar{u}(s))\, ds.$$

The proof is exactly the same as for the corresponding result for the controlled stopping problem or for the controlled discounted problem.

Theorem 9.6.3 *Fix* x *and assume the conditions of Theorem* 9.6.2, *and let* h *index any convergent subsequence. Let* $\bar{u}(\cdot)$ *be the control referred to in Theorem* 9.6.2. *Assume Case I or II. Then*

$$\limsup_h E_x \rho_h < \infty, \tag{6.12}$$

where ρ_h *is the first time that* $\xi^h(\cdot)$ *leaves* G_h *under the optimal control* $u^h(\cdot)$. *Then* $E_x \tau(\bar{u}) < \infty$. *Assume that*

$$P_x\{\tau(\bar{u}) = \tau'(\bar{u})\} = 1 \tag{6.13}$$

(*or assume* A9.5.1). *Then*

$$\lim_{h\to 0} V^h(x) = R(x, \bar{u}) \le R(x, u) \qquad \text{all} \quad u \in \hat{\mathscr{C}}_R^+(x, G). \tag{6.14}$$

If $f(\cdot, \cdot)$ *and* $\sigma(\cdot)$ *satisfy a uniform Lipschitz condition in* x, *then* $\bar{u} \in \overline{\mathscr{C}}_R(x)$.

Proof The result $E_x \tau(\bar{u}) < \infty$ follows from (6.12) and Fatou's lemma. Let $R_\lambda(x, u)$ denote the discounted cost with constant discount factor $\lambda(x) \equiv \lambda$ and for the process $X(\cdot)$ under a control $u(\cdot)$ and let $u_\lambda^h(\cdot)$ denote the optimal pure Markov stationary control for the discounted problem on the chain $\{\xi_n^h\}$. Equation (6.12) follows from the argument in Theorem 9.6.1, which also implies that

$$\limsup_{\substack{h \to 0 \\ \lambda \to 0}} E_x^h \rho_\lambda^h < \infty.$$

For each $\varepsilon > 0$, (6.12) and the last estimate imply that there is a $\lambda_\varepsilon > 0$ such that, for $\lambda \le \lambda_\varepsilon$,

$$R^h(x, u^h) \le R_\lambda^h(x, u_\lambda^h) + \varepsilon. \tag{6.15}$$

Now, let $u \in \hat{\mathscr{C}}_{\mathrm{R}}^+(x, G)$. By Theorem 9.5.2 (when A9.5.1 is dropped from Theorem 9.5.2, $R(x, \bar{u})$ is replaced by $\overline{\lim}_h V^h(x)$).

$$\overline{\lim_{h \to 0}}\, R_\lambda^h(x, u_\lambda^h) \le R_\lambda(x, u). \tag{6.16}$$

Also, as $\lambda \to 0$,

$$R_\lambda(x, u) \to R(x, u). \tag{6.17}$$

Equations (6.15)–(6.17), Theorem 9.6.2, and the arbitrariness of ε imply the theorem. Q.E.D.

REMARK Suppose that $k(\cdot, \cdot) \equiv 0$ and $b(\cdot) \le 0$. Let the cost be $R(x, u) = E_x^u b(X(\tau)) I_{\{\tau < \infty\}}$. Then the discrete problem is always well formulated, and Theorem 9.6.2 continues to hold. Theorem 9.6.3 holds without (6.12), and if the $\hat{\mathscr{C}}_{\mathrm{R}}^+(x, G)$ in (6.14) is replaced by $\hat{\mathscr{C}}_{\mathrm{R}}(x, G)$. Cases I and II are not needed.

9.7 The Impulsive Control Problem

We are able to treat the impulsive control problem when there is also a continuously acting control. Only a sketch will be given. First we define a class of controls $\tilde{\mathscr{A}}_{\mathrm{R}}(x)$. We say that $(u(\cdot), \{\tau_i, u_i\}) \in \tilde{\mathscr{A}}_{\mathrm{R}}(x)$, if $u(\cdot)$ is a $\mathscr{U}$-valued measurable process, if the u_i are $\mathscr{V}$-valued random variables and (all i) $\tau_{i+1} \ge \tau_i \ge 0$, and if there is a Wiener process $w(\cdot)$ such that the functions $u(\cdot), \tau_i(\cdot), u_i(\cdot)$, (where $\tau_i(t) = I_{\{\tau_i \le t\}}$, $u_i(t) = u_i I_{\{\tau_i \le t\}}$) are nonanticipative with respect to $w(\cdot)$, and the equation

$$\begin{aligned} X(t) &= X(\tau_i^-) + u_i + \int_{\tau_i}^t f(X(s), u(s))\, ds + \int_0^t \sigma(X(s))\, dw(s), \\ X(0) &= x, \qquad t \in [\tau_i, \tau_{i+1}), \end{aligned} \tag{7.1}$$

has a unique (and nonanticipative) solution, in the sense that the law of $(X(\cdot), u(\cdot), \{\tau_i, u_i\})$ depends only on the law of $(w(\cdot), u(\cdot), \{\tau_i, u_i\})$.

Define the cost

$$R(x, u, \{\tau_i, u_i\}) = E_x\left[\sum_{i=1}^{\infty}(\exp - \lambda\tau_i)p(X(\tau_i^-), u_i) + \int_0^\infty (\exp - \lambda s)k(X(s), u(s))\, ds\right],$$

with

$$V(x) = \inf R(x, u, \{\tau_i, u_i\}),$$

where the inf is over the class $\tilde{\mathscr{A}}_R(x)$. There is an optimal pure Markov stationary policy for the discretized problem. Let $\{\xi^h(\cdot), F^h(\cdot), K^h(\cdot), W^h(\cdot), \{\tau_i^h, u_i^h\}\}$ denote the interpolated process and optimal jump times and jumps, and let $(\xi(\cdot), F(\cdot), K(\cdot), W(\cdot), \{\rho_i, \bar{u}_i\})$ denote the limit of a convergent subsequence. Define $\rho_i(\cdot), \bar{u}_i(\cdot)$, by $\rho_i(t) = I_{\{\rho_i \le t\}}, \bar{u}_i(t) = \bar{u}_i I_{\{\rho_i \le t\}}$. Using the techniques of Sections 9.1–9.5, we can show that $\{\rho_i(\cdot), \bar{u}_i(\cdot)\}$ are nonanticipative and that there is a measurable nonanticipative $\mathscr{U}$-valued function $\bar{u}(\cdot)$ such that

$$\begin{aligned} \xi(t) &= \xi(\rho_i^-) + \bar{u}_i + \int_0^t f(\xi(s), \bar{u}(s))\, ds + \int_0^t \sigma(\xi(s))\, dW(s), \\ K(t) &= \int_0^t (\exp - \lambda s)k(\xi(s), \bar{u}(s))\, ds. \end{aligned} \tag{7.2}$$

Under the assumptions A8.4.1, A9.1.1–A9.1.3, and a natural analog of A9.1.4, we can also show that

$$R(x, \bar{u}, \{\rho_i, \bar{u}_i\}) \le R(x, u, \{\rho_i, v_i\}) \tag{7.3}$$

for any $u, \{\rho_i, v_i\} \in \tilde{\mathscr{A}}_R(x)$. Under a Lipschitz condition on $f(\cdot, \cdot)$ and $\sigma(\cdot)$, we have the uniqueness under $\bar{u}, \{\rho_i, \bar{u}_i\}$; hence $R(x, \bar{u}, \{\rho_i, \bar{u}_i\}) = V(x)$.

The problem of multiple simultaneous jumps of $X(\cdot)$ or $\xi(\cdot)$ can be handled as indicated in Chapter 8.

9.8 Numerical Results

Figures 9.1–9.3 illustrate some numerical data for an optimal stopping and control problem for the system

$$dX_1 = X_2\, dt, \qquad dX_2 = -(X_1 + 3X_2)\, dt + u\, dt + \sqrt{2}\, dw,$$

where

$$R(x, \tau, u) = E_x^u \int_0^\tau [1 + 0.25|u(t)|]\, dt + E_x^u(X_1^2(\tau) + X_2^2(\tau)).$$

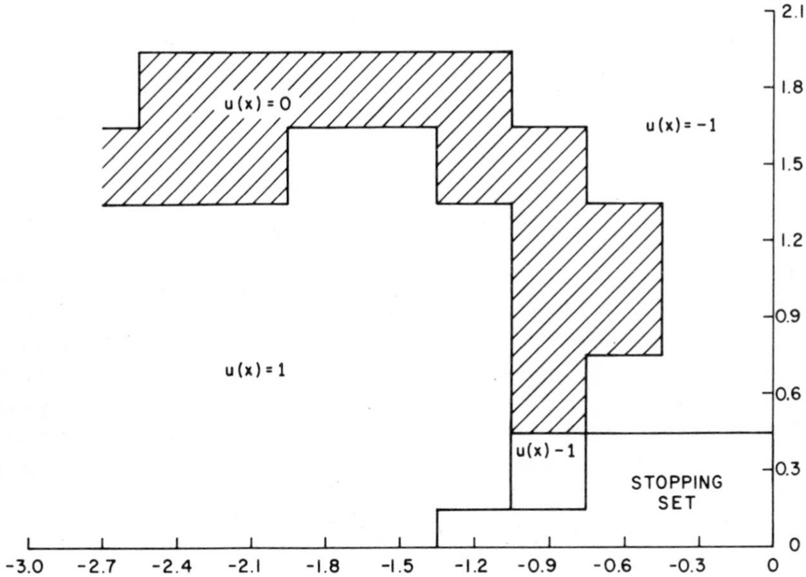

FIG. 9.1 Optimal stopping and control, decision sets, $h = 0.3$.

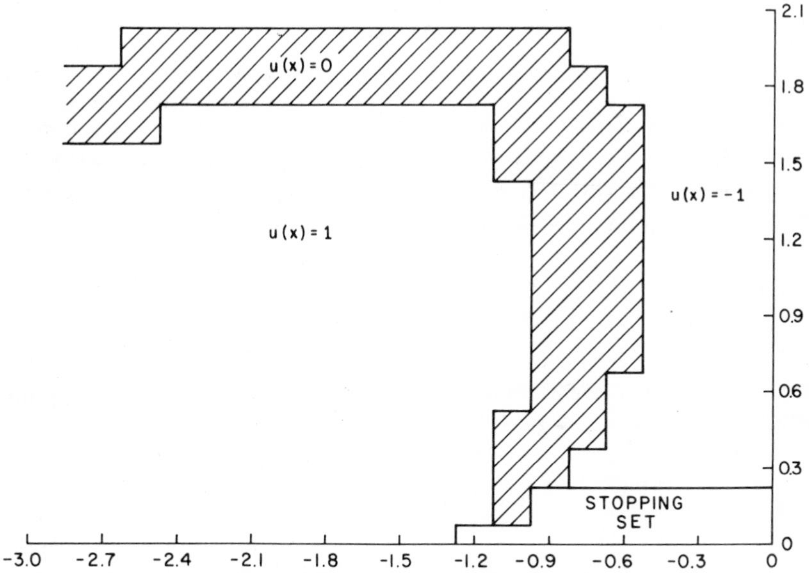

FIG. 9.2 Optimal stopping and control, decision sets, $h = 0.15$.

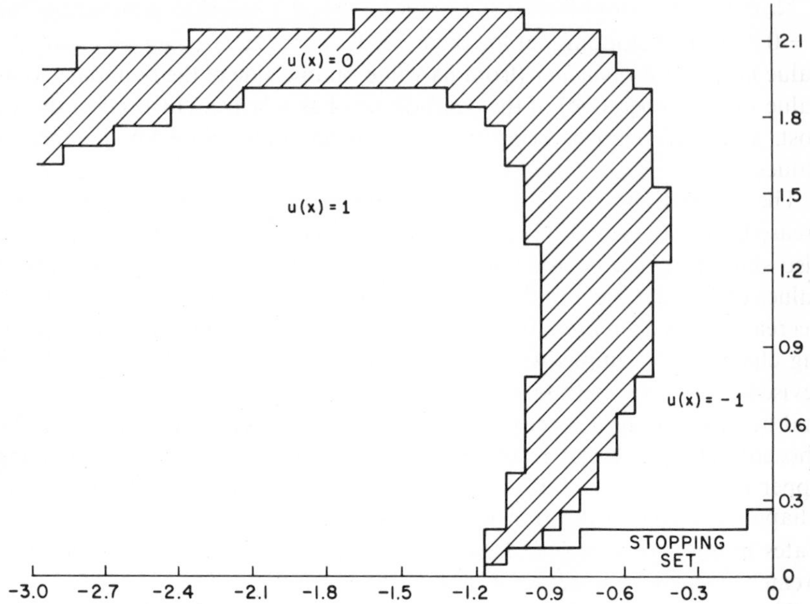

FIG. 9.3 Optimal stopping and control, decision sets, $h = 0.075$.

The figures give data for the revised version of method 1 (Section 9.1) where $\bar{f}_1(x) = f_1(x) = x_2$, $\bar{f}_2(x) = -(x_1 + 3x_2)$, and $\tilde{f}_2(x) = u$. There was enforced stopping on first exist from $G = \{x: |x_1| < 3, |x_2| < 3\}$, unless the process was stopped earlier.

The sets where $u(x) = 0, \pm 1$, or where we stop, are plotted for the upper left-hand quadrant only.

Since the algorithm yields values of $u(x)$ for x on the grid G_h only, we used the following naive rule to interpolate the decision sets from G_h to G: if the value of $u(x + e_i h)$ was different from that of $u(x)$, we assumed that the control for the continuous process would change values at $x + e_i h/2$ and similarly for $-h$ replacing h. The scheme is too simple. It is quite possible that improved decision sets would be obtained by first interpolating the $V^h(\cdot)$, $x \in G_h$, into a smooth function on G and then (assuming that the interpolation is optimal for the continuous time procedure or for a fixed time interval discretization of it) use the principle of optimality for the diffusion (or for a fixed time interval discretization of the diffusion) to get the decision sets. Also, a reasonable "smoothing" of the decision sets in the figures would bring them even closer to each other. In fact, such a smoothing would be used for an actual continuous time control.

Note that the stopping set is largest for $h = 0.3$, but that set is within the $\varepsilon = 0.2$ optimal stopping set for the $h = 0.075$ problem. In fact, changing the value $u(\cdot)$ on the *boundary* (from its value on one side of the boundary to its value on the other side) of the decision sets has relatively little effect on the cost, and such changes bring the decision sets quite close for the various values of h.

The simulations raised more questions than they answered. As h decreased, the values of $V^h(\cdot)$ decreased. The revised version of method 3 was also simulated. The decision sets were close to those in the figures, but the values of V^h were higher (about 20% or so), although the absolute difference decreased as h decreased. It is far from clear how to choose an approximating chain. For example, which of the methods is preferable (although the revised version of method 1 seems to work best) and why? Should $f(\cdot, \cdot)$ and $\sigma(\cdot)$ be averaged locally before discretization? How should the grid be chosen? The numerical results in our runs suggest that we are obtaining upper bounds to $V(\cdot)$, but we have not been able to prove that. Here, as in Chapter 8, physical intuition is helpful in selecting useful orderings of the states in the computations. This intuition can probably be used to a much greater extent in connection with the choice of algorithm.

The data indicate that the procedure is useful and reliable but much remains to be done before a reasonably good understanding of the numerical properties and best choices is available. It should be pointed out, however, that (at least at present) such a warning should be attached to any of the known numerical techniques, whether or not there are associated convergence theorems.

CHAPTER 10

Approximations to Stochastic Delay Equations and to Diffusions and Partial Differential Equations with Reflecting Boundaries

The chapter treats extensions of the concepts of Chapters 6 and 7 to two interesting classes of processes. The first class consists of processes which are solutions to stochastic differential delay equations. The discretizations and approximations of functionals are discussed.

The second class of processes may be considered to be diffusions which are reflected from a boundary. These processes are defined to be the solutions to the submartingale problem of Strook and Varadhan [S4], and are related to solutions of elliptic and parabolic equations with Neumann or mixed boundary conditions in the same way that the stochastic differential equation of Chapter 6 was related to solutions of elliptic and parabolic equations with Dirichlet and Cauchy boundary conditions. The discretizations are Markov chains which are reflected from a boundary. The solutions to the discretized equations are approximations to the solutions of the partial differential equations. The ideas are outlined and a convergence theorem stated. Details of the proof are not given but can be found in Kushner [K7]. Finally, several extensions to the approximation problem when there are boundary reflections are discussed.

10.1 Approximations to Stochastic Differential Delay Equations. Introduction

The ideas of Chapter 7 carry over to the case where the functional dependence of the $f(X(t), t)$ and $\sigma(X(t), t)$ of A7.1.1 on $X(t)$ is replaced by a smooth dependence on the solution over an interval of time, say over $[t-1, t]$. We will treat a simple one-dimensional problem. The higher-dimensional forms are treated similarly, but the state space becomes so large that the method will probably be impractical. For any α, β, $\alpha < \beta$, let $C[\alpha, \beta]$ denote $C^1[\alpha, \beta]$.

Let μ denote a finite measure on the interval $[-1, 0]$. Assume that

A10.1.1 *$f(\cdot, \cdot)$, $\sigma(\cdot, \cdot)$, $k(\cdot, \cdot)$, $b_1(\cdot, \cdot)$, and $b_T(\cdot)$ are bounded continuous real-valued functions on $C[-1, 0] \times [0, T]$ and (for $b_T(\cdot)$ only) on $C[-1, 0]$, resp. Assume*

$$|f(x(\cdot), t) - f(y(\cdot), t)| + |\sigma(x(\cdot), t) - \sigma(y(\cdot), t)| \leq \int_{-1}^{0} |x(t) - y(t)| \mu(dt), \tag{1.1}$$

$$\inf_{x(\cdot),\, 0 \leq t \leq T} \sigma(x(\cdot), t) > 0,$$

where $x(\cdot)$, $y(\cdot)$ range over $C[-1, 0]$ and $x(t)$ is the value of the function $x(\cdot)$ at time $t \in [-1, 0]$.

For any continuous function $Y(\cdot)$ on $[-1, T]$, let Y_t denote the element of $C[-1, 0]$ with values $Y_t(s) = Y(t + s)$, $s \in [-1, 0]$, $t \geq 0$.

The equation

$$X(t) = x_0(0) + \int_0^t f(X_s, s)\, ds + \int_0^t \sigma(X_s, s)\, dw(s) \tag{1.2}$$

has a unique, continuous, nonanticipative solution for each $w(\cdot)$ process and initial condition $x_0 \in C[-1, 0]$. [The condition (1.1) plays the same role that the uniform Lipschitz condition plays in the ordinary case.] See Itô and Nisio [I2]. Let $b < c$ denote real numbers and let G denote an open set in $C[-1, 0]$ such that $G \subset \{y(s) : y(0) \varepsilon (b, c)\}$. *Henceforth let x_0 be fixed.* If τ and τ' denote the first exit times from G and $\bar{G}$, resp., suppose that

$$P_{x_0}\{\tau = \tau' | \tau \leq T\} = 1, \qquad P_{x_0}\{\tau = T\} = 0. \tag{1.3}$$

Equation (1.3) is satisfied, for example, by any set of the form

$$G = \{y(\cdot)\,;\, y(-a_1) \in A_1, \ldots, y(-a_s) \in A_s\},$$

where $a_i \in [-1, 0]$ and the A_i are finite unions of open intervals.

Define the functional, for some $T > 0$,

$$R(x_0) = E_{x_0}\left[\int_0^T k(X_s, s)\,ds + b_1(X_\tau, \tau)I_{\{\tau \le T\}} + b_T(X_T)I_{\{\tau > T\}}\right]. \quad (1.4)$$

The solution $X(\cdot)$ to (1.2) is not a Markov process but the sequence $\{X_t, t \ge 0\}$ is when we define the probability space appropriately. Since the state space of X_t is infinite-dimensional, we cannot expect to be able to approximate the solution to (1.2) by an interpolated Markov chain whose variables are real-valued. The dimensionality of the approximating chain must grow as the finite difference intervals go to zero. There is an abstract operator which is related to functionals of (1.2) in the same way that the operator $(\partial/\partial t + \mathscr{L})$ was related to the functionals of (7.1.1). However, we will proceed in a simpler way by using the technique of approximation which was developed in Chapter 7, with finite difference intervals h and Δ.

Definitions Assume that $1/\Delta$ is an integer and define $L_\Delta = 1 + 1/\Delta$. Let x_0 denote the fixed initial condition for (1.2). Define $\xi_0^{h,\Delta} \equiv (x_0(0), x_0(-\Delta), \ldots, x_0(-1)) \equiv (\xi^{h,\Delta}(0), \xi^{h,\Delta}(-1), \ldots, \xi^{h,\Delta}(1 - L_\Delta))$, where the equality defines $\xi^{h,\Delta}(i)$, $i \le 0$. The number L_Δ is the dimension of the state space of the approximating Markov chain. We will shortly define the sequence $\xi^{h,\Delta}(n)$ for all integers $n > 0$. With this in mind, define $\tilde{\xi}^{h,\Delta}(\cdot)$ to be the *piecewise linear interpolation* of $\{\xi^{h,\Delta}(n), n \ge 1 - L_\Delta\}$, *with constant interpolation intervals* Δ. Define the vector $\xi_n^{h,\Delta} = (\xi^{h,\Delta}(n), \ldots, \xi^{h,\Delta}(n - L_\Delta + 1))$. We use a piecewise linear (rather than piecewise constant) interpolation since $f(\cdot, t)$, etc., are only defined on $C[-1, 0]$ and it seems to be the simplest procedure in the present case. The interpolations are elements of $C[-1, T]$.

The Process $\{\xi^{h,\Delta}(n)\}$

For $n > 0$, define the random variable $\xi^{h,\Delta}(n)$ via the transition probability (7.2.6a, b) as follows:

$$\begin{aligned} P\{\xi^{h,\Delta}(n+1) &= \xi^{h,\Delta}(n) \mid \tilde{\xi}_{n\Delta}^{h,\Delta}\} \\ &= 1 - (\Delta/h^2)[h\,|f(\tilde{\xi}_{n\Delta}^{h,\Delta}, n\Delta)| + \sigma^2(\tilde{\xi}_{n\Delta}^{h,\Delta}, n\Delta)] \\ P\{\xi^{h,\Delta}(n+1) &= \xi^{h,\Delta}(n) \pm h \mid \tilde{\xi}_{n\Delta}^{h,\Delta}\} \\ &= (\Delta/h^2)[\sigma^2(\tilde{\xi}_{n\Delta}^{h,\Delta}, n\Delta)/2 + hf^{\pm}(\tilde{\xi}_{n\Delta}^{h,\Delta}, n\Delta)]. \end{aligned} \quad (1.5)$$

We assume, as in Chapter 7, that the first term of (1.5) is nonnegative.

If $N\Delta = T$ and $N(h, \Delta)$ denotes the first value of n for which $\tilde{\xi}_{n\Delta}^{h, \Delta} \notin G$ and $\tau(h, \Delta) = N(h, \Delta)\Delta$, then we define the approximation

$$R^{h, \Delta}(x_0) = E_{x_0}\left[\sum_{i=0}^{(N \cap N(h, \Delta))-1} k(\tilde{\xi}_{i\Delta}^{h, \Delta}, i\Delta)\Delta + b_1(\tilde{\xi}_{\tau(h, \Delta)}^{h, \Delta}, \tau(h, \Delta))I_{\{N(h, \Delta)\le N\}}\right.$$

$$\left. + b_T(\tilde{\xi}_T^{h, \Delta})I_{\{N(h, \Delta)>N\}}\right]. \tag{1.6}$$

The Markov chain $\tilde{\xi}_n^{h, \Delta}$ approximates $X_{n\Delta}$ and the law of the chain can be used to compute (1.6) recursively.

For each value of $\xi_n^{h, \Delta}$, there are at most three values of $\xi_{n+1}^{h, \Delta}$. Furthermore, if $n > 0$, then $\xi^{h, \Delta}(n) - \xi^{h, \Delta}(n - 1)$ can differ by at most h. If this is true for $n = -L_\Delta + 2, \ldots, 0$ also, then there are at most

$$[(b - c)/h - 1]3^{1/\Delta}$$

states in the state space of $\{\xi_n^{h, \Delta}\}$ which need to be used.

Tightness and Convergence

We can write

$$R^{h, \Delta}(x_0) = E_{x_0}\left[\int_0^{T \cap \tau(h, \Delta)} k(\tilde{\xi}_s^{h, \Delta}, s)\, ds + b_1(\tilde{\xi}_{\tau(h, \Delta)}^{h, \Delta}, \tau(h, \Delta))I_{\{\tau(h, \Delta)\le T\}}\right.$$

$$\left. + b_T(\tilde{\xi}_T^{h, \Delta}, T)I_{\{\tau(h, \Delta)>T\}}\right] + \tilde{\varepsilon}(h, \Delta),$$

where $\tilde{\varepsilon}(h, \Delta)$ is an error due to the approximation of the sum by the integral and goes to zero as $h, \Delta \to 0$. Write, in obvious notation,

$$\tilde{\xi}^{h, \Delta}(i\Delta) = x_0(0) + \sum_{j<i} f(\tilde{\xi}_{j\Delta}^{h, \Delta}, j\Delta)\Delta + \sum_{j<i}\beta^{h, \Delta}(j), \tag{1.7}$$

$$\tilde{\xi}^{h, \Delta}(t) = x_0(0) + \int_0^t f(\tilde{\xi}_s^{h, \Delta}, s)\, ds + B^{h, \Delta}(t) + \varepsilon^{h, \Delta}(t)$$

$$= x_0(0) + F^{h, \Delta}(t) + B^{h, \Delta}(t) + \varepsilon^{h, \Delta}(t), \tag{1.8}$$

where $B^{h, \Delta}(t)$ is a linear interpolation of the function of t which is defined by $\sum_{j\Delta<t}\beta^{h, \Delta}(j)$ and the error $\varepsilon^{h, \Delta}(\cdot)$ is due to the use of the interpolation and to the approximation of the sum of the f terms by the integral and goes to zero as $h, \Delta \to 0$.

The sequence $\{\tilde{\xi}^{h, \Delta}(\cdot), B^{h, \Delta}(\cdot), \varepsilon^{h, \Delta}(\cdot)\}$ is tight on $C^3[-1, T]$. Let h, Δ index a convergent subsequence with limit $(\xi(\cdot), B(\cdot), 0)$. Note that, if $\tilde{\xi}^{h, \Delta}(\cdot)$ converges to $\xi(\cdot)$ w.p. 1 on $[-1, T]$, then $\tilde{\xi}_t^{h, \Delta} \to \xi_t$, w.p. 1, $t \in [0, T]$ as

$h, \Delta \to 0$. There is a Wiener process $W(\cdot)$ such that $\xi(\cdot)$ and $B(\cdot)$ are nonanticipative with respect to $W(\cdot)$ and (1.2) holds for $\xi(\cdot) = X(\cdot)$ and $W(\cdot) = w(\cdot)$. By (1.3) and the uniqueness of the solution to (1.2),

$$R^{h,\Delta}(x_0) \to R(x_0), \qquad \text{as } h, \Delta \to 0.$$

The technique of discretization of Chapter 6 is more difficult (than that of Chapter 7) for the delay problem since the interpolation intervals Δt_i^h would be random and the minimal number of terms of $\{\xi^h(n)\}$ which are needed to represent the process on each interval $[t-1, t]$ would be random and would vary with t.

10.2 Approximations to Elliptic and Parabolic Equations with Neumann Boundary Conditions. Formulation

The Martingale Problem

An alternative formulation of the problem of existence and uniqueness of the solution of the stochastic differential equation

$$X(t) = x + \int_0^t f(X(s))\, ds + \int_0^t \sigma(X(s))\, dw(s) \tag{2.1}$$

is the following, where we suppose that $f(\cdot)$ and $\sigma(\cdot)$ do not depend on time, purely for notational simplicity. Let $q(\cdot, \cdot)$ be a bounded function in $C^{2,1}[R^r \times [0, \infty)]$ and define the functional $F_q(\cdot, \cdot)$ on $C^r[0, \infty) \times [0, \infty)$ by

$$F_q(x(\cdot), t) = q(x(t), t) - q(x(0), 0) - \int_0^t [\partial/\partial s + \mathscr{L}]q(x(s), s)\, ds. \tag{2.2}$$

Suppose that, for each $x \in R^r$, there is a measure P_x on $C^r[0, \infty)$ such that $P_x\{x(0) = x\} = 1$ and, for each $q(\cdot, \cdot)$, the process with values† $M_q(\omega, t) \equiv F_q(x(\cdot), t)$ is a martingale (with respect to the sequence $\mathscr{B}_t$, $0 \le t < \infty$, of sub-σ-algebras of $C^r[0, \infty)$ which are induced by the projections $x(s)$, $s \le t$). Then $\{P_x, x \in R^r\}$ is said to be a solution to the martingale problem.

If $a(\cdot)$ is bounded, continuous, and strictly positive definite and $f(\cdot)$ bounded and measurable, then there is a unique solution to the martingale problem (Strook and Varadhan [S3]). There is a unique solution to the martingale problem if and only if there is a unique solution (in the sense of probability law) to (2.1) for each $x \in R^r$ [K1]. Thus, the martingale problem has a unique solution if $f(\cdot)$ and $\sigma(\cdot)$ satisfy a uniform Lipschitz condition. Note that, if $X(\cdot)$ is a solution to (2.1) with $X(0) = x$, then by Itô's Lemma

$$F_q(X(\cdot), t) = \int_0^t q_x'(X(s), s)\sigma(X(s))\, dw(s) \tag{2.3}$$

† Here $\omega = x(\cdot)$ is the generic variable of the probability space $C^r[0, \infty)$.

is a martingale under P_x, the measure induced by the solution (2.1), and the appropriate sequence of σ-algebras.

Thus, the solution to the martingale problem is intimately related to solutions of elliptic and parabolic equations with Cauchy or Dirichlet types of boundary conditions and to diffusion processes for which $\mathscr{L}$ is the differential generator. There is a very nice extension of the idea of "martingale problem," due to Strook and Varadhan [S4], which relates a class of reflecting diffusions to solutions of the partial differential equations with Neumann boundary conditions.

The Submartingale Problem

Let G denote a bounded open set, $\gamma(\cdot)$ a bounded continuous real-valued, and $\phi(\cdot)$ a twice continuously differentiable real-valued function such that $G = \{x : \phi(x) > 0\}$, $\partial G = \{x : \phi(x) = 0\}$, $|\phi_x(x)| \geq 1$ on ∂G, and $\inf_{x \in \partial G} \phi_x'(x)\gamma(x) > 0$. Let $\rho = 0$ or 1. Suppose that, for each $x \in \bar{G}$, there is a measure P_x on $\bar{C}_G^r[0, \infty) \equiv C^r[0, \infty) \cap \{x(\cdot) : x(t) \in \bar{G} \text{ all } t < \infty\}$, such that $P_x\{x(0) = x\} = 1$ and, for each $q(\cdot, \cdot) \in C^{2,1}[\bar{G} \times [0, \infty)]$, for which

$$\rho q_t(x, t) + \gamma'(x)q_x(x, t) \geq 0 \quad \text{on} \quad \partial G, \qquad t \geq 0, \tag{2.4}$$

the process defined by

$$F_q(x(\cdot), t) = q(x(t), t) - q(x(0), 0) - \int_0^t [\partial/\partial s + \mathscr{L}]q(x(s), s)I_G(x(s))\, ds \tag{2.5}$$

is a submartingale, with respect to $\{\mathscr{B}_t\}$, where $\mathscr{B}_t$ is the σ-algebra induced by the projections $x(s)$, $s \leq t$. Then $\{P_x, x \in \bar{G}\}$ is said to be a *solution of the* submartingale problem (Strook and Varadhan [S4]).

Let $a(\cdot)$ be strictly positive definite and $a(\cdot)$ and $f(\cdot)$ bounded and continuous. Then [S4] there is a unique solution to the submartingale problem. Furthermore [S4], there is a unique nondecreasing scalar-valued process $\mu(\cdot)$, which increases only when $x(t) \in \partial G$, such that

$$F_q(x(\cdot), t) - \int_0^t (\rho q_s(x(s), s) + \gamma'(x(s))q_x(x(s), s))\, d\mu(s) \tag{2.6}$$

is a martingale (with respect to P_x and $\{\mathscr{B}_t\}$). The measure P_x induces a process which satisfies

$$X(t) = x + \int_0^t f(X(s))I_G(X(s))\, ds + \int_0^t \sigma(X(s))I_G(X(s))\, dw(s) + \int_0^t \gamma(X(s))\, d\mu(s) \tag{2.7}$$

for some Wiener process $w(\cdot)$. The last integral in (2.7) accounts for the reflection into G from ∂G. It is the term which "pushes" $X(t)$ away from ∂G in the direction $\gamma(X(t))$ when $X(t) \in \partial G$. This will be made clear when we develop the Markov chain approximation.

Let $\lambda(\cdot) \geq 0$, $\beta(\cdot) \geq 0$, $g(\cdot)$, $b_T(\cdot)$, $k(\cdot)$, and $b(\cdot)$ be real-valued, bounded continuous functions on some neighborhood N_1 of $\bar{G}$, with $\lambda(x) \geq \lambda_0 > 0$, where λ_0 is a real number, and let $b_1(\cdot, \cdot)$ be a bounded continuous function on $N_1 \times [0, \infty)$. Consider the two sets of equations (2.8) and (2.9).

$$\mathscr{L}V(x) + k(x) - \lambda(x)V(x) = 0, \qquad x \in G, \tag{2.8a}$$

$$\gamma'(x)V_x(x) - \beta(x)V(x) + g(x) = 0, \qquad x \in \partial G, \tag{2.8b}$$

$$V_t(x, t) + \mathscr{L}V(x, t) + k(x) - \lambda(x)V(x) = 0, \qquad x \in G \times [0, T), \tag{2.9a}$$

$$\rho V_t(x, t) + \gamma'(x)V_x(x) - \beta(x)V(x) + g(x) = 0,$$
$$x \in \partial G \times [0, T), \quad \rho = 0 \text{ or } 1, \tag{2.9b}$$
$$V(x, T) = b_T(x), \qquad x \in \bar{G}.$$

Define

$$B(s, t) = \exp - \int_s^t \beta(X(v))\, d\mu(v), \qquad B(t) = B(0, t),$$

$$A(s, t) = \exp - \int_s^t \lambda(X(v))I_G(X(v))\, dv, \qquad A(t) = A(0, t),$$

$$D(s, t) = A(s, t)B(s, t), \qquad D(t) = A(t)B(t),$$

$$R(x) = E_x \int_0^\infty D(s)k(X(s))I_G(X(s))\, ds + E_x \int_0^\infty D(s)g(X(s))\, d\mu(s), \tag{2.10}$$

$$\begin{aligned} R(x, t) = {} & E_{x,t} \int_t^T D(t, s)k(X(s))I_G(X(s))\, ds \\ & + E_{x,t} \int_t^T D(t, s)g(X(s))\, d\mu(s) \\ & + E_{x,t} D(t, T)b_T(X(T)). \end{aligned} \tag{2.11}$$

The functions (2.10) and (2.11) have the same relationship to Eqs. (2.8) and (2.9) as Eqs. (4.2.2) and (4.2.4b) had to (4.5.2) and discounted (4.5.4), resp. Boundary costs $b(\cdot)$ and $b_1(\cdot, \cdot)$ and hitting time τ do not appear since the boundary is reflecting. The functions defined by (2.10) and (2.11) are weak solutions to (2.8) and (2.9), where the underlying process $X(\cdot)$ is the unique solution to the appropriate submartingale problem.

Weakened Conditions; $a(\cdot)$ Not Necessarily Positive Definite

The positive definiteness of $a(\cdot)$ was used in Strook [S4] in several ways. Indeed, it was basic to the proof of uniqueness. More important for our purposes, it implied the estimate (2.12). Let N_ε denote the ε-neighborhood of ∂G. Then [S4, Lemma 3.3] for each real $T < \infty$, there is a real K_T such that

$$E_x \int_0^T I_{N_\varepsilon}(X(s)) I_G(X(s))\, ds \le K_T \varepsilon. \tag{2.12}$$

The estimate (2.12) says that $X(\cdot)$ spends "little" time near ∂G (unless it is on ∂G). It is particularly useful in the approximation problem. For example, if $\xi^h(\cdot)$ is the "approximating" process, then to show convergence of the approximations to the functionals (2.10) or (2.11) (or even to the paths (2.7)), we need to show that (for an appropriate definition of G_h, the finite difference grid on G)

$$I_{G_h}(\xi^h(s)) \to I_G(X(s)), \qquad \text{w.p. 1, almost all } s \tag{2.13}$$

(using Skorokhod imbedding) as $h \to 0$. If the paths $X(\cdot)$ lie in a compact set contained in G, then (2.13) is obvious. The estimate (2.12) (and a similar estimate for $\xi^h(\cdot)$) allows us to act *almost* as if the paths $X(\cdot)$ were contained in such a compact set, except for the actual time spent on the boundary. So even if we weaken the conditions, we will still need some condition which guarantees (2.12).

In Chapters 6 and 7 the assumptions were of two types. One type concerned the continuity of various functions and the second type concerned uniqueness (in the sense of the probability law) of the solution to (2.1) for each initial condition. We will do the same thing here. The properties of the functions listed above (2.8) will always be assumed to hold.

Let us split ∂G into two disjoint parts, ∂G_R and ∂G_A, where ∂G_A is relatively open, the process $X(\cdot)$ being reflected on the first and absorbed on the second. Replace the boundary condition (2.8b) by

$$\begin{aligned} \gamma'(x)V_x(x) - \beta(x)V(x) + g(x) &= 0, \qquad x \in \partial G_R \\ V(x) &= b(x), \qquad x \in \partial G_A, \end{aligned} \tag{2.8b$'$}$$

and (2.9b) by

$$\begin{aligned} \rho V_t(x, t) + \gamma'(x)V_x(x) - \beta(x)V(x) + g(x) &= 0, \qquad x \in \partial G_R \times [0, T), \\ V(x, t) &= b_1(x, t), \qquad x \in \partial G_A \times [0, T), \\ V(x, T) &= b_T(x), \qquad x \in \bar{G}. \end{aligned} \tag{2.9b$'$}$$

We make the following assumptions on the process $X(\cdot)$. There is a (unique) solution to the submartingale problem for $t \le \tau$, where $\tau =$ first hitting time of ∂G_A. This assumption holds if $a(\cdot)$ is positive definite on a neighborhood of ∂G_R, and the stochastic differential equation (2.1) has a unique solution (in the sense of probability law) on $[0, \infty)$ for any initial condition which is independent of $w(\cdot)$.

Assume that there is a real $K > 0$ such that

$$\phi'_x(x)a(x)\phi_x(x) \geq K|\phi_x(x)|^2 \quad \text{on} \quad \partial G_R. \tag{2.14}$$

The estimate (2.14) is used to prove (2.12) (with T replaced by $T \cap \tau$) and the related estimate (2.15) [K7], where ρ_h is the first time that $\xi^h(\cdot)$ hits the discretization of ∂G_A and $\xi^h(\cdot)$ is the interpolation of the approximating chain.

$$E_x \int_0^{T \cap \rho_h} I_{N_\varepsilon}(\xi^h(s))I_{G_h}(\xi^h(s))\, ds \leq K_T \varepsilon \qquad \text{for small} \quad h. \tag{2.15}$$

Assumption (2.14) states, essentially, that if the diffusion is degenerate at some $x \in \partial G_R$, then it cannot be (locally) concentrated in the tangent plane to ∂G_R at x. Indeed, the assumption (2.14) together with $\gamma'(x)\phi_x(x) > 0$ insure that the reflection is well defined; i.e., that the process does not "chatter" near the boundary but eventually drifts away, even though it may hit the boundary infinitely often in some small interval of time.

In order to avoid an ambiguity in the boundary condition on $\overline{\partial G_A} \cap \overline{\partial G_R}$, we assume

$$P_x\{X(\tau) \in \overline{\partial G_A} \cap \overline{\partial G_R}\} = 0. \tag{2.16}$$

Also, suppose that

$$\begin{gathered}\tau \text{ is continuous w.p. 1 relative to the measure on} \\ C^r[0, \infty) \text{ or on } D^r[0, \infty) \text{ induced by } X(\cdot),\end{gathered} \tag{2.17}$$

and, in the case of the parabolic equation, that

$$P_{x,t}\{\tau = T\} = 0. \tag{2.18}$$

We now define (2.19) to be the weak-sense solution to (2.8a) and (2.8b′), and (2.20) to be the weak-sense solution to (2.9a) and (2.9b′). Actually, there do not seem to be specific references which we can cite to justify the definition. Yet the development in [S4] strongly suggests that (2.19) and (2.20) are the weak solutions. In any case, the approximations converge to (2.19) and (2.20).

$$\begin{aligned} R(x) = E_x \int_0^\tau D(s)k(X(s))I_G(X(s))\, ds + E_x D(\tau)b(X(\tau)) \\ + E_x \int_0^\tau D(s)g(X(s))\, d\mu(s), \end{aligned} \tag{2.19}$$

$$R(x, t) = E_{x,t} \int_t^{\tau \cap T} D(t, s)k(X(s))I_G(X(s))\, ds + E_{x,t} D(t, \tau)b_1(X(\tau), \tau)I_{\{\tau < T\}}$$
$$+ E_{x,t} \int_t^{\tau \cap T} D(t, s)g(X(s))\, d\mu(s) + E_{x,t} D(t, \tau)b_T(X(\tau))I_{\{\tau \geq T\}}. \tag{2.20}$$

Owing to the splitting of the boundary, Eq. (2.7) is replaced by

$$X(t) = x + \int_0^{t \cap \tau} f(X(s))I_G(X(s))\, ds + \int_0^{t \cap \tau} \sigma(X(s))I_G(X(s))\, dw(s)$$
$$+ \int_0^{t \cap \tau} \gamma(X(s))I_{\partial G_R}(X(s))\, d\mu(s). \tag{2.21}$$

10.3 The Finite Difference Approximations

We will concentrate on the elliptic case (2.8a) and (2.8b′) and its solution (2.19). The problem for the parabolic case involves only minor changes (see Kushner [K7]). By applying a finite difference method similar to the one used in Chapter 6, the process (2.21) will be approximated by a "reflected and absorbed" chain. The procedure is much the same as that which we used previously; the only new difficulties concern the behavior near ∂G_R.

Define the set ∂G_R^h in $\bar{G} \cap R_h^r$ as follows: $x \in \partial G_R^h$ if $x \in \bar{G} \cap R_h^r$ and either $x \in \partial G_R$ or the line connecting x to one of the neighboring grid points $x \pm e_i h$ or $x \pm e_i h \pm e_j h$ or $x \pm e_i h \mp e_j h$ touches ∂G_R. The set ∂G_R^h approximates ∂G_R from the "inside." Define $G_h = G \cap R_h^r - \partial G_R^h$, the discretized interior of G, and $\partial G_A^h = R_h^r - G_h - \partial G_R^h$, the discretized stopping set.

To discretize (2.8a), use (6.2.1)–(6.2.3) and assume (6.2.4). Then (6.2.5) gives the transition probability of the approximating chain $\{\xi_n^h\}$ in G_h. Suppose that the approximating chain *stops* on first contact with ∂G_A^h. To complete the construction of the approximating chain, we need to discretize the first line of (2.8b′). To do this we use

$$V_{x_i}(x) \to \begin{cases} [V(x + e_i h) - V(x)]/h & \text{if } \gamma_i(x) \geq 0, \\ [V(x) - V(x - e_i h)]/h & \text{if } \gamma_i(x) < 0,\ x \in \partial G_R^h. \end{cases} \tag{3.1}$$

Define $|\gamma(x)| = \sum_{i=1}^r |\gamma_i(x)|$ and define the transition probabilities on ∂G_R^h by

$$p^h(x, x \pm e_i h) = \gamma_i^{\pm}(x)/|\gamma(x)|. \tag{3.2}$$

Note that

$$E_x(\xi_{n+1}^h - \xi_n^h \mid \xi_n^h = y \in \partial G_R^h) = \gamma(y)h/|\gamma(y)|, \tag{3.3}$$

which is consistent with the fact that the reflection from a point $y \in \partial G_R$ is in the direction $\gamma(y)$. In order for (3.2) to make sense, we must require that

$$x + e_i h \text{ sign } \gamma_i(x) \in \bar{G} \qquad \text{for small } h \text{ and } x \in \partial G_R^h (\text{sign } 0 = 0). \quad (3.4)$$

Condition (3.4) is not essential. It can be dropped if we modify (3.1) and such modifications (still yielding a Markov chain approximation) will always exist. We stay with (3.1) and (3.4) purely for notational convenience. Thus we have constructed a Markov chain which is reflected from ∂G_R^h, and stopped on ∂G_A^h.

Now, let us return to the discretization of (2.8a). The discretization (for $x \in G_h$) is done exactly as it was done for the discounted case in Section 6.5. For notational definiteness, assume that $\sup_x \Delta t^h(x) \to 0$ as $h \to 0$ and use (6.5.3). Thus, letting $V^h(\cdot)$ denote the solution to the finite difference equation, we have

$$V^h(x) = \exp -\lambda(x)\, \Delta t^h(x)[E_x V^h(\xi_1^h) + k(x)\, \Delta t^h(x)], \qquad x \in G_h. \quad (3.5)$$

Constructions similar to those used in Section 6.5 can be used at $x \in \partial G_R^h$. A direct approximation of the first line of (2.8b′) yields

$$V^h(x) = \sum_{i,\pm} p^h(x, x \pm e_i h) V^h(x \pm e_i h) + \frac{hg(x)}{|\gamma(x)|} - \frac{h\beta(x)}{|\gamma(x)|} V^h(x),$$
$$x \in \partial G_R^h. \quad (3.6)$$

Define

$$d\mu^h(x) = h/|\gamma(x)|, \qquad d\mu_i^h = d\mu^h(\xi_i^h) I_{\partial G_R^h}(\xi_i^h)$$
$$\mu_n^h = \mu_{n-1}^h + d\mu_{n-1}^h, \qquad \mu_0^h = 0.$$

The following approximations to (3.6) can be used:

$$V^h(x) = (1 - \beta(x)\, d\mu^h(x))[E_x V^h(\xi_1^h) + g(x)\, d\mu^h(x)], \quad (3.7a)$$

$$V^h(x) = (\exp -\beta(x)\, d\mu^h(x))[E_x V^h(\xi_1^h) + g(x)\, d\mu^h(x)], \quad (3.7b)$$

$$V^h(x) = (\exp -\beta(x)\, d\mu^h(x)) E_x V^h(\xi_1^h) + \frac{(1 - \exp -\beta(x)\, d\mu^h(x))}{\beta(x)} g(x). \quad (3.7c)$$

If $d\mu^h(x) \to 0$ as $h \to 0$ uniformly for $x \in \partial G_R$, then (3.7a–c) can all be used. (This is the case here since $\inf |\gamma(x)| > 0$ on ∂G_R.) If not, then we will have to use (3.7c), as we had to use (6.5.4) when $\Delta t^h(x) \not\to 0$ uniformly in x as $h \to 0$. It is possible to handle both absorbing and reflecting boundaries simultaneously if we set $b(x) = g(x)/\beta(x)$, let $\gamma(x) = 0$ on ∂G_A, and use (3.7c). But, for simplicity, we use (3.7b), and continue to use the fact $\inf_{x \in \partial G_R} |\gamma(x)| > 0$.

Finally, we must have the boundary condition

$$V^h(x) = b(x), \qquad x \in \partial G_A^h. \tag{3.8}$$

Equations (3.5), (3.7b), and (3.8) will constitute the discretization of (2.8a), (2.8b′). The form (3.7b) is selected for simplicity of notation.

Define

$$A_n^h = \prod_{i=0}^{n} (\exp -\lambda(\xi_i^h)\, \Delta t_i^h I_{G_h}(\xi_i^h)),$$

$$C_n^h = \prod_{i=0}^{n} (\exp -\beta(\xi_i^h)\, d\mu_i^h), \qquad D_n^h = A_n^h C_n^h.$$

Then the unique solution to (3.5), (3.7b), (3.8) and the discrete approximation to (2.19) is

$$V^h(x) = E_x \sum_{i=0}^{N_h-1} D_i^h k(\xi_i^h)\, \Delta t_i^h I_{G_h}(\xi_i^h) + E_x D_{N_h-1}^h b(\xi_{N_h}^h) + E_x \sum_{i=0}^{N_h-1} D_i^h g(\xi_i^h)\, d\mu_i^h, \tag{3.9}$$

where $N_h = \min\{n : \xi_n^h \in \partial G_A^h\}$.

10.4 Continuous Time Interpolations and Convergence

In order to be able to use weak convergence theory to relate (3.9) to (2.19), we must interpolate $\{\xi_n^h\}$ into a continuous parameter process $\xi^h(\cdot)$. We will again take the interpolations to be piecewise constant. If $\xi_n^h = y \in G_h$, then the natural interpolation interval is $\Delta t^h(y)$ as in Chapter 6. However, if $y \in \partial G_R^h$, then there are several choices for the interval. We will concentrate on two of them (although intermediate cases are possible), to be called *scaling* 0 *and* 1 (which will ultimately correspond to $\rho = 0$ or 1, resp., and which will determine the actual amount of time that the process $\xi^h(\cdot)$ and limit $\xi(\cdot)$ spend on the reflecting boundary).

Note that

$$E_x(\xi_{n+1}^h - \xi_n^h \mid \xi_n^h = y \in \partial G_R^h) = \gamma(y)\, d\mu^h(y). \tag{4.1}$$

For scaling 0, set

$$\Delta t^h(x) = h^2/|\gamma(x)| = h\, d\mu^h(x).$$

For scaling 1, set

$$\Delta t^h(x) = h/|\gamma(x)| = d\mu^h(x).$$

Define $\Delta t_i^h = \Delta t^h(\xi_i^h)$ for $\xi_i^h \in \partial G_R^h$ also. Define $\rho_h = t_{N_h}^h$ and

$$t_n^h = \sum_0^{n-1} \Delta t_i^h,$$

and let $\xi^h(\cdot)$ denote the usual piecewise constant interpolation of $\{\xi_n^h\}$ with interpolation intervals $\{\Delta t_n^h\}$. Define $n_t = \max\{n : t_n^h \le t\}$ and $D^h(t) = D_{n_t}^h$, $D^h(0^-) = I$. Then

$$\begin{aligned} V^h(x) = E_x \int_0^{\rho_h} D^h(s) k(\xi^h(s)) I_{G_h}(\xi^h(s))\, ds + E_x D^h(\rho_h^-) b(\xi^h(\rho_h)) \\ + E_x \int_0^{\rho_h} D^h(s) g(\xi^h(s))\, d\mu^h(s), \end{aligned} \tag{4.2}$$

where

$$\begin{aligned} d\mu^h(s) &= ds\, I_{\partial G_{R^h}}(\xi^h(s))/h, \qquad \text{scaling } 0, \\ d\mu^h(s) &= ds\, I_{\partial G_{R^h}}(\xi^h(s)), \qquad \text{scaling } 1, \\ \mu^h(0) &= 0. \end{aligned} \tag{4.3}$$

Define

$$I_i^h = I_{\{\xi_i^h \in G_h,\, i < N_h\}}, \qquad I_{i,R}^h = I_{\{\xi_i^h \in \partial G_R^h,\, i < N_h\}}.$$

Then

$$\begin{aligned} \xi_n^h = x + \sum_{i=0}^{n-1} f(\xi_i^h)\, \Delta t_i^h I_i^h + \sum_{i=0}^{n-1} \beta_i^h I_i^h + \sum_{i=0}^{n-1} \gamma(\xi_i^h)\, d\mu_i^h I_{i,R}^h \\ + \sum_{i=0}^{n-1} [\delta \xi_i^h - \gamma(\xi_i^h)\, d\mu_i^h] I_{i,R}^h, \end{aligned} \tag{4.4}$$

where β_i^h is defined, as usual, to be $[\xi_{i+1}^h - \xi_i^h - f(\xi_i^h)\, \Delta t_i^h] I_i^h$. Let $F^h(\cdot)$, $B^h(\cdot)$, $\Gamma^h(\cdot)$, and $H^h(\cdot)$ denote the piecewise constant interpolations (with interpolation intervals $\{\Delta t_i^h\}$) of the series defined by the sums on the right-hand side of (4.4), each taking the value 0 at $t = 0$.

The following result is proved in Kushner [K7].

Theorem 10.4.1 *Fix x. The sequence $\{\xi^h(\cdot), F^h(\cdot), B^h(\cdot), \Gamma^h(\cdot), H^h(\cdot), \mu^h(\cdot)\}$ is tight in $D^{5r+1}[0, \infty)$ under each scaling. Let h denote a convergent subsequence with limit $(\xi(\cdot), F(\cdot), B(\cdot), \Gamma(\cdot), H(\cdot), \mu(\cdot))$. The process $\xi(\cdot)$ satisfies, for some Wiener process† $W(\cdot)$,*

† As in Chapter 6, if $a(\cdot)$ is not strictly positive definite, then to get $W(\cdot)$, we may need to augment the probability space by adding an independent Wiener process.

$$\begin{aligned}\xi(t) &= x + F(t) + B(t) + \Gamma(t) \\ &= x + \int_0^{t \cap \tau} f(\xi(s)) I_G(\xi(s))\, ds \\ &\quad + \int_0^{t \cap \tau} \sigma(\xi(s)) I_G(\xi(s))\, dW(s) \\ &\quad + \int_0^{t \cap \tau} \gamma(\xi(s)) I_{\partial G_R}(\xi(s))\, d\mu(s), \qquad H(t) \equiv 0, \end{aligned} \tag{4.5}$$

where $\tau = \inf\{t : \xi(t) \in \partial G_A\}$. *The functions* $\mu(\cdot)$ *and* $\xi(\cdot)$ *are nonanticipative with respect to* $W(\cdot)$.

Furthermore, if $q(\cdot, \cdot)$ *in* $C^{2,1}[R^r \times [0, \infty)]$ *satisfies* (2.4) *on* ∂G_R, *where* $\rho = 0$ *under scaling* 0 *and* $\rho = 1$ *under scaling* 1, *then*

$$q(\xi(t), t) - q(x, 0) - \int_0^{t \cap \tau} [\partial/\partial s + \mathscr{L}] q(\xi(s), s) I_G(\xi(s))\, ds \tag{4.6}$$

is a submartingale. Also

$$V^h(x) \to R(x) \qquad \text{(given by (2.19)) as } h \to 0. \tag{4.7}$$

We will not give the details of the proof. Several long sequences of estimates are involved, and the details appear in Kushner [K7]. It is interesting that the discrete approximations $\{\xi_n^h\}$ are the same irrespective of the value of ρ but, owing to the different time scalings on ∂G_R^h, the limiting processes are different—however, they are only different insofar as their time scale is concerned.

10.5 Extensions of the Reflection Problem

The problem of calculating invariant measures can be handled as in Section 6.8. There are also extensions to the various optimal control problems—optimal stopping, impulsive control, and continuously acting control. In the last case, we can have an internally acting control (in G), a boundary control (affecting the cost for time spent on the boundary, and the direction of reflection) or both. The appropriate dynamic programming equations (for minimizing the controlled analogue of (2.19)) are

$$\begin{aligned} &\min_{\alpha \in \mathscr{U}} [\mathscr{L}^\alpha V(x) + k(x, \alpha) - \lambda(x) V(x)] = 0, && x \in G, \\ &\min_{\alpha \in \mathscr{U}_1} [V_x'(x)\gamma(x, \alpha) + g(x, \alpha) - \beta(x) V(x)] = 0, && x \in \partial G_R, \\ &V(x) = b(x), && x \in \partial G_A, \end{aligned} \tag{5.1}$$

where $\mathscr{U}$ and $\mathscr{U}_1$ are appropriate action sets.

It is far from clear whether the submartingale model, as discussed in Section 10.2, is useful or not for problems in control theory. Of course, there are undoubtedly mathematical problems of interest in controlled reflections, but it is hard to find realistic applications to control theory at the present time. For this reason, we will not develop the discretization of (5.1) here. We note, however, that the pair $\mu(\cdot)$, $w(\cdot)$ has the same role that $w(\cdot)$ would have in the discussion of classes of control strategies and comparison of "discretizable" controls in Chapters 8 and 9.

There are physical problems where there are "hard" reflections. An example is when the velocity of a particle changes sign when the particle hits a wall. Such problems are simpler to treat than the submartingale model since the change of state is instantaneous and the particle does not linger near the boundary from which it is reflected, owing to the nature of the reflection.

We will now give a simple illustration of a "chattering" type of problem, on which more work is required. Define

$$\dot{X}_1 = \dot{X}_2 = 1, \qquad X(0) = 0,$$

$$G = \{x : x_2 < 1\}, \qquad \partial G_R = \{x : x_2 = 1\},$$

and refer to Figure 10.1. Let $h = 1/\text{integer}$, and $\partial G_R^h = \{(0, 1), (h, 1), \ldots\}$. Define the boundary scaling by $\Delta t^h(x) = h\rho$ for $x \in \partial G_R^h$, $\rho \geq 0$.

For $x \in G_h$,

$$p^h(x + e_1 h) = p^h(x + e_2 h) = \tfrac{1}{2}, \qquad \Delta t^h(x) = h/2.$$

Define the direction of reflection on ∂G_R by $\gamma_1(x) = 0$, $\gamma_2(x) = -1$ (the reflection is "straight down").

After hitting line 1 for the first time, the path $\{\xi_n^h\}$ spends an average of $\frac{1}{3}$ of the steps on line 0 and the rest on line 1. The interpolation intervals on lines 0 and 1 are ρh and $h/2$, resp. If x is on line 1, then the average velocity of $\xi^h(\cdot)$ to the right can be calculated from the facts that $p^h(x, x + e_1 h) = \frac{1}{2}$ and $\Delta t^h(x) = h/2$ and is 1. Thus, the average velocity of $\xi^h(\cdot)$ to the right is $1/(1 + \rho)$, after first reaching line 1. The sequence $\{\xi^h(\cdot)\}$ is tight on $D^r[0, \infty)$ and the limit $\xi(\cdot)$ satisfies

$$\dot{\xi}_1(t) = \dot{\xi}_2(t) = 1, \qquad t < 1,$$

$$\dot{\xi}_1(t) = 1/(1 + \rho), \qquad \dot{\xi}_2(t) = 0, \qquad \xi_2(t) = 1, \qquad t \geq 1.$$

Eventually, the chain $\{\xi_h^h\}$ bounces back and forth between line 0 and line 1 and it is not true that

$$I_{G_h}(\xi^h(s)) \to I_G(\xi(s)) \qquad \text{as } h \to 0$$

w.p. 1. Approximations to the cost functionals do not necessarily converge.

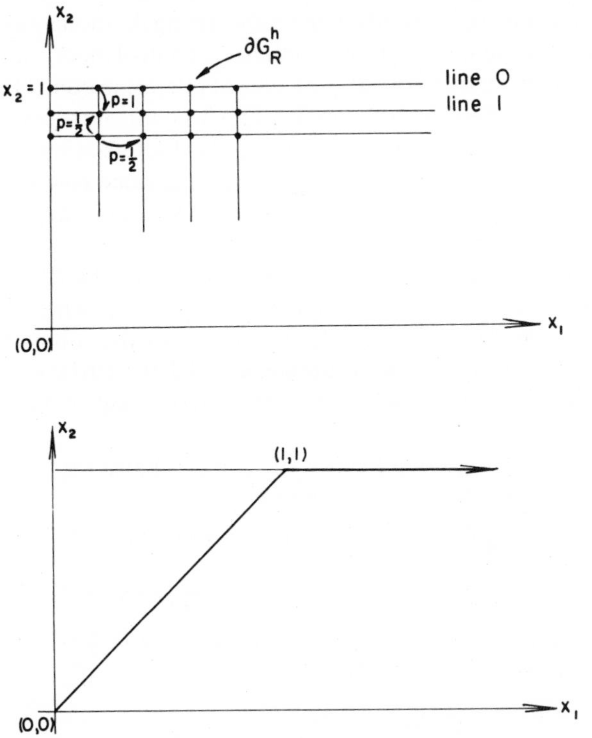

FIG. 10.1 An example where the discretization chatters near ∂G_R^h.

Of course, this particular problem can easily be handled by slightly modifying the method of calculating the cost functionals near the boundary. In particular, if there is chattering and if it is clear that part of the limit process will be on the boundary, then calculate the approximate cost as if the approximation were on the boundary. But it is not so easy to see how to treat the general case. This type of model sometimes appears in applications when we artificially truncate a state space by introducing a reflecting boundary.

CHAPTER 11

The Separation Theorem of Optimal Stochastic Control Theory

The stochastic control problem, where the system dynamics are linear in the state variable [i.e., $f(x, u) = Ax + d(u)$ for a matrix A and function $d(\cdot)$, and $\sigma(x) = \text{constant}$] is one of the most useful and important models in control theory. Frequently in applications, one observes not the state variable itself, but only some linear function of the state plus additive "white" noise. Thus the control function can depend only on the state via the noise corrupted observations. The separation principle is a fundamental idea in the control of such systems. Loosely speaking, it states that the optimal control can be obtained in two steps. The first step is the derivation of a dynamical (Itô-like) equation for the conditional mean and the second step is the computation of the optimal control—assuming that the dynamical equation which is satisfied by the conditional mean is the true dynamical equation of interest. The exact idea will be made clear below.

In the discrete time problem, the proof is rather obvious and almost classical [J1, G3, K5]. The continuous time parameter problem has been treated by a number of authors, a few among them being Kushner [K15],

Wonham [W4], Balakrishnan [B5], and Fleming and Rishel [F1]. This case is much more subtle and many questions center around the choice of an admissible control and even about an admissible probability space. Wonham's results [W4] depend heavily on the strict positive definiteness of $\sigma\sigma'$ or on the explicit solvability of the Bellman equation by a smooth ($C^{2,1}$) function. The first condition is rarely met in practice and the Bellman equation is usually only solvable by numerical methods. Balakrishnan [B5] treats a narrow class of cost functionals and it is not entirely clear whether there always exist solutions to the equations corresponding to the arbitrary observation dependent controls.

Here, we take an approach that is somewhat different from those used previously. The point of view toward the admissible control is different and is more natural in certain respects. In a sense, a large part of the solution lies in that definition. We can treat problems where control stops at a random (stopping or hitting) time or where there are state space constraints (dependent on the observations or conditional means or covariance).

The problem is defined and the class of admissible controls is described in Section 11.1 and an existence theorem for an optimal observation dependent control appears in Section 11.2. The proof of the theorem uses many of the ideas of weak convergence that we have developed in the preceding chapters. Even the idea for the class of admissible controls arose from the work of Chapters 8 and 9.

The point of view taken here (deliberately) avoids a number of problems. Although our class of admissible controls seems to be rather natural, it is still important to understand the filtering and control problem, when the admissible controls are defined as suitable functions of the observations only, and directly. Unfortunately, except for the rare case where $E(\cdot)$ is invertable for all s, the problems of uniqueness and meaning have not yet been resolved. This is one of the motivations for our point of view.

11.1 Assumptions and the System Model

First, we will collect some assumptions, then the basic control model will be defined, and the optimization problem stated.

Let $w(\cdot)$ and $z(\cdot)$ denote independent Wiener processes with values in R^r and $R^{r'}$, resp., and let $A(\cdot)$, $E(\cdot)$, and $C(\cdot)$ denote bounded measurable matrix-valued functions, defined on $[0, \infty)$, with $C(\cdot)$ and $E(\cdot)$ being continuous. In the theorem, the control values will be constrained to lie in a compact convex set $\mathscr{U} \subset R^m$. However, as mentioned after the proof, the general case can also be treated. Theorem 11.2.1 will deal with a control problem defined on a finite interval. It should be clear that the formulation is readily extended to problems on unbounded or random time intervals. We suppose until mentioned otherwise, that the control is defined on $[0, \infty)$, and we will consider it either as a sequence of random variables (which form a separable and measurable process), or as an element of one of the L_2 spaces $L_{2,T}^m$ or $L_{2,l}^m$, depending on the situation, as we have done in past chapters. In Theorem 11.2.1, we will use the assumptions

A11.1.1 *$k(\cdot, \cdot)$ and $b(\cdot)$ are bounded continuous real-valued functions on $R^r \times \mathscr{U}$ and R^r, resp.*

A11.1.2 *$d(\cdot)$ is an R^r-valued continuous bounded function on $\mathscr{U}$. The set $\{d(\alpha), k(x, \alpha), \alpha \in \mathscr{U}\} \equiv (d(\mathscr{U}), k(x, \mathscr{U}))$ is convex for each $x \in R^r$.* The boundedness assumptions on $k(\cdot, \cdot)$, $b(\cdot)$, $d(\cdot)$, and on $\mathscr{U}$ can be weakened; they have nothing to do with the separation theorem *per se* but only with the question of existence of an optimal control.

The class of admissible controls will be defined below. If $u(\cdot)$ is admissible, then the *uncontrolled* and *controlled state equations* are (1.1) and (1.2), resp., where $X(0)$ is a normally distributed random variable with mean $m(0)$ and covariance $P(0)$, and which is independent of $w(\cdot)$ and $z(\cdot)$.

$$X^0(t) = X(0) + \int_0^t A(s)X^0(s)\, ds + \int_0^t E(s)\, dw(s), \tag{1.1}$$

$$X(t) = X(0) + \int_0^t (A(s)X(s) + d(u(s)))\, ds + \int_0^t E(s)\, dw(s). \tag{1.2}$$

Let $P(\cdot)$ denote the solution to the Ricatti equation (1.3), with initial condition $P(0)$, define $L(t) \equiv P(t)C'(t)$, and let $\Phi(\cdot, \cdot)$ denote the fundamental matrix of $\dot{x} = Ax$.

$$\dot{P} = AP + PA' - PC'CP + EE'. \tag{1.3}$$

The *uncontrolled and controlled observation processes* are defined by (1.4) and (1.5), resp.

$$Y^0(t) = \int_0^t C(s)X^0(s)\, ds + z(t) \tag{1.4}$$

$$\begin{aligned} Y(t) &= \int_0^t C(s)X(s)\, ds + z(t) \\ &= Y^0(t) + \int_0^t C(s)\, ds \int_0^s \Phi(s, v)\, d(u(v))\, dv. \end{aligned} \tag{1.5}$$

Define the *innovations process* $\nu(\cdot)$ and processes $m(\cdot)$ and $m^0(\cdot)$ by

$$d\nu(t) = C(t)(X(t) - m(t))\,dt + dz$$
$$= C(t)(X^0(t) - m^0(t))\,dt + dz,$$
$$dm^0(t) = A(t)m^0(t)\,dt + L(t)\,d\nu(t), \qquad m^0(0) = m(0) = EX(0), \tag{1.6}$$
$$dm(t) = (A(t)m(t) + d(u(t)))\,dt + L(t)\,d\nu(t). \tag{1.7}$$

It is well known (see, e.g., Fleming and Rishel [F1]) that

$$m^0(t) = E[X^0(t) \mid Y^0(s), s \le t], \qquad \text{w.p. 1.}$$

Furthermore, if $u(\cdot)$ is a separable and measurable control for which the corresponding process $X(\cdot)$ exists and $E|X(t)| < \infty$ for each t, and where each $u(t)$ is measurable over $\mathscr{G}_t$, the smallest σ-algebra which measures $\nu(s)$, $s \le t$, then it is also true that

$$m(t) = E[X(t) \mid Y(s), u(s), s \le t]$$
$$= m^0(t) + \int_0^t \Phi(t, v)\, d(u(v))\, dv, \qquad \text{w.p. 1}$$
$$= E[X(t) \mid \nu(s), u(s), s \le t]. \tag{1.8}$$

The choice of the admissible class of controls is one of the crucial points in formulating the separation theorem problem. An admissible control cannot depend directly on $X(\cdot)$, but must depend on it only via the noise corrupted observations $Y(\cdot)$. Wonham [W4] supposed that $u(t)$ was a function of the solution to (1.7) over $[0, t]$ and that it satisfied a type of Lipschitz condition. The Lipschitz condition guaranteed that $u(t)$ was $\mathscr{G}_t$ measurable. Balakrishnan [B5] and Fleming and Rishel [F1] take the admissible controls to be the functions with values in the desired set and where $u(t)$ is $\mathscr{G}_t$ measurable for each $t \ge 0$. In general, of course, we *cannot* let $u(t)$ be an arbitrary function of $Y(s)$, $s \le t$, for it may then make no sense. There may not be a solution to (1.2) for such a general control and so the $Y(\cdot)$ process may not exist. So we must consider controls for which, at least, (1.2) and (1.7) have well-defined solutions. Also, we would like (1.8) to continue to hold.

The following definition of admissibility uses a somewhat broader class of control functions, but it is a very natural class, and is well suited for use in a general proof of an existence or separation theorem. *The control $u(\cdot)$ is said to be admissible if it is separable, measurable, $\mathscr{U}$-valued for almost all ω, t, nonanticipative with respect to $\nu(\cdot)$, and if for each t, $u(t)$ is conditionally independent of $w(\cdot)$ and $z(\cdot)$, given $\nu(s)$, $s \le t$.* The conditional independence and the fact that $\nu(\cdot)$ is nonanticipative with respect to $w(\cdot)$ and $z(\cdot)$ implies that $u(\cdot)$ is nonanticipative with respect to $w(\cdot)$, $z(\cdot)$. Again, where useful, we suppose that the paths of $u(\cdot)$ are points in an appropriate L_2 space.

Our class of admissible controls includes the classes of controls considered by the other cited authors. It includes (nonanticipative) controls which are $v(\cdot)$ functions, and also feedback functions (of $m(\cdot)$) which yield unique nonanticipative solutions to (1.7). The conditional independence guarantees that, if $v(s)$, $s \leq t$, is available, then we can learn nothing more about $u(t)$ by knowing $z(s)$ or $w(s)$, $s < \infty$. Thus, the condition truly implies that the control does not use any more information about $X(\cdot)$, $w(\cdot)$, or $z(\cdot)$ than is contained in the observations and in past values of the control. Of course, it does allow randomizations. For $u(\cdot)$ admissible, the corresponding solutions to (1.2) and (1.7) are unique, both pathwise and in the sense of probability law. Also, $v(\cdot)$ is a Wiener process and (1.8) holds.

Theorem 11.2.1 will treat the cost functional

$$R(u) = E^u\left[\int_0^T k(X(s), u(s))\, ds + b(X(t))\right]. \tag{1.9}$$

Remarks on our point of view The probability space on which the processes are defined is not important. Only the probability law is important. We do not need to distinguish between strong sense and weak sense solutions to the differential equations (1.2) or (1.7). For convenience in what follows, suppose that (whatever the probability space) the control is an element of an $L^m_{2,l}$ space (for the simple reason that the space, being separable and complete, will allow us to use a regular conditional probability).

Suppose that there is a probability space $(\tilde{\Omega}, \tilde{\mathscr{B}}, \tilde{P})$ supporting processes $(\tilde{X}(\cdot), \tilde{m}(\cdot), \tilde{Y}(\cdot), \tilde{z}(\cdot), \tilde{w}(\cdot), \tilde{v}(\cdot), \tilde{u}(\cdot), \tilde{X}^0(\cdot), \tilde{m}^0(\cdot), \tilde{Y}^0(\cdot))$, where $(\tilde{X}^0(\cdot), \tilde{w}(\cdot))$, $(\tilde{X}(\cdot), \tilde{u}(\cdot), \tilde{w}(\cdot))$, $(\tilde{m}^0(\cdot), \tilde{v}(\cdot))$, $(\tilde{m}(\cdot), \tilde{u}(\cdot), \tilde{v}(\cdot))$, $(\tilde{Y}^0(\cdot), \tilde{X}^0(\cdot), \tilde{z}(\cdot))$, and $(\tilde{Y}(\cdot), \tilde{X}(\cdot), \tilde{z}(\cdot))$ solve (1.1), (1.2), (1.6), (1.7), (1.4), and (1.5), resp., and $\tilde{u}(t)$ is $\mathscr{U}$-valued. Let all the processes (other than $\tilde{w}(\cdot)$, $\tilde{z}(\cdot)$, $\tilde{v}(\cdot)$) be nonanticipative with respect to $\tilde{w}(\cdot)$, $\tilde{z}(\cdot)$, $\tilde{v}(\cdot)$, let $\tilde{w}(\cdot)$ and $\tilde{z}(\cdot)$ be independent Wiener processes, and let any measurable function of $\tilde{u}(s)$, $s < t$, be conditionally independent of $\tilde{z}(\cdot)$, $\tilde{w}(\cdot)$, given $\tilde{v}(s)$, $s \leq t$. Let $\tilde{P}_{\tilde{v}}(d\tilde{u})$ denote the regular conditional distribution function of $\tilde{u}(\cdot)$, given $\tilde{v}(\cdot)$. For each measurable set $A \in L^m_{2,l}$, we can suppose that there is a measurable function $Q(A, \cdot)$ on $C^r[0, \infty)$ such that†

$$Q(A, y(\cdot)) = \tilde{P}_{\tilde{v}}(A)$$

for all $\tilde{\omega}$ such that $\tilde{v}(\tilde{\omega}, \cdot) = y(\cdot) \in C^r[0, \infty)$.

† Strictly speaking, we should delete the $y(\cdot)$ in a $C^r[0, \infty)$ measurable set N, where $\tilde{P}\{\tilde{v}(\tilde{\omega}, \cdot) \in N\} = 0$. For $y(\cdot) \in N$, set $Q(A, y(\cdot)) = \tilde{Q}(A)$, where $\tilde{Q}(\cdot)$ is an arbitrary distribution.

Now, let $(\Omega, \mathscr{B}, P)$ denote the probability space on which are defined the actual uncontrolled processes of concern in the application, namely, $z(\cdot)$, $w(\cdot)$, $m^0(\cdot)$, X^0, $Y^0(\cdot)$, and hence the innovation $v(\cdot)$. The measure P is that determined by $z(\cdot)$, $w(\cdot)$ but we write it as $P(dv\ dz\ dw)$. We allow ourselves to control the process in any way at all, provided only that we do not use "unavailable" information, and that the control take values in $\mathscr{U}$ for almost all ω, t. There is no reason why we cannot "randomize" or let $u(\cdot)$ depend also on quantities that have nothing at all to do with the processes as defined on $(\Omega, \mathscr{B}, P)$. In this sense, the probability space itself could also depend on the control $u(\cdot)$. We will define $u(\cdot)$ via an augmentation procedure.

Define $\Omega' = \Omega \times L^m_{2,\,t}$ and let $\mathscr{B}'$ denote the associated σ-algebra of subsets of Ω'. A measure P' will be defined on $(\Omega', \mathscr{B}')$ via a conditional independence assumption: Let ω' denote the generic variable of Ω', but continue to write $v(\cdot)$, $z(\cdot)$, $w(\cdot)$ as ω functions. Define

$$P'(du\ dv\ dw\ dz) = P(dv\ dz\ dw)Q(du,\ v(\omega,\ \cdot)) \tag{1.10}$$

(with a slight abuse of notation regarding the conditional probability). The marginal measures of $z(\cdot)$, $w(\cdot)$, $v(\cdot)$, $X^0(\cdot)$, $Y^0(\cdot)$ are unaffected. But we can now generate a control according to the new law and define the processes $X(\cdot)$, $m(\cdot)$, and $Y(\cdot)$. The process $u(\cdot)$ is nonanticipative with respect to $v(\cdot)$, $w(\cdot)$, and $z(\cdot)$ and is an admissible control. The cost functionals (say, of the form (1.9)) are the same for $(\tilde{X}(\cdot), \tilde{u}(\cdot))$ and $(X(\cdot), u(\cdot))$ since their laws are the same. Thus it makes no difference on which space we do our optimization. The main problem with our formulation is that it is not always clear how to generate the values of $u(t)$ recursively from the past values and the innovations for an arbitrary admissible $u(\cdot)$. This is not a serious difficulty. Note the remarks that we make below in connection with Theorem 11.2.1.

11.2 The Separation and Optimality Theorems

Define the functions $\hat{k}(\cdot, \cdot, \cdot)$ and $\hat{b}(\cdot)$ by

$$\hat{k}(x, \alpha, t) = \int k(x + \xi, \alpha)\ dN_t(\xi)$$

$$\hat{b}(x) = \int b(x + \xi)\ dN_T(\xi),$$

where $N_t(\cdot)$ is the distribution of $X^0(t) - m^0(t)$ (normal with mean zero and covariance $P(t)$). Both functions are continuous in their arguments and (except for the t-dependence of $\hat{k}$, which is continuous in all its arguments) satisfy A11.1.1 and A11.1.2. If $u(\cdot)$ is admissible with corresponding solutions $X(\cdot)$, $m(\cdot)$ then

$$E^u b(X(T)) = E^u[E^u b(X(T)) \mid v(s), u(s), s \leq T]$$
$$= E^u \int b(m(T) + \xi)\, dN_T(\xi) = E^u \hat{b}(m(T)),$$

and a similar relation holds for $k(\cdot, \cdot)$. Thus

$$R(u) = E^u \int_0^T \hat{k}(m(s), u(s), s)\, ds + \hat{b}(m(T))]. \tag{2.1}$$

Theorem 11.2.1 *Assume* A11.1.1 *and* A11.1.2, *then there is an admissible control which minimizes* (2.1).

REMARK To get the optimal control, we only need look for the control which minimizes (2.1) with system (1.7) over all $\mathscr{U}$-valued controls nonanticipative with respect to $v(\cdot)$, where $v(\cdot)$ is assumed to be an arbitrary Wiener process with the correct covariance function. The regular conditional probability of the optimal control, given $v(\cdot)$, yields the optimal law for the original problem, when the actual innovations are substituted for the arbitrary Wiener process $v(\cdot)$.

If the optimal control is representable as a $v(\cdot)$ function, then there is a rule which maps the observed values of the innovations directly into values of the control. This is true whether $\mathscr{U}$, $k(\cdot, \cdot)$, $b(\cdot)$, or $d(\cdot)$ are bounded or not. Theorem 11.2.1 is an existence theorem; it does not provide a rule for getting the *values* of $u(t)$ from the values of $u(s)$, $s < t$, and $v(s)$, $s \leq t$, in a dynamic manner. It is not always clear how we can generate a measurable process $u(\cdot)$ (with the correct joint law with $v(\cdot)$) recursively from the $v(\cdot)$ data as $v(\cdot)$ evolves. It says simply that there *is* an optimal admissible control.

It is not satisfactory that the existence theorem does not provide a rule. But, under an additional convexity condition, $u(\cdot)$ can be *assumed* to be a $v(\cdot)$ function. See the remarks after the proof. In any case, it follows from the implicit function theorem used in the proof that the minimizing $u(t)$ is a measurable function of $v(s)$, $s \leq t$, and the corresponding solution $m(t)$. But, when the control is used in this feedback form, we cannot guarantee that the solution to (1.7) is unique.

By the proof of Theorem 9.3.2, for any admissible control $u(\cdot)$ and $\varepsilon > 0$, there is an admissible $v(\cdot)$ dependent control $\tilde{u}(\cdot)$, which depends on $v(\cdot)$ at only finitely many time points, which is piecewise constant, and

$$R(\tilde{u}) \leq R(u) + \varepsilon.$$

Thus, the optimum in the class of admissible $v(\cdot)$ dependent controls (if it exists) is as good as the optimum admissible control. The technique of proof of Theorem 9.5.2 yields a constructive procedure for getting a (possibly randomized) admissible control $\hat{u}(\cdot)$ which takes finitely many values and is piecewise constant and is such that

$$R(\hat{u}) \leq R(u) + \varepsilon.$$

The theorem to be proved says, basically, that if we have a sequence of admissible controls, each for a filtering problem, then in the limit we get an optimal control for a filtering problem. The "filtering" structure is not lost, as we go to the limit.

PROOF There is at least one admissible control. Let $\{u^n(\cdot)\}$ denote a minimizing sequence. If we wish, we can suppose that the probability space is different for each n. Our notation will take this possibility into account. Let $(w^n(\cdot), z^n(\cdot), X^{0,n}(\cdot), m^{0,n}(\cdot), Y^{0,n}(\cdot), v^n(\cdot), X^n(\cdot), m^n(\cdot), Y^n(\cdot)) \equiv (S^{0,n}(\cdot), X^n(\cdot), m^n(\cdot), Y^n(\cdot)) \equiv S^n(\cdot)$ denote the relevant quantities, where $\{u^n(\cdot)\}$ is the minimizing sequence of admissible controls. Define

$$K^n(t) = \int_0^t \bar{k}(m^n(s), u^n(s), s)\, ds$$

$$D^n(t) = \int_0^t d(u^n(s))\, ds.$$

The set $\{S^n(\cdot), K^n(\cdot), D^n(\cdot)\}$ is tight on $C^\alpha[0, T]$ for some α. Let n index a convergent subsequence, suppose that the Skorokhod imbedding is used and denote the limit by $(\bar{S}(\cdot), \bar{K}(\cdot), \bar{D}(\cdot))$. From the weak convergence results of Chapters 6 and 8, we can show that $\bar{w}(\cdot)$, $\bar{z}(\cdot)$ are independent Wiener processes, $\bar{v}(\cdot)$ is a Wiener process which is nonanticipative with respect to $\bar{w}(\cdot)$ and $\bar{z}(\cdot)$, that all the other processes in $\bar{S}(\cdot)$, $\bar{K}(\cdot)$, $\bar{D}(\cdot)$ are nonanticipative with respect to $\bar{w}(\cdot)$, $\bar{z}(\cdot)$, $\bar{v}(\cdot)$, and (2.2)–(2.8) hold, where $\bar{X}(0)$ and $X(0)$ have the same distribution. The covariance of $\bar{v}(\cdot)$ is that of $v^n(\cdot)$.

$$\bar{X}(t) = \bar{X}(0) + \int_0^t A(s)\bar{X}(s)\, ds + \bar{D}(t) + \int_0^t E(s)\, d\bar{w}(s), \tag{2.2}$$

$$\bar{m}(t) = m(0) + \int_0^t A(s)\bar{m}(s) + \bar{D}(t) + \int_0^t L(s)\, d\bar{v}(s), \tag{2.3}$$

$$\bar{Y}(t) = \int_0^t C(s)\bar{X}(s) + \bar{z}(t), \tag{2.4}$$

$$\bar{X}^0(t) = \bar{X}(0) + \int_0^t A(s)\bar{X}^0(s)\, ds + \int_0^t E(s)\, d\bar{w}(s), \tag{2.5}$$

$$\bar{m}^0(t) = m(0) + \int_0^t A(s)\bar{m}^0(s)\,ds + \int_0^t L(s)\,d\bar{v}(s), \tag{2.6}$$

$$\bar{Y}^0(t) = \int_0^t C(s)\bar{X}^0(s)\,ds + \bar{z}(t), \tag{2.7}$$

$$\bar{v}(t) = \int_0^t C(s)[\bar{X}^0(s) - \bar{m}^0(s)]\,ds + \bar{z}(t). \tag{2.8}$$

It is clear that (w.p. 1)

$$\bar{m}^0(t) = E[\bar{X}^0(t) \mid \bar{Y}^0(s), s \le t] = E[\bar{X}^0(t) \mid \bar{v}(s), s \le t] \tag{2.9}$$

$$\hat{b}(m^n(T)) \to \hat{b}(\bar{m}(T)), \qquad K^n(t) \to \bar{K}(t), \qquad t \le T \quad \text{as} \quad n \to \infty. \tag{2.10}$$

We need to represent $\bar{K}(\cdot)$ and $\bar{D}(\cdot)$ in terms of an admissible control. But first we will show (for each t) that $\bar{K}(s)$ and $\bar{D}(s)$, $s \le t$, are conditionally independent of $\bar{w}(\cdot)$, $\bar{z}(\cdot)$, given $\bar{v}(s)$, $s \le t$. Let q be an arbitrary integer, let $t_1, \ldots, t_q$ be arbitrary positive scalars and $\le t$, let $s_1, \ldots, s_q$ be arbitrary positive scalars, and let $\alpha_j, \beta_j, \rho_j, \gamma_j$ be arbitrary vectors. If (2.11) holds for all such sequences, then we have the conditional independence

$$\begin{aligned} &E \exp i \sum_j (\alpha_j' \bar{D}(t_j) + \rho_j' \bar{K}(t_j) + \beta_j' \bar{z}(s_j) + \gamma_j' \bar{w}(s_j)) \\ &= E\left\{ E\left[\exp i \sum_j (\alpha_j' \bar{D}(t_j) + \rho_j' \bar{K}(t_j)) \mid \bar{v}(s), s \le t\right]\right. \\ &\qquad \left. \cdot E\left[\exp i \sum_j (\beta_j' \bar{z}(s_j) + \gamma_j' \bar{w}(s_j)) \mid \bar{v}(s), s \le t\right]\right\}. \end{aligned} \tag{2.11}$$

By the admissibility of $u^n(\cdot)$, (2.11) holds if the bar is replaced by the superscript n. Let us examine the factor

$$E\left[\exp i \sum_j (\beta_j' z^n(s_j) + \gamma_j' w^n(s_j)) \mid v^n(s), s \le t\right]. \tag{2.12}$$

The conditional expectation in (2.12) is independent of n. In fact, there are bounded continuous functions $\delta_j(\cdot)$, $\varepsilon_j(\cdot)$, $Q(\cdot)$ such that (2.12) can be written in the form (w.p. 1)

$$\left[\exp i \int_0^t \delta(s)\,dv^n(s)\right] Q(t) \equiv F(t),$$

where

$$\delta(s) = \sum_j (\beta_j' \delta_j(s) + \gamma_j' \varepsilon_j(s)).$$

The function $Q(\cdot)$ depends only on the conditional covariances of $w^n(\cdot)$, $z^n(\cdot)$, given $v^n(\cdot)$, and is a nonrandom quantity and the integral $\int_0^t \delta_j(s)\,dv^n(s)$

is the conditional mean of $z^n(s_j)$ given $v^n(s)$, $s \le t$, etc. This representation of (2.12) and the weak convergence imply (2.11). By the conditional independence,

$$\begin{aligned} \bar{m}(t) &= E[\bar{X}(t) \mid \bar{v}(s), \bar{D}(s), s \le t] \\ &= E[\bar{X}(t) \mid \bar{Y}^0(s), \bar{D}(s), s \le t] \\ &= E[\bar{X}(t) \mid \bar{Y}(s), \bar{D}(s), s \le t], \qquad \text{w.p. } 1. \end{aligned} \tag{2.13}$$

By the results in Section 9.2, there is an admissible control $u(\cdot)$ such that

$$\bar{D}(t) = \int_0^t d(u(s))\, ds,$$

$$\bar{K}(t) = \int_0^t \hat{k}(\bar{m}(s), u(s), s)\, ds, \tag{2.14}$$

and (2.13) holds with $u(s)$ replacing $\bar{D}(s)$. By minimality and admissibility, $u(\cdot)$ is an optimal admissible control. Q.E.D.

Remarks and extensions The method can be extended to problems where T is replaced by a random stopping or hitting time, provided that these times have the appropriate conditional independence properties. The following extensions can also be handled.

Let $k(x, \alpha) = k_0(x) + k_1(\alpha)$, where $k_0(x) \ge 0$. Let $b(x) \ge 0$, let $(k_1(\mathscr{U}), d(\mathscr{U}))$ be bounded and convex as before and suppose that $b(\cdot)$ and $k_0(\cdot)$ are convex. Both $k_0(\cdot)$ and $b(\cdot)$ may be unbounded, but suppose that they grow no faster than a polynomial as $|x| \to \infty$. All the above functions are assumed to be continuous. Then, the proof of Theorem 11.2.1 implies that there is an admissible control $u(\cdot)$ such that (2.2)–(2.9) and (2.13) hold and that

$$\begin{aligned} \lim_{n\to\infty} E^{u^n}[K^n(T) + \hat{b}(m^n(T))] &\ge E^u[\bar{K}(T) + \hat{b}(\bar{m}(T))] \\ &= E^u\left[\int_0^T \hat{k}(\bar{m}(t), u(t), t)\, dt + \hat{b}(\bar{m}(T))\right]. \end{aligned} \tag{2.15}$$

Since $\{u^n(\cdot)\}$ is minimizing, there must be equality in (2.15).

Now, note that there is a measurable nonanticipative and $\bar{v}(\cdot)$ dependent process $\tilde{u}(\cdot)$ such that, for each $t \le T$,

$$\left.\begin{aligned} E[d(u(t)) \mid \bar{v}(s), s \le t] &= d(\tilde{u}(t)) \\ E[k_1(u(t)) \mid \bar{v}(s), s \le t] &= k_1(\tilde{u}(t)) \end{aligned}\right\} \text{w.p. } 1.$$

In fact $\tilde{u}(\cdot)$ can be supposed to be admissible.

Define $\tilde{m}(t) = E[\bar{m}(t) \mid \bar{v}(s), s \le t]$. Then

$$\tilde{m}(t) = m(0) + \int_0^t A(s)\tilde{m}(s)\, ds + \int_0^t d(\tilde{u}(s))\, ds + \int_0^t L(s)\, d\bar{v}(s) \qquad (2.16)$$

and $\tilde{m}(\cdot)$ is the conditional expectation, given $\tilde{u}(s)$, $s \le t$, $\bar{v}(s)$, $s \le t$. Also, by convexity and Jensens' inequality

$$\begin{aligned} E\hat{k}_0(\bar{m}(t), t) &\ge E\hat{k}_0(\tilde{m}(t), t), \\ E\hat{b}(\bar{m}(t)) &\ge E\hat{b}(\tilde{m}(T)). \end{aligned} \qquad (2.17)$$

These calculations imply that the optimal control can be assumed to be a function of the innovations. Thus, there is an explicit rule for calculating it.

The only problem in allowing $k_1(\cdot)$, $d(\cdot)$, or $\mathscr{U}$ to be unbounded is in showing that $\{K^n(t)\}$ and $\{D^n(t)\}$ are tight and converge to absolutely continuous functions. Various additional assumptions can be used to guarantee this but we will not pursue the question further.

References

[B1] Bensoussan, A., and Lions, J. L., Nouvelle formulation de problems de control impulsionnel et applications, *C. R. Acad. Sci. Paris, Sér. A* **276**, 1189–1192, 1279–1284, 1333–1338 (1973).

[B2] Bensoussan, A., and Lions, J. L., Sur le contrôle impulsionnel et les inequations quasi-variationnelles d'evolution, *C. R. Acad. Sci. Paris, Sér. A* **280**, 1049–1053 (1975).

[B3] Billingsley, P., *Convergence of Probability Measures.* Wiley, New York, 1968.

[B4] Breiman, L., *Probability.* Addison-Wesley, Reading, Massachusetts, 1968.

[B5] Balakrishnan, A. V., A note on the structure of optimal stochastic controls, *Appl. Math. Optimization* **1**, 87–94 (1974).

[C1] Chung, K. L., *Markov Chains with Stationary Transition Probabilities.* Springer-Verlag, Berlin, 1960.

[D1] Derman, C., *Finite State Markovian Decision Processes.* Academic Press, New York, 1970.

[D2] Doob, J. L., *Stochastic Processes.* Wiley, New York, 1953.

[D3] Dudley, R. M., Distances of probability measures and random variables, *Ann. Math. Statist.* **39**, 1563–1572 (1968).

[D4] Dynkin, E. B., *Markov Processes.* Springer-Verlag, Berlin, 1965 (English translation of Russian original).

[F1] Fleming, W. H., and Rishel, R. W., *Deterministic and Stochastic Optimal Control.* Springer-Verlag, Berlin, 1975.

[F2] Forsythe, G., and Wasow, W., *Finite Difference Methods for Partial Differential Equations*. Wiley, New York, 1960.

[F3] Friedman, A., *Stochastic Differential Equations and Applications*. Academic Press, New York, 1975.

[G1] Garabedian, P. R., *Partial Differential Equations*. Wiley, New York, 1964.

[G2] Gikhman, I. I., and Skorokhod, A. V., *Introduction to the Theory of Random Processes*. Saunders, Philadelphia, 1969 (Translation of Russian original).

[G3] Gunckel, T. L., and Franklin, G. F., "A generalized solution for linear sampled data control," *J. Basic Engrg.* **85**, 197–201 (1963).

[H1] Hillier, F. S., and Lieberman, G. J., *Operations Research* (2nd ed.). Holden-Day, San Francisco, 1974.

[H2] Howard, R. A., *Dynamic Programming and Markov Chains*. Technology Press, M.I.T., Cambridge, Massachusetts, 1960.

[I1] Iglehart, D. E., Diffusion approximations in applied probability, *Mathematics of the Decision Sciences, Part II, Lectures in Applied Math* Vol. 12. Amer. Math. Soc., Providence, Rhode Island, 1968.

[I2] Itô, K., and Nisio, M., On stationary solutions of a stochastic differential equation, *Math. J. Kyoto Univ.* **15**, 777–794 (1966).

[J1] Joseph, P. D., and Tou, J. F., On linear control theory, *Trans. AIEE* (part II) **80**, 193–196 (1961).

[K1] Karoui, N., and Reinhard, H., Processus de diffusion dans R^n, *Lecture Notes in Mathematics*, No. 321. Springer-Verlag, Berlin, 95–116 (1973).

[K2] Khazminskii, R. Z., Necessary and sufficient conditions for the asymptotic stability of linear stochastic systems, *Theor. Probability Appl.* **1**, 144–147 (1967).

[K3] Kushner, H. J., "On the differential equations satisfied by conditional probability densities of Markov processes, *SIAM J. Control* **2**, 106–119 (1964).

[K4] Kushner, H. J., Dynamical equations for optimal nonlinear filtering, *J. Differential Equations* **2**, 179–190 (1967).

[K5] Kushner, H. J., *Introduction to Stochastic Control Theory*. Holt, New York, 1971.

[K6] Kushner, H. J., Finite difference methods for the weak solutions of the Kolmogorov equations for the density of both diffusion and conditional diffusion processes, *J. Math. Anal. Appl.* **52**, 251–265 (1976).

[K7] Kushner, H. J.: Probabilistic methods for finite difference approximations to degenerate elliptic and parabolic equations with Neumann and Dirichlet boundary conditions, *J. Math. Anal. Appl.* **53**, 644–668 (1976).

[K8] Kushner, H. J., and Chen, C. H., Decomposition of systems governed by Markov chains, *IEEE Trans. Automatic Control.* **AC-19**, 501–507 (1974).

[K9] Kushner, H. J., and Kleinman, A. J., "Accelerated procedures for the solution of discrete Markov control problems," *IEEE Trans. Automatic Control* **AC-16**, 147–152 (1971).

[K10] Kushner, H. J., and Kleinman, A. J., Mathematical programming and the control of Markov chains, *Internat. J. Control* **13**, 801–820 (1971).

[K11] Kushner, H. J., and Yu, C. F., Probability methods for the convergence of finite difference approximations to partial differential equations, *J. Math. Analysis Appl.* **43**, 603–625 (1973).

[K12] Kushner, H. J., and Yu, C. F., Probability methods for the convergence of finite difference approximations to partial differential integral equations, *J. Math. Analysis Appl.* **45**, 54–72 (1974).

[K13] Kushner, H. J., and Yu, C. F., Approximations, existence and numerical procedures for optimal stochastic controls, *J. Math. Analysis Appl.* **45**, 563–587 (1974).

[K14] Kunita, H., and Watanabe, S., On square integrable martingales, *Nagoya Math. J.* **30**, 209–245 (1967).

[K15] Kushner, H. J., *Stochastic Stability and Control.* Academic Press, New York, 1967.

[K16] Kushner, H. J., "Approximations for Functionals and Optimal Control Problems on Jump-Diffusion Processes," to appear.

[L1] Lindvall, T., Weak convergence of probability measures and random functions in the function space $D[0, \infty)$, *J. Appl. Probability* **10**, 109–121 (1973).

[L2] Lo, J. T. H., and Willsky, A. S., Estimation for rotational processes with one degree of freedom, Part I, *IEEE Trans. Automatic Control* **AC-20**, 10–21 (1975).

[L3] Loève, M., *Probability Theory*, 3rd ed. Van Nostrand-Reinhold, Princeton, New Jersey, 1963.

[M1] McShane, E. J., and Warfield, R. B., On Fillipov's implicit function lemma, *Proc. Amer. Math. Soc.* **18**, 41–47 (1967).

[M2] McShane, E. J., *Stochastic Calculus and Stochastic Models.* Academic Press, New York, 1974.

[M3] Meyer, P. A., *Probability and Potentials.* Blaisdell, Waltham, Massachusetts, 1966.

[M4] Mitchell, R. R., and Kozin, F., Sample stability of second-order linear differential equations with wide band noise coefficients, *SIAM J. Appl. Math.* **27**, 571–605 (1974).

[M5] Mortenson, R. E., Optimal control of continuous time stochastic systems, Ph.D. Thesis. Dept. Electrical Engrg., Univ. California, Berkeley, California, 1966.

[N1] Neveu, J., *Mathematical Foundations of the Calculus of Probability.* Holden-Day, San Francisco, 1965.

[P1] Pinsky, M., A note on degenerate diffusion processes, *Theor. Probability Appl.* **14**, 502–506 (1969).

[R1] Roxin, E., The existence of optimal controls, *Michigan Math. J.* **9**, 109–119 (1962).

[S1] Shiryaev, A. N., *Statistical Sequential Analysis.* Translations of mathematical monographs, Vol. 38. Amer. Math. Soc., Providence, Rhode Island, 1973.

[S2] Skorokhod, A. V., Limit theorems for stochastic processes, *Theor. Probability Appl.* **1**, 262–290 (1956).

[S3] Strook, D. W., and Varadhan, S. R. S., Diffusion processes with continuous coefficients, I, II, *Comm. Pure Appl. Math.* **22**, 345–400; 479–530 (1969).

[S4] Strook, D. W., and Varadhan, S. R. S., Diffusion process with boundary conditions, *Comm. Pure Appl. Math.* **24**, 1971.

[S5] Strook, D. W., and Varadhan, S. R. S., On degenerate elliptic and parabolic operators of second order and their associated diffusions, *Comm. Pure Appl. Math.* **25**, 651–713 (1972).

[V1] Varga, R. S., *Matrix Iterative Analysis.* Prentice-Hall, Englewood Cliffs, New Jersey, 1962.

[W1] Wagner, H. M., *Principles of Operations Research*, 2nd ed. Prentice-Hall, Englewood Cliffs, New Jersey, 1975.

[W2] Wong, E., *Stochastic Processes in Information and Dynamical Systems.* McGraw-Hill, New York, 1971.

[W3] Wong, E., and Zakai, M., On the convergence of ordinary integrals to stochastic integrals, *Ann. Math. Statist.* **36**, 1560–1564 (1965).

[W4] Wonham, W. M., On the separation theorem of stochastic control, *SIAM J. Control* **6**, 312–326 (1968).

[Y1] Yamada, T., and Watanabe, S., On the uniqueness of solutions of stochastic differential equations, *J. Math. Kyoto Univ.* **11**, 155–167 (1971).

[Z1] Zakai, M., On the optimal filtering of diffusion processes, *Z. Wahrscheinlichkeitstheorie und Verw. Gebiete* **11**, 230–243 (1969).

Index of Selected Symbols

Index

P

Q

R

S

T

W

A
B 7
C 8
D 9
E 0
F 1
G 2
H 3
I 4
J 5